169
Structure and Bonding

Aims and Scope

The series *Structure and Bonding* publishes critical reviews on topics of research concerned with chemical structure and bonding. The scope of the series spans the entire Periodic Table and addresses structure and bonding issues associated with all of the elements. It also focuses attention on new and developing areas of modern structural and theoretical chemistry such as nanostructures, molecular electronics, designed molecular solids, surfaces, metal clusters and supramolecular structures. Physical and spectroscopic techniques used to determine, examine and model structures fall within the purview of *Structure and Bonding* to the extent that the focus is on the scientific results obtained and not on specialist information concerning the techniques themselves. Issues associated with the development of bonding models and generalizations that illuminate the reactivity pathways and rates of chemical processes are also relevant.

The individual volumes in the series are thematic. The goal of each volume is to give the reader, whether at a university or in industry, a comprehensive overview of an area where new insights are emerging that are of interest to a larger scientific audience. Thus each review within the volume critically surveys one aspect of that topic and places it within the context of the volume as a whole. The most significant developments of the last 5 to 10 years should be presented using selected examples to illustrate the principles discussed. A description of the physical basis of the experimental techniques that have been used to provide the primary data may also be appropriate, if it has not been covered in detail elsewhere. The coverage need not be exhaustive in data, but should rather be conceptual, concentrating on the new principles being developed that will allow the reader, who is not a specialist in the area covered, to understand the data presented. Discussion of possible future research directions in the area is welcomed.

Review articles for the individual volumes are invited by the volume editors.

In references *Structure and Bonding* is abbreviated *Struct Bond* and is cited as a journal.

More information about this series at http://www.springer.com/series/430

D. Michael P. Mingos
Editor

The Chemical Bond I

100 Years Old and Getting Stronger

With contributions by

V. Arcisauskaite · W.-J. Chen · S. Ding · G. Frenking ·
J.M. Goicoechea · J.-F. Halet · M.B. Hall · M. Hermann ·
X. Jin · Z. Lin · J.E. McGrady · D.M.P. Mingos ·
J.-Y. Saillard · F.K. Sheong · D. Stalke

Springer

Editor
D. Michael P. Mingos
Inorganic Chemistry Laboratory
University of Oxford
Oxford, United Kingdom

ISSN 0081-5993 ISSN 1616-8550 (electronic)
Structure and Bonding
ISBN 978-3-319-33541-4 ISBN 978-3-319-33543-8 (eBook)
DOI 10.1007/978-3-319-33543-8

Library of Congress Control Number: 2016940193

Printed on acid-free paper

This Springer imprint is published by Springer Nature
The registered company is Springer International Publishing AG Switzerland

Preface

These three volumes of *Structure and Bonding* celebrate the 100th anniversary of the seminal papers by Lewis and Kossel. These papers, which formed the basis of the current view of the chemical bond, were published independently in 1916 and have greatly influenced the development of theoretical chemistry during the last century. Their essential ideas, which were initially formulated within classical Newtonian framework, have withstood many experimental tests and proved to be sufficiently flexible to incorporate the newer quantum mechanical ideas, which emerged in the 1920s and 1930s. Most importantly, Lewis' description of the covalent bond provided a graphical notation and a language for experimental chemists, which enabled generations of chemists to constructively discuss and predict the structures of molecules and graphically represent the course of chemical reactions. The Lewis and Kossel descriptions of chemical bonding are cornerstones of the undergraduate curriculum. They have achieved this pre-eminent distinction by evolving and incorporating a flexible view of chemical bonding, based on the symmetry characteristics and radial distribution functions of atomic orbitals. The development of a universally accepted notation for representing the bonds in inorganic and organic molecules has been particularly significant. Spectroscopic and structural results, which emerged as chemistry incorporated quantum mechanical concepts, provided detailed information concerning the structures of molecules not only in the solid state but also in the liquid and gas phases. These have provided increasingly rigorous tests of the bonding models, which emerged from the quantum mechanical description of the chemical bond.

The idea to celebrate this important anniversary in chemical evolution struck a chord with leading figures in the area of theoretical chemistry and resulted in the submission of 18 chapters, and it became necessary to produce three separate volumes of *Structure and Bonding* to satisfactorily account for the enormous influence Lewis and Kossel's seminal ideas had on modern chemistry. Following a historical introduction by myself, Volume 1 contains chapters by Dietar Stalke, Zhenyang Lin, Gernot Frenking, Jean-Francois Halet, Jen-Yves Saillard, José M. Goicoechea, John McGrady and Michael Hall covering a variety of experimental and theoretical studies of topical chemical bonding issues. Examples include the implications of experimentally determined electron densities on Lewis bond structures, the Lewis description of lone pairs in transition metal complexes, dative Lewis bonds, the bonding patterns in large metal clusters and the role of carbonyl ligands in stabilising such clusters and the electronic properties of endohedral metal clusters.

Volume 2 starts with a detailed account of Lewis and Kossel's legacy in defining the bonding in ionic and covalent compounds of main group elements and addresses the thermochemical and bond length implications of the Lewis and Kossel models. The subsequent chapters by Paul Poppelier, Miroslav Kohout, Sason Shaik, Philippe Hiberty and Bernard Silvi use highly accurate theoretical calculations to address and explore the fundamental nature of the covalent bond. Discussions of quantum chemical topology, the definition of electron pairs in positional space, provide a deeper insight into the nature of the chemical bond and the relevance of the ELF topological approach to the Lewis bond model and the evolution of electron pair bonding in covalent, ionic and charge shift bonds. The Lewis description of the chemical bond was limited to single, double and triple bonds, but in recent years compounds with bond orders greater than three have become commonplace, and the final chapter by Santiago Alvarez compares the electronic characteristics of Cr–Cr quadruple and quintuple bonds.

In Volume 3, the implications of the Lewis bonding ideas for modern inorganic, organic and organometallic chemistry are discussed by Douglas Stephen, Philip Miller, Robert Crabtree, Malcolm Green, Ged Parkin, Didier Bourissou and Ghenwa Bouhadir. These fascinating articles demonstrate how non-conventional Lewis acids and bases have been used to develop new chemistry based on frustrated Lewis pairs and describe the modern coordination chemistry of triphosphine ligands and its catalytic implications. Lewis developed the concept that bases function by donating non-bonding electron pairs, but Crabtree recounts how this view has had to be modified by the discovery of complexes where π-bonds and σ-bonds act as donors. Green and Parkin extend the basic Lewis concepts to organometallic complexes with three-centre two-electron bonds. Bourissou and Bouhadir describe compounds where the lone pairs on transition metals are able to function as Lewis bases – a field which has grown enormously in recent years.

This brief summary provides an indication of how the basic ideas introduced by Lewis and Kossel have blossomed over the last century as a result of the nourishment provided by quantum theory and the love and attention bestowed on them by successive generations of chemists. We hope that the quality and depth of the many contributions in these three volumes will convince the reader that the sentiment expressed in the title of this series "The Chemical Bond 100 Years Old and Getting Stronger" is appropriate.

Oxford, UK
April 2016

D. Michael P. Mingos

Contents

Struct Bond (2016) 169: 1–56
DOI: 10.1007/430_2015_203

Published online: 24 April 2016

The Chemical Bond: Lewis and Kossel's Landmark Contribution

D. Michael P. Mingos

Abstract The seminal papers of Lewis and Kossel in 1916 are put into a historical perspective. Mendeleev's periodic table, Thompson's discovery of the electron, Ramsay and Raleigh's discovery of the noble gases, Rutherford's model of the atom and Bohr's description of the stationary orbitals for the electrons in atoms all paid an important role in providing the background for Lewis and Kossel's proposal that the chemical bond originated either from the transfer of electrons or the sharing of electron pairs. These insights depended on the attainment of inert gas configurations by the atoms either directly by electron transfer or electron-pair sharing. The model incorporated an evolutionary gene which has enabled it to survive and grow by incorporating subsequent developments in quantum physics. The simplicity of the model has resulted in the development of a notation, which is universally used by chemists and has evolved to plot the course of chemical reactions and predict their regioselectivities. Its initial limitations are discussed, and the way in which they have been overcome by an orbitally based model is recounted. The model has been repeatedly enriched by quantum mechanically based theoretical studies.

Keywords Chemical bond • Covalent bond • Dative bonds • Effective atomic number rule • Hyper-valent • Hypo-valent • Ionic bond • Lewis structures

Contents

D.M.P. Mingos (✉)
Inorganic Chemistry Laboratory, University of Oxford, South Parks Road, Oxford OX1 3QR, UK
e-mail: Michael.mingos@seh.ox.ac.uk

Abbreviations

ccp	Cubic close packed
DFT	Density functional theory
EAN	Effective atomic number rule
Et	Ethyl
hcp	Hexagonal close packed
HOMO	Highest occupied molecular orbital
LCAO	Linear combination of atomic orbitals
LUMO	Lowest unoccupied molecular orbital
Me	Methyl
MO	Molecular orbital
Ph	Phenyl
VB	Valence bond
VSEPR	Valence shell electron-pair repulsion theory
XRD	X-ray diffraction

1 Introduction

These volumes of *Structure and Bonding* celebrate the 100th anniversary of the seminal papers by Lewis and Kossel [1–4] on the chemical bond and their influence on the development of chemical theory during the last century. Spectroscopic and structural results, which provided detailed information concerning the structures of molecules and the distribution of electron density in molecules, have provided increasingly rigorous tests of their bonding models. Their essential ideas, which were formulated in a classical Newtonian framework, have withstood many tests and proved to be sufficiently flexible to incorporate the newer quantum mechanical ideas. Most importantly it provided a graphical notation and a language for experimental chemists, which enabled them to constructively discuss and predict the structures of molecules and graphically represent the course of chemical reactions. Although the Lewis and Kossel descriptions of chemical bonding are cornerstones of the undergraduate curriculum, they have achieved this distinction by evolving and incorporating a more flexible view of chemical bonds and the development of a universally accepted notation – in Newton's modest words, progress in science is achieved by standing "on shoulders of others".

2 Historical Development of the Lewis/Kossel Model

2.1 *The Periodic Table*

The Victorian age was characterised by an obsession with the classification of the natural world, and animals, rocks and indeed everything were collected, classified and put on display in museums. The study of minerals and the animal kingdom had begun to yield great insights which had begun to undermine the traditional biblical view of the origins and age of the earth. By 1863, 56 chemical elements had been isolated and characterised as unique on the basis of their atomic weights and valencies – a sufficient number to develop a system of classification [5, 6]. In 1864, John Newlands [7, 8] noted that recurring similarities in their chemical properties could be emphasised if the elements were ordered according to their relative atomic weights. A repeating pattern occurred for groups of eight elements, in a way that was reminiscent of musical octets and therefore described by him as the *Law of Octaves* [7, 8]. Gaps in these octaves suggested other elements, which may be discovered in the future, but he lacked the self-confidence to make firm predictions. Lothar Meyer showed a similar diffidence when in 1864 he failed to predict any new elements, when he developed his periodic table based on the valencies of 28 elements [9, 10]. Unaware of Newlands and Meyer's earlier work, Mendeleev began to classify the elements according to their chemical properties while writing the two volumes of the textbook *Principles of Chemistry* (1868–1870). At an early stage, he recognised the following relationships based on atomic weights for elements which had similar chemical properties [11–13] ([14] and reference [5] page 156):

Cl 35.5	K 39	Ca 40
Br 80	Rb 85	Sr 88
I 127	Cs 133	Ba 137

He then developed an extended version of the periodic table by incorporating additional elements which followed a similar pattern. Mendeleev made a formal presentation *The Dependence between the Properties of the Atomic Weights of the Elements* to the Russian Chemical Society on 6th March 1869 [11–13]. The resulting table classified the elements on the basis of their atomic weight and valency. Mendeleev took the important step of predicting several new elements in the gaps which were present in his table and underlined the table's usefulness by predicting very specific physical and chemical properties for these elements. His predictions were based on interpolations between the established physical and chemical properties of elements, which belonged to the same column in his table. A few months later, Meyer published a virtually identical table. Meyer and Mendeleev were therefore codiscoverers of the periodic table, but Mendeleev's decision to accurately predict the properties of ekasilicon (germanium), ekaaluminium (gallium) and ekaboron (scandium) resulted in him being regarded as the more important contributor by the chemical community. The award of the Nobel Prize in 1904 to Sir William Ramsay helped to cement his premier position for future generations.

He established that the elements, if arranged according to their atomic weight, exhibit an apparent periodicity of properties and his conclusions were summarised as follows:

1. Elements which are similar regarding their chemical properties have atomic weights which are either of nearly the same value (e.g. Pt, Ir, Os) or which increase regularly (e.g. K, Rb, Cs).
2. The arrangement of the elements in groups of elements according to their atomic weights (with some exceptions) highlights the common valencies and their distinctive chemical properties. The lightest elements of these groups are Li, Be, B, C, N, O and F.
3. The elements which are the most widely diffused have small atomic weights.
4. The atomic weight of an element may sometimes be amended by a knowledge of those of its contiguous elements. Thus, the atomic weight of tellurium must lie between 123 and 126 and cannot be 128. (*Tellurium's atomic mass is 127.6, and Mendeleev was incorrect in his assumption that atomic mass must increase with position within a period.*)
5. Certain characteristic properties of elements can be predicted from their position in the periodic table.
6. He was puzzled about where to put the known lanthanides and predicted the existence of another row in the table for them and the actinides.

Mendeleev based the regularities in the table primarily on the atomic weights of the elements rather than their valencies, because it had been established that some elements were capable of exhibiting more than one valency. Lothar Meyer noted that the saturation capacity of elements (the valency) rises and falls regularly and evenly in both intervals [9, 10], e.g.:

Valency	1	2	3	4	3	2	1
	Li	Be	B	C	N	O	F
	Na	Mg	Al	Si	P	S	Cl

As Russell has noted [15], "Thus out of a study of the periodic dependence of general chemical behaviour on atomic weights there emerged a new set of valency relationships that for the first two periods at least, revealed an underlying simplicity that was to prompt still more fundamental questions".

2.2 *Discovery of Inert (Noble) Gases*

Although Mendeleev's periodic table led to many predictions, it completely failed to anticipate the existence of a whole group of monatomic gases. The first of the noble gases to be discovered by Lord Raleigh and William Ramsey in 1894 was argon [16]. Besides not being predicted, physical measurements on argon suggested that it was monatomic, a property which had only been observed previously for mercury vapour. Since valency and atomic weight were the two important parameters for the periodic table, the atomic weight depended on the atomicity of the new element. This problem was exacerbated when it was realised that the sample of argon had not been obtained in a pure form. Since the gas was completely inert, it was necessary to determine its atomic weight from specific heat measurements, and a valency of zero was unprecedented. In 1895 at a meeting at the Royal Society, Raleigh and Ramsey suggested that the new element, if a pure gas, would have an atomic weight of 39.9, which would not fit in with the periodic table. However, if it were a mixture of two gases with atomic weights of 37 (93.3%) and 82 (6.7%), the two elements would neatly fit in positions between chlorine and potassium and bromine and rubidium. Recognising that this new group of elements may represent a serious threat to his periodic classification, Mendeleev published his alternative interpretation [14]. He dismissed the possibility that it was monoatomic on the grounds that there was no room in the periodic table for such an element. Furthermore, it would be necessary to have a group of eight in the third series between chlorine and potassium. Indeed he concluded that the new gas was a triatomic form of nitrogen. In 1897 terrestrial helium was discovered and in 1900 krypton, neon and xenon and thereby confirming the presence of a completely new family of elements which had not been predicted by Mendeleev or anyone else. Ramsay proposed that their atomic weights placed them between the halogens and the alkali metals, i.e. extending Mendeleev's table by extending each period by one element on the right. This removed the threat which he feared, and he was able to celebrate in the following terms "for me it is a glorious confirmation of the general applicability of the periodic law". This "magnificent survival" of the periodic system after a "critical test" had resulted [14]. The incorporation of the noble gases into the periodic table provided an important component for the development of the chemical bonding principles proposed by Lewis and Kossel in 1916.

2.3 Valency

Chemists and alchemists before them had recognised for centuries that the behaviour of chemical species was governed by a type of chemical affinity, which resulted from specific chemical bonds. In 1704, Sir Isaac Newton famously outlined his atomic bonding theory, in "Query 31" of his Opticks, whereby atoms attach to each other by some "force". He acknowledged previous theories of how atoms were thought to attach to each other, i.e. "hooked atoms", "glued together by rest" or "stuck together by conspiring motions", but favoured the view that the cohesion whereby "particles attract one another by some force, which in immediate contact is exceedingly strong, at small distances performs the chemical operations, and reaches not far from the particles with any sensible effect".

The development of valency arose from Berzelius' theory of chemical combination which stressed [17, 18] the electronegative and electropositive character of combining atoms. In the mid-nineteenth century, Frankland, Kekulé, Couper, Butlerov and Kolbe [19–26], building on the theory of radicals, developed the theory of valency in which elements in compounds were joined by an attraction of positive and negative poles. The concept of valency preceded the discovery of the electron and the planetary view of the atom and may be traced to the 1850 paper by Frankland [19, 24]. He combined the older theories of free radicals and "type theory" and demonstrated that elements have the tendency to combine with other elements to form compounds containing an integer number of attached elements, e.g. in the three attached atoms NH_3, NI_3, four attached atoms in CH_4 and five attached atoms in PCl_5. Based on these examples and postulates, Frankland articulated the truism:

"A tendency or law prevails (here), and that no matter what the characters of the uniting atoms may be, the combining power of the attracting element, if I may be allowed the term, is always satisfied by the same number of atoms". The convention that pairs of atoms are held together by a force which was described as a bond was first used by Couper [21] and Crum–Brown [27] around 1860. Representing a bond by a line eventually became a graphical convention of great importance to chemists, but of course has no direct physical reality.

Chemistry has a knack of using terms such as valency, electronegativity and bonding which have a multiplicity of meanings. In its broadest sense, valency has been used to describe the ability of elements to combine with others. Russell's book provides a thorough analysis of the history of valency [15]. A chemical bond is more precisely defined as the force which holds two chemical entities together, but the definition encompasses a duality which at its extremes is based on either electrostatic (ionic) or covalent bonding and in between a variable amount of covalent and ionic character.

This "combining power" was subsequently described as quantivalence or valency. The International Union of Pure and Applied Chemistry (IUPAC) has made several attempts to arrive at an unambiguous definition of valence. The current version, adopted in 1994 [28]:

> The maximum number of univalent atoms (originally hydrogen or chlorine atoms) that may combine with an atom of the element under consideration, or with a fragment, or for which an atom of this element can be substituted

Although Frankland's definition worked well for a wide range of inorganic and organic molecules, it was less effective in the classification of salts. In these compounds, it was more convenient to consider the number of electrons which are transferred between the atoms. The "oxidation state" of an atom in a molecule gives the number of valence electrons it has gained or lost. In contrast to the *valency number*, the *oxidation state* can be positive (for an electropositive atom) or negative (for an electronegative atom). For example, the oxidation states of the metals in NaCl, $MgCl_2$ and $AlCl_3$ is +1, +2 and +3, and the chloride has a charge of −1. In Na_2O, MgO and Al_2O_3, the oxidation states of the metals are identical to those in the chlorides because the oxide bears a charge of −2.

Mendeleev and Meyer's periodic classification highlighted the relationship between an element's valency and its position in a particular group of the periodic table. In 1904 Abegg [29, 30] expanded the concept into a generalisation which he described as the group of 8. Drude [31] clearly summarised Abegg's group of 8 as follows: "An elements' positive *valency number v* signifies the number of loosely attached negative electrons in the atom"; his *negative valency number v′* means that the atom has the power of removing v′ negative electrons from other atoms, or at least of attaching them more firmly to itself. The prospect of electrons being related to the valencies of atoms followed soon after the discovery of the electron by Thomson [32, 33], who speculated that valency must be associated with the transfer of electrons between atoms. In crystalline solids, it was speculated that the forces holding the ions together involved electrostatic attraction between opposite charges, but these concepts could not be readily adapted to non-polar molecular solids. Rutherford's study [34] of the scattering of alpha particles by metal foils in 1911 showed that although the majority of particles passed directly through the foil, a small number were reflected by large angles. These experiments led Rutherford to propose a model of the atoms based on a localisation of the nucleus in 1/10,000 the volume occupied by the much lighter electrons occupying the large volume of the atom. Moseley's study in 1913 [35, 36] of the X-ray spectral lines of atoms showed that their position depended primarily on the atomic number of the atom, i.e. the number of electrons or protons in a neutral atom. These observations established that Mendeleev's periodic classification depended primarily on atomic number rather than atomic weight. In addition it provided an important insight into his use of valency as a parameter and suggested that atomic number must be related to the number of electrons in an atom of an element. Bohr [37, 38] developed in 1913 a planetary view of the atom, which restricted the electron to specific orbits based on the quantisation of the electron's angular momentum according to Planck's condition. Bohr also recognised that the shell structure which resulted from his quantum restrictions had implications for understanding the electronic structures of molecules and the periodic table. The model was extended to heavier atoms by Sommerfeld [39–41] who developed a model based on elliptical orbitals, which required two quantum integers.

2.4 Lewis/Kossel Papers

The modern view of valency can be traced primarily to two papers published by Lewis and Kossel in 1916. Their independent analyses both associated the stabilities of chemical compounds of the lighter elements to the attainment of eight electrons in their outer electron shells, i.e. the attainment of inert gas electronic configurations.

Kossel [4] focussed attention on the strongly electropositive character of elements succeeding the inert gases and the electronegative character of the elements preceding the inert gases. He proposed that when the atoms of these elements combine, they lose or gain sufficient electrons to achieve the closed shells associated with the inert gas atoms. The resulting positive and negative ions experience classical electrostatic attractive forces and more than recoup the energy expended in forming the ions especially if they form a crystalline solid. The ionic charges which results when the electrons are lost or gained may be associated with the valencies of the atoms. Kossel therefore may be considered as the co-originator of the octet rule, but he failed to recognise the possibility that octets may also be achieved by sharing rather than electron transfer. Lewis proposed a similar analysis but also provided a description of the chemical bonds in molecular organic and inorganic compounds. He proposed that an inert gas configuration may also be achieved in a molecule such as H_2 if the pair of electrons was shared equally by both atoms, thereby achieving the same closed shell configuration as He [1]. In the fluorine molecule F_2, the sharing of a pair of electrons would similarly result in both achieving the same electron configuration as a neon atom. To Lewis a "shared" electron pair resulted in a single pair of electrons occupying the valence shells of both bonded atoms. He postulated that in an element-hydrogen bond, the hydrogen achieved a doublet and the element to which it was bonded an octet by sharing an electron pair. Langmuir [42–45], who had been a student of Lewis', and did much to popularise the model, introduced the term *covalent bond* to describe the sharing of electron pairs in such molecules to distinguish it from the *ionic* or *electrovalent bond* found in salts such as Na^+Cl^-.

Lewis was unable to explain why two electrons favoured forming localised electron-pair bonds, although they would be expected to repel each other. Indeed to resolve this contradiction, he proposed (wrongly) that Coulomb's law may not be valid at the short interelectron distances found in bonds. He also recognised the disparity between his static view of the electrons in atoms and the planetary model which Bohr had developed in 1913. In 1923 Lewis proposed [2] that if the electron orbits had a fixed spatial orientation, then the average position of the electrons coincided with the fixed position of his static electron pair. The discovery that the electron had a spin in 1925 [46, 47] and the development of the Pauli exclusion principle [48] led to the recognition that a pair of electrons with the same spin keep as far apart as possible, whereas a pair of electrons with opposite spin experience reduced electron repulsion. The importance of these charge and spin correlation effects was not fully appreciated until the 1950s as a result of the work of Lennard–Jones [49] and Linnett [50].

2.5 Representation of Lewis Structures

Lewis and Kossel's proposals coincided with the shell structure of atoms which resulted from the hybrid classical/quantum model for the hydrogen atom developed by Bohr [37, 38] and subsequently extended by Sommerfeld [39–41] to other atoms. They did not fully appreciate the physical implications of a quantum model. Specifically Lewis based his model on the following postulates:

1. Kernel electrons (or core electrons in closed shells) remain unaltered in all ordinary chemical changes.
2. An atom in a molecule tends to hold an even number of electrons in its valence shells.
3. Electrons in shells which lie outside the kernel are mutually interpenetrable, and their pairing leads to the formation of a covalent bond.

Lewis and Kossel both suggested the electrons in molecules and ions form concentric groups of either two or eight electrons, although they represented them in quite different ways. Lewis preferred to represent them using a cubic model (his static representation of the electrons led to a symmetrical arrangement if they were located at the vertices of a cube), whereas Kossel preferred to use concentric rings to illustrate the successive shells. The different representations are summarised for neon in Fig. 1.

Lewis and Kossel both concluded that the stable electronic configurations in molecules resemble the two and eight electrons found in the inert gases and noted

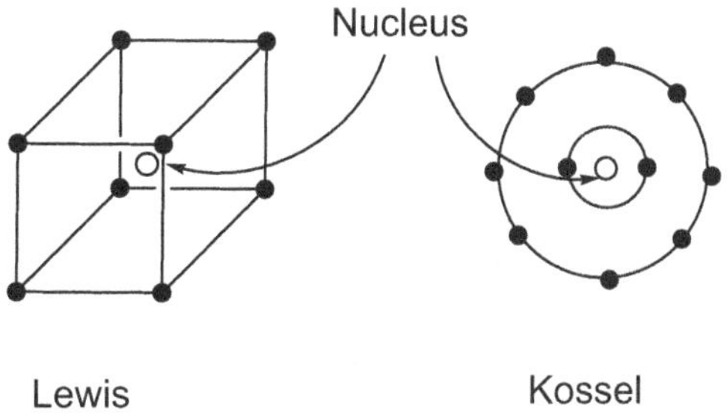

Fig. 1 Kossel and Lewis' representations of electrons in atoms [1, 2, 4]

that the attainment of these configurations in molecules either by sharing electrons or transferring electrons provides the driving force for chemical bonding.

Lewis had lectured on his ideas in undergraduate courses from 1902, i.e. in pre-quantum times, but was discouraged from publishing the work because he was uncomfortable with the duality of his chemical bond theory. He also found it problematic to apply his ideas to hydrocarbons and especially those with multiple bonds. As he has noted "I could not bring myself to believe in two distinct kinds of chemical union". Eventually in 1916 Lewis made the important extension to add a "rule of 2" to his "rule of 8". He recognised that with minor exceptions such as NO, NO_2 and ClO_2, the great majority of molecules, known at that time, had even numbers of electrons [3]. Thus, he established the importance of the electron-pair bond and recognised that it no longer belonged to either atom exclusively, but was shared between them. He extended his ideas to multiple bonds and initially represented these electron-pair bonds graphically using his cubes as shown at the top of Fig. 2 [1, 2].

Lewis noted his representations for the hydrogen and the fluorine molecules and molecules with double bonds, e.g. ethene. He could not represent the carbon–carbon triple bonds found in alkynes using the cubic notation, and this led him to modify the cube to a tetrahedron, in which pairs of electrons have been attracted together (see bottom of Fig. 2). The model thereby combined two important ideas – a pair of electrons was responsible for each covalent bond, and molecules with single, double and triple bonds were represented by a pair of tetrahedra sharing vertices, edges or faces. The latter incorporated the stereochemical implications of the tetrahedral carbon atom established earlier by van't Hoff and leBel [51, 52]. In later publications, Lewis abandoned cubic representations and used colons to represent electron-pair bonds and preferred the dot structures shown at the top of Fig. 3. Pedagogically these dot structures which emphasise the attainment of the octet of electrons around the central atom and doublets at hydrogen are still used to introduce basic bonding concepts. To emphasise the valencies of the atoms, the

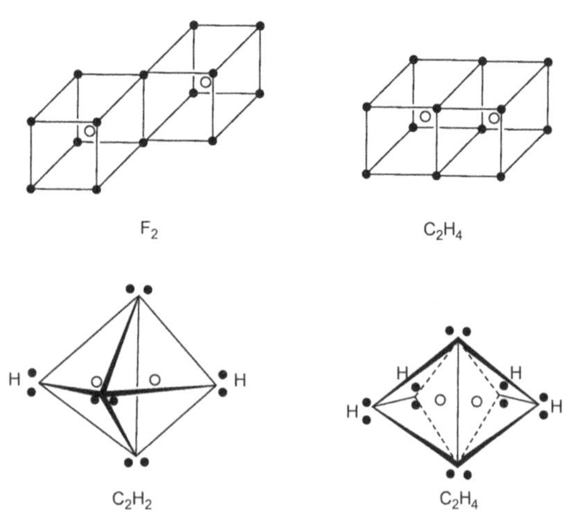

Fig. 2 Lewis' description of covalent bonds in F_2 and C_2H_4 based on the sharing of electrons from two cubes which leads to single and double bonds respectively. The model could not be adapted to C_2H_2, but the alternative description based on four-electron pairs at the vertices of a tetrahedron could result in the sharing of three electron pairs required for the triple bond in C_2H_2

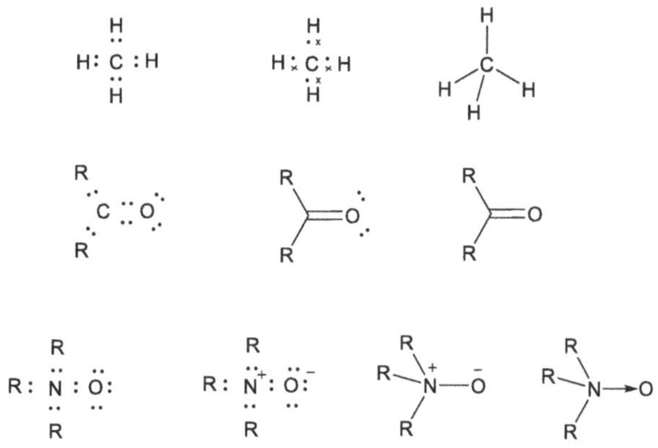

Fig. 3 Representation of Lewis structures based on the attainment of closed shells by electron-pair sharing. The *initial dot structures* have been progressively replaced by line structures to represent the two-electron two-centre bonds. *Dots* are only retained when they have stereochemical consequences or are required to represent organic reactions using the curly arrow notation (see Fig. 4)

origins of the electrons are sometimes indicated by using the dots and crosses shown in Fig. 3. As the concepts become familiar, then the structures are represented by line structures. In organic chemistry, this also carries with it implications incorporating the stereochemistries of the carbon, nitrogen and oxygen atoms.

2.6 Lewis Acids/Bases: Dative Bond Representations

In 1923, Lewis provided an important general definition of acids and bases: "An acid substance is one which can employ an electron lone pair from another molecule in completing the stable group of one of its own atoms" [3]. The Brønsted–Lowry acid–base theory was published in the same year. The two theories are distinct but complementary. Nevertheless, Lewis suggested that an electron-pair donor may be classified as a base and an electron-pair acceptor be classified as acid. Langmuir recognised that Me_3BNH_3 and Me_3CCH_3 were isoelectronic and consequently the B–N and C–C bonds at their centres must be closely related since they were both based on the sharing of an electron pair. Sidgwick proposed that when both electrons come from one of the atoms, it could be described as a dative covalent bond or coordinate bond [53, 54]. The distinction was not universally accepted, and Pauling, for example, rarely used the terms coordinate or dative bonds in his publications and books [55–62]. The alternative representations of dative covalent bonds are shown at the bottom of Fig. 3. The Lewis acid/base theory has had an important impact on understanding the reactions of organic molecules and was extended by Sidgwick to the transition metal coordination compounds studied by Werner [53, 54]. To represent organic reactions as a series of Lewis acid/base steps, it is common to indicate the lone pairs in organic molecules as shown at the bottom of Fig. 3. Ingold and Robinson [63–69] were primarily responsible for showing how the Lewis acid/base ideas and the

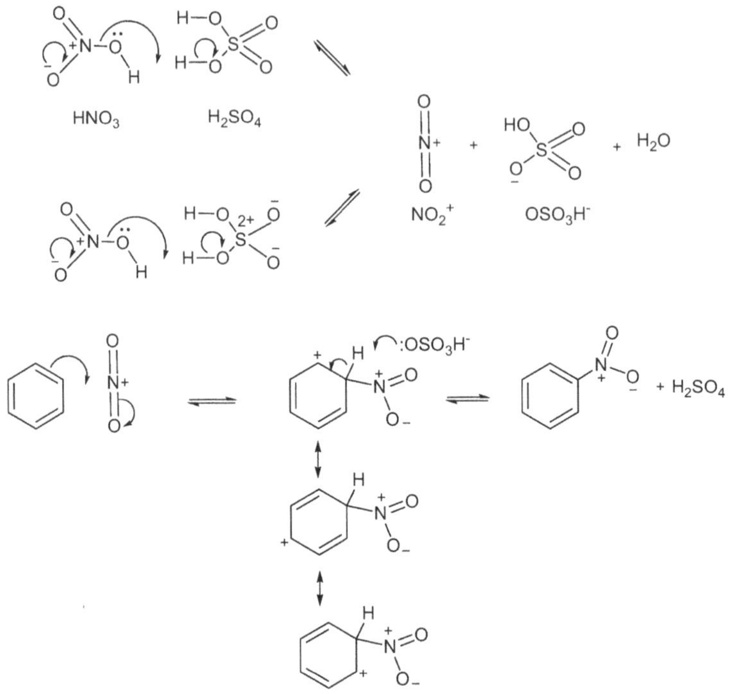

Fig. 4 An example of the use of the *curly arrow notation* to represent the course of organic reactions. The resonance structures shown in the middle of the figure suggest that electron-releasing groups in the *ortho-* and *para-*substituents of the benzene ring will favour the substitution process. Although, the sulphur-containing reagents are drawn with multiple bonds, the *top* of the figure shows that the *curly arrow notation* works equally well if single-bonded octet structures are drawn for these compounds

Lewis structures could be used to represent organic reactions, and the resulting curly arrow representation, which may be viewed as an extension of the Sidgwick dative bond, is universally accepted and used to describe the mechanistic pathways of organic reactions. Figure 4 gives some specific examples of the notation as it is used in organic chemistry today.

Robinson, Lapworth, Ingold, Pauling and Wheland [58, 59, 62–72] extended these basic concepts to describe the inductive and mesomeric effects of substituents in organic molecules and provided a very widely accepted methodology. This accounted for the preferred locations of substitution reactions in aromatic rings and the relative rates of these substitution reactions. A specific example is shown in Fig. 4. The curly arrow notation provided a convenient way of describing the distribution of charges in organic molecules and transition states. The adherence of the octet rule ensures that a movement of an electron pair from one atom (or bond) is only permitted if an electron-pair hole is simultaneously created to accept it – this defines the pathway across the molecule. This convenient notation was underpinned by the valence bond model developed by Pauling and particularly the concept of resonance [55, 60, 61]. Sidgwick and Sutton [73–75] provided experimental evidence for these inductive and mesomeric effects by measuring the dipole moments of a wide range of molecules and interpreted the data using the bonding models developed by Pauling.

2.7 Summary

Lewis has gained more recognition for developing a coherent bonding model than Kossel, but like Mendeleev, this was not recognised by the award of a Noble Prize, although he was nominated more than 35 times! In 1923 Lewis developed the concepts which had been presented in the 1916 *Journal of the American Chemical Society* in his book "Valence and the Structures of Atoms and Molecules" [3]. Pauling recognised his enormous contribution by dedicating his classic "Nature of the Chemical Bond" to him in 1938. In summary, his theory incorporated the following basic ideas:

1. The description of the chemical bond depends on making a distinction between valence electrons, which contribute to the chemical bond, and core electrons, which do not participate significantly in chemical bonding.
2. A covalent chemical bond results from the sharing of pairs of electrons.
3. An ionic bond results from the transfer of electrons from the electropositive atom to the electronegative atom. The number of electrons transferred is dictated by the achievement of an inert gas configuration.
4. The Lewis–Kossel description provided a consistent description of chemical bonding, which depends on the attainment of the inert gas rule either by sharing or transfer of electrons.
5. Covalent molecules may have electron pairs involved in covalent chemical bonds and also electron pairs which do not contribute to the chemical bond. For example, F_2 has one covalent bond holding the fluorine atoms together and three non-bonding electron pairs on each fluorine atom.
6. Although homonuclear molecules such as Cl_2 and F_2 are non-polar, NaCl and KCl are highly polar. It emphasised the similarity between many Brønsted acids, with elimination of molecular compounds and a distinction between primary and secondary affinities.
7. It provided an effective notation of the electronic structures of inorganic and organic molecules. Initially this was based on the representation of electron pairs as colons, but subsequently developed so that covalent bonds were represented by lines joining the atoms and non-bonding electron pairs as colons.
8. It anticipated electronegativity as a way of describing polarised bonds, which bridged the gap between the extreme forms of covalent and ionic bonding.

9. It provided a general way of accounting for the reactivities of unsaturated compounds and the effect of substituents on the regioselectivities of many organic reactions.
10. The definition of the chemical bond as a shared electron pair could be extended to describe the dative bond and the elaboration of Lewis acid/base interactions.

What is remarkable is the success and widespread use of a model which stern critics would argue owes more to numerology than modern physics and was not based solidly on quantum or even Newtonian physics. In a contradictory manner, it defines the chemical bond in terms of a classical electrostatic interaction between oppositely charged ions (the ionic bond) and the pairing of negatively charged electron sharing a small region of the molecule (the covalent chemical bond). It is hardly surprising that this contradiction made Lewis delay publication from 1902 when he first introduced the basic ideas to undergraduates in his lectures. The modern description of the chemical bond is based on a quantum mechanical description of atoms and molecules which depends on defining the electron in an atom not as a particle but a wave and whose properties depend on four quantum numbers, three of which define the radial and nodal characteristics of the wave and the fourth the spin of the electron. The resulting orbital picture of chemical bonding has not only encouraged the development of pictorial representations which explain the occurrence of bonds with bond orders which exceed the triple bonds described by Lewis but has also provided great insights into the three-dimensional geometries of molecules and their reactivity patterns. Lewis and Kossel's generalisations did not assist in defining these fundamental questions of physics, but they did emphasise the importance of electron pair in a chemical bond and the importance of attaining inert gas configurations in ions and molecules. Most importantly it provided a very effective means of communicating in the chemical community the valency, the stereochemistry of atoms in molecules and a way of auditing the movement of electron pairs between reactants and products in chemical reactions.

Chemists recognised from an early stage that the Lewis–Kossel approach provided alternative molecular structures for molecules with the same number of valence electrons. This ambiguity was even apparent for the elements belonging to the same group of the periodic table. For example, although N_2 and O_2 are diatomic molecules having strong multiple bonds, the related elements phosphorus and sulphur have allotropic forms, which are based on single bonds between the elements. The number of covalent bonds formed by each atom is identical, but the lighter elements show a great preference for forming multiple bonds as shown in Fig. 5.

The ability of the first long row of elements to form strong multiple bonds is an important general characteristic of the periodic table, but the classical Lewis description of ethene has to be modified for the analogous compounds of the heavier Group 14 elements. As shown in Fig. 6, the planar structure characteristic of ethene is no longer maintained and the molecules show a folded structure, and the fold angle increases with the atomic number of the element. It is noteworthy that the resulting structure may be described as singlet "carbenoid" structures which

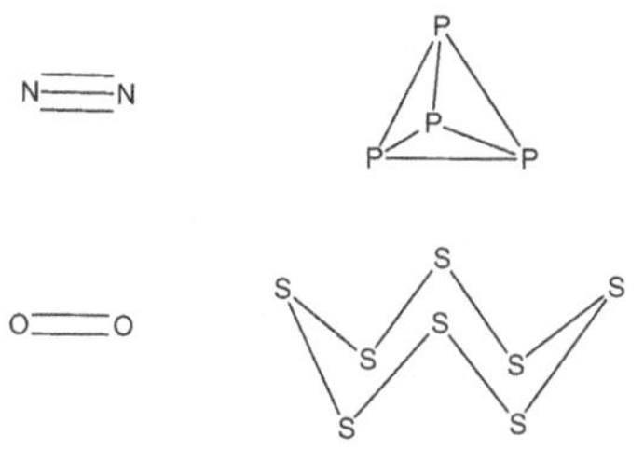

Fig. 5 Lewis structures giving rise to multiply bonded dimmers or polyhedral and ring compounds for elements belonging to the same group of the periodic table

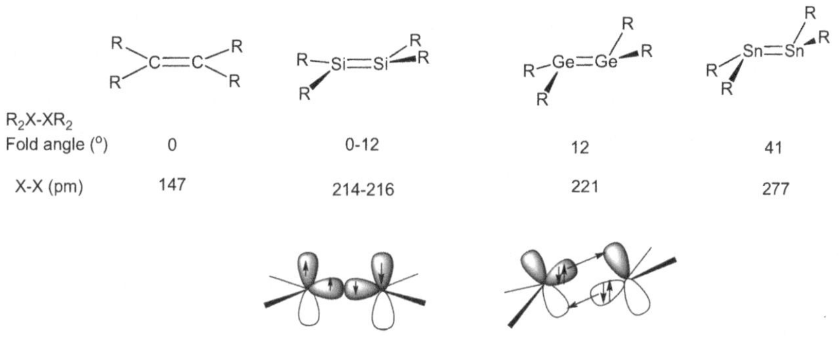

Fig. 6 Two pairs of electrons either forming a double bond in ethene using the classical Lewis description or two dative bonds. The latter is observed in analogues of ethene for the heavier group 14 elements. The geometric consequences of the different bonding modes result in the progressive folding of the molecule [76]

may be considered to interact more weakly through dative bonds as shown in Fig. 6 [76–78].

3 Extensions of the Lewis/Kossel Model

3.1 Generalisations of the Lewis Structures

The discussion above has indicated some of the limitations of the original Lewis/Kossel description of chemical bonding and the manner in which it has been adapted to assimilate the multitude of new compounds being reported from chemical laboratories during the last century. Central to the model is the definition of the chemical bond as a pair of electrons and the adherence to the octet rule.

The relevance of completed electronic shells associated with the inert gases was extended by Langmuir [42–45], who developed specific formulae relating the covalence of the central atom to the number of valence electrons in the inert gases. Since the atomic numbers of the inert gases are 2, 10, 18, 36, 54 and

86, these numbers were identified with the completion of stable electronic configurations. If the core electrons are excluded, these configurations are associated with 2, 8, 8, 18, 18, 32 valence electrons. Bury [79] clarified the Langmuir proposal by suggesting that the maximum numbers of electrons in the various shells are 2, 8, 18 and 32. Bury noted that in transition metal and lanthanide atoms, inner building occurs, i.e. the filling up of inner electronic shells, while the outermost ones remain constant. These developments led chemists to use the octet rule for organic and main group molecules, and Blanchard [80] applied the 18-electron rule to transition metal carbonyl complexes such as $Ni(CO)_4$. An alternative electron-counting procedure, based on the electron shell structures proposed by Bohr and Bury, was introduced by Sidgwick in 1923 [81]. The effective atomic number (EAN) rule, focussed not just on the valence shell electron count but on the total atom electron count. Attainment of an octet or an 18-electron outer configuration was equivalent to attaining the total electron count (or atomic number) of the nearest noble gas. Sidgwick's EAN rule was first applied to the burgeoning number of transition metal carbonyls and nitrosyls by Reiff in 1931 [82], and in 1934 Sidgwick extended its use to complexes with bridging, carbonyls [83]. Sidgwick and Blanchard popularised the rule in the 1940s. In the 1960s [84], there was a reversion to electron-counting procedures based solely on the valence electrons, because the main group molecules could be referred to the octet rule, and the three rows of transition metals could be referred to the 18-electron rule. Sidgwick's EAN rule, which includes the chemically inactive core electrons, results in a separate electron count for each row of the main group and transition metal blocks. The octet and 18-electron rules are subject to many exceptions, but they, nevertheless, proved very useful as a pedagogical tool in organometallic and inorganic chemistry [85].

It was noted above that the initial octet rule was extended to an 18-electron rule for transition metal compounds, and the dative bond notation introduced by Sidgwick was used very widely for describing coordination compounds and organometallic compounds. The duality arising from the formal description of the bonding in such compounds in terms of formal oxidation states of the central metal ion or a covalent model based on the valency of the metal has presented certain issues, which have been discussed at some length in the reviews of Green and Parkin [86, 87]. The increasing number of organometallic compounds since 1950 and their importance as intermediates in catalytic processes led to a detailed study of alkene and carbonyl complexes of transition metals in low oxidation states. This revealed that the dative bonding in such compounds could proceed simultaneously in both directions, i.e. from a ligand lone pair to the metal and from a filled d orbital on the metal to an empty orbital on the ligand. This synergic bonding model (discussed more fully in Sect. 6.3) represents one of the most important outcomes of the Lewis electron-pair model, and Green and Parkin have introduced a convenient and flexible notation for classifying such compounds.

In this review, attention will be directed towards some important differences in the way in which octet and 18-electron compounds are commonly represented in the literature to describe structures and reactions. Figure 7 compares the representations for typical main group and transition metal compounds which conform to

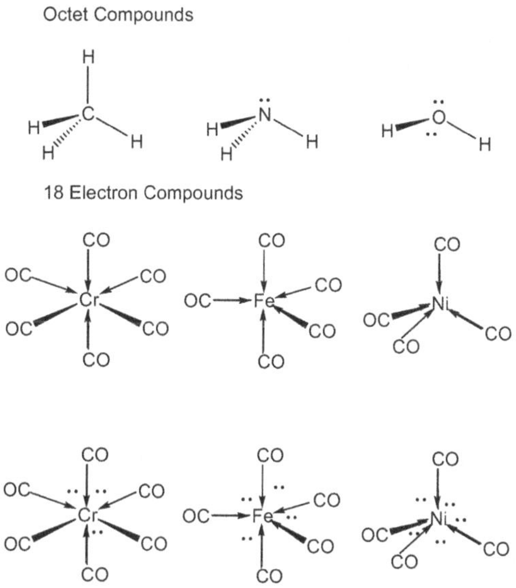

Fig. 7 Comparison of Lewis structures for typical main group and transition metal compounds

the octet and 18-electron rules and emphasises the dative bond notation introduced by Sidgwick. The ubiquitous presence of CO as a two-electron donor ligand resulted in a simplification so that dative bond arrow is commonly replaced by a single bond line, although this may be misleading to newcomers to the field, who have been introduced to Lewis acid–base reactions represented by dative bond arrows. The other important omission concerns the lone pairs. In the octet compounds, the lone pairs are clearly shown and are important for the use of these Lewis formulae for describing reactions of these molecules through the curly arrow representations. It also has structural implications because these lone pairs are stereochemically active and occupy space as if they were covalent bonds. Thus, all three main group molecules in Fig. 7 may be related to the parent tetrahedron with lone pairs successively replacing bonds. The stereochemical importance of lone pairs in main group molecules was recognised by Sidgwick and Powell and reviewed in 1940 [88]. This stereochemical generalisation which was described as valence shell electron-pair theory was subsequently amplified by Gillespie and Nyholm [89–92] and is discussed in more detail in Sect. 3.4. For the transition metal carbonyls, the metals also have pairs of electrons which are not used in the metal–ligand sigma bonds, but are not generally shown in the Lewis/Sidgwick representations. Specifically Cr, Fe and Ni have 6, 8 and 10 electrons paired on the metal, i.e. 3, 4 and 5 electron pairs, which are omitted (see Fig. 7). This difference may initially have arisen for printing and aesthetic reasons but also reflected the current view that the d valence electrons belong to an inner shell. Showing all these electron pairs can lead to rather cluttered representations as shown at the bottom of Fig. 7, and more significantly the electron pairs are not stereochemically active in the way that has been described above for the octet compounds. This significant difference has been interpreted using a quantum mechanically based free-electron

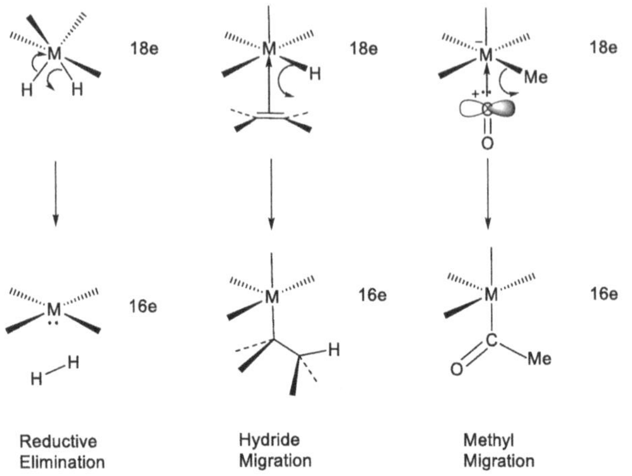

Fig. 8 *Curly arrow representations* of archetypical organometallic transformations

model described as *The Complementary Spherical Electron Density Model* [93–96]. The ligand and metal orbitals in an 18-electron compound are related to those of an inert gas, and their wave function representations provide complete and complementary orthogonal sets. The nodal characteristics of the electron pairs localised on the metal make them orthogonal to the metal–ligand orbitals, and consequently they are not stereochemically active. Zhenyang Lin has discussed this aspect of transition metal coordination chemistry in a separate chapter of this volume [97].

The other intriguing question which results from the extension of the Lewis notation to transition metal chemistry is why the curly arrow notation so commonly used in organic chemistry has not been used more widely in organometallic and coordination chemistry [98–100]. Figure 8 illustrates the primary transformations of metal complexes, viz., oxidative addition, hydrogen migration and methyl migration using curly arrow notation which have analogues in organic chemistry. It is noteworthy that all the transformations and their reverse reactions involve changes in electron counts from 18 to 16 or vice versa, and consequently the electron book keeping is slightly more complex than those of organic reactions where the octet rule is maintained. The negligible use of this notation in contrast to organic chemistry may result, because, in contrast to organic reactions, a small number of centres are involved. Nevertheless, Ghosh and Berg have recently shown how the curly arrow notation may be used to systematise main group chemistry [100]. The concerted movements of electron pairs along many centres and around a ring of atoms are much less common in transition metal reactions. The representation of π-bonded ligands with 3–8 carbon ligands in organometallic complexes provides an additional complication for this type of representation. For simple organometallics, these problems can be overcome using canonical forms such as those illustrated in Fig. 9.

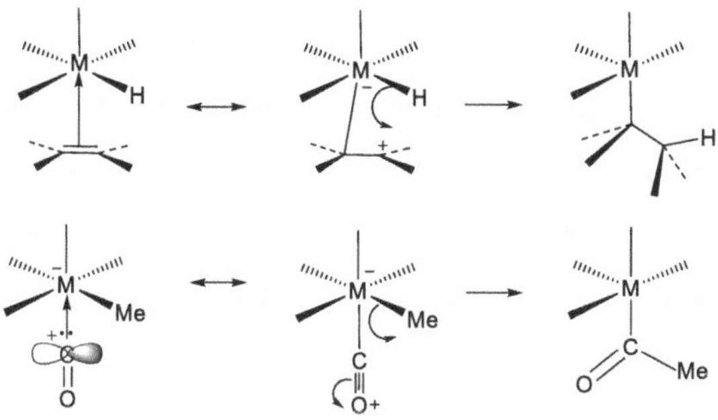

Fig. 9 Canonical forms which illustrate how the *curly arrow notation* may be used for organometallic compounds

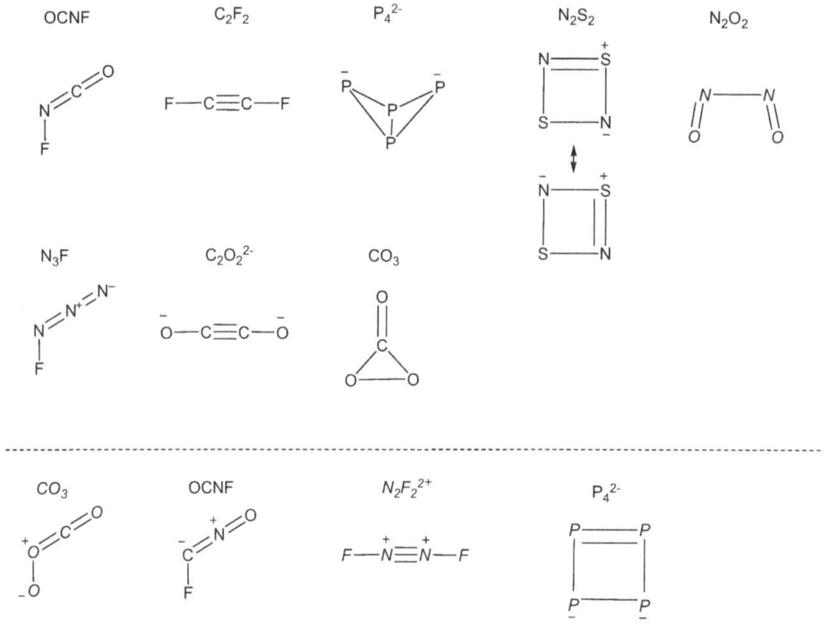

Fig. 10 Alternative structures for molecules and ions with 4 atoms and 22 valence electrons. Since the Lewis structures are based on the attainment of the inert gas configuration at each atom, all the structures have five bonds. The *structures below the line* are generally less favoured because of the dipolar nature of the structures or the occurrence of identical charges on adjacent atoms

3.2 *Isosteric and Isoelectronic Relationships*

The electron-pair/octet rule formalism shows significant limitations when applied to more complex molecules, because alternative isomeric structures, all of which are consistent with the Lewis assumptions, may be written. For example, molecules and ions with four atoms and a total of 22 valence electrons are consistent with all the Lewis structures shown in Fig. 10. Linear, angular, ring and butterfly structures occur, although each structure is associated with five covalent two-electron bonds. Consequently additional criteria are required in order to establish which possibilities are more stable. The following criteria provide a preliminary way of understanding why certain structures are preferred [98].

1. The less electronegative atom is generally located in central locations and the more electronegative atoms on the outside.
2. The heavier elements favour ring structures rather linear structures with multiple bonds.
3. Structures with uncharged atoms are generally preferred relative to charged structures, and positively charged atoms are generally disfavoured for very electronegative atoms such as F and O.
4. Structures with the same charge on adjacent atoms are disfavoured.

The structures shown above the dotted line illustrate the implications of these criteria and suggest more stable structures. In addition although the structure shown for N_2O_2 is consistent with the Lewis formalism, the weak N–N single bond means that the structure is only observed at low temperatures. The dissociation energy of the NO dimer is only 8.3 kJ mol^{-1}, and it represents an example of a molecule which is not adequately represented by Hartree–Fock molecular orbital calculations [76]. The structures below the dotted line are disfavoured because of they are dipolar or have identical charges on adjacent atoms.

The number of bonds, x, associated with a Lewis structure may be summarised by the following relationship, where the total number of valence electrons, t, and h and m represent the number of hydrogen and main group atoms:

$$x = [2h + 8m - t]/2$$

where h is the number of hydrogen atoms (EAN = 2, corresponding to the closed shell for He) and m is the number of p block atoms (which attain the eight electrons associated with the adjacent inert gas). Clearly the examples in Fig. 10 share five bonds in common because of this relationship. Some other examples of this relationship are given in Table 1 [101], and Fig. 11 illustrates how x varies systematically as the number of valence electrons t is varied. As t increases, the number of covalent bonds decreases. The table gives examples of electron precise molecules which obey the octet rule and also molecules which are apparently hypo-valent (i.e. do not achieve the octet rule) and hyper-valent (i.e. exceed the octet rule).

3.3 Hypo-valent and Hyper-valent Main Group Molecules

Table 1 gives examples of electron precise molecules where the inert gas rule is obeyed. The number of bonds is consistent with the determined structures. These molecules may have multiple bonds or single bonds. The table also gives examples

Table 1 Examples of simple molecules adhering to the effective atomic number rule (EAN)

Molecule	h	m number of main group atoms	$[2h+8m]$	t number of valence electrons	x number of bonds
Electron precise molecules (obey EAN rule)					
CH_4	4	1	16	8	4
NH_3	3	1	14	8	3
NF_3	0	4	32	26	3
NF_3O		5	40	32	4
C_2H_2	2	2	20	10	5
C_2H_4	4	2	24	12	6
C_2H_6	6	2	28	14	7
Hypo-valent molecules (short of electron pair at central atom)					
BF_3	0	4	32	24	4
B_2H_6	6	2	22	12	8
Hyper-valent molecules (excess electron pairs at central atom)					
XeF_2	0	3	24	22	1
XeF_4	0	5	40	36	2
XeF_6	0	7	56	50	3

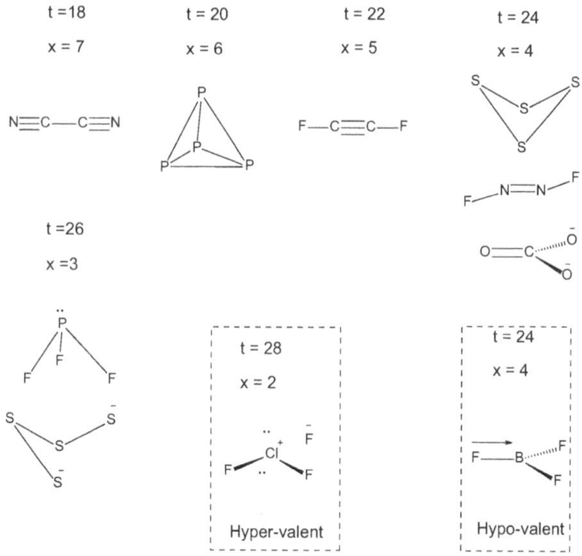

Fig. 11 Examples of molecules with the same number of atoms, but with differing numbers of valence electrons, *t*, which result in a variation in the number of bonds, *x*, to satisfy the requirements of the octet rule

of molecules which do not obey the octet rule, and the predicted number of bonds is not consistent with x number of σ-bonds. Such molecules are described as hypo-valent or hyper-valent according to whether the octet rule is not met or exceeded [102–105]. In these molecules, if the value x exceeds the number of σ-bonds, it indicates the number of additional bonds required to make the compound conform to the inert gas rule. For example, in BF_3 and BH_3, $x=4$ if the inert gas rule is

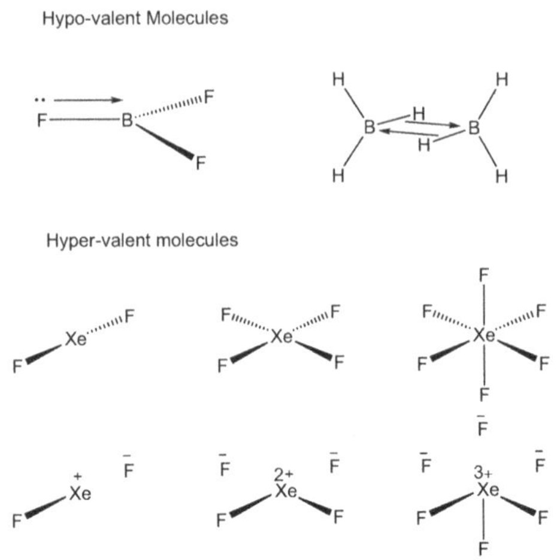

Fig. 12 The extension of the Lewis ideas for hypo-valent and hyper-valent molecules. In the former, the electron deficiency is relieved by the formation of dative bonds either from lone pairs or B–H bonds. In the latter, the hyper-valency is brought into line with the EAN rule by using Lewis bond structures where the central atom bears positive charges which compensate the negative charges on the fluorines. The results in Table 1 suggest that *x* is associated with the number of three-centre four-electron Xe–F bonds in these structures

obeyed, although there are only three B–F or B–H bonds. This deficiency may be made up by adding a dative bond as shown in Fig. 12. In BF_3 a mesomeric interaction involving a p_π orbital on F perpendicular to the plane and the empty B 2p orbital has been proposed. In BH_3 this is precluded, but an intermolecular dative bond from a B–H bond to the boron 2p orbital helps to resolve the hypo-valency. Resonance between the symmetry-related canonical forms leads to the observed structure of B_2H_6 which is more commonly described in terms of a pair of three-centre two-electron B–H–B bonds [104, 105]. Of course steric constraints which prevent polymerisation means that this way of achieving the octet rule may not be realisable.

The closed shell inert gas configuration ns^2np^6 for the noble gases is associated with a high ionisation potential and unfavourable electron affinity. Their closed electron configurations result in total spin and orbital angular momenta to be zero and therefore reduce the possibility of electron-pair interactions with electrons on other atoms. Therefore, the noble gases should provide robust examples of the octet rule by not forming compounds. Indeed it could be argued that the Lewis ideas may have hindered the discovery of noble gas compounds. Bartlett's pioneering work on xenon compounds in the early 1960s [102, 103] showed that the heavier noble gases do indeed form compounds with electronegative atoms such as F and O. Their valence electrons become involved in electron-pair bonds with electronegative atoms such as F and O when the ionisation energy of the inert gas is sufficiently low. For these hyper-valent compounds (XeF_2, XeF_4 and XeF_6), $x = 1$, 2 and 3 according to the formula given above, i.e. half the number of Xe–F bonds.

Initially these hyper-valent compounds were thought to result from the promotion of electrons from xenon valence orbitals into higher-lying d orbitals and subsequent formation of two-centre two-electron bonds with O or F, i.e. they exceed the octet rule. Contemporary interpretations of the bonding in these compounds favour the formation of three-centre four-electron bonds. This interpretation results in the ionic canonical forms illustrated in Fig. 12, which when in resonance reproduce the observed symmetric geometries. In these molecules, x represents the number of three-centre four-electron bonds in the molecule [101].

These examples illustrate the way in which chemists have modified the formal two-centre two-electron Lewis bond representations in order to extend the description to hypo- and hyper-valent compounds. It involves an extension of the ideas to encompass sharing of electron pairs between three rather than two atoms and the use of resonance in order to match the canonical forms to the symmetry of the molecule. For hypo- and hyper-valent compounds, the octet rule may be attained by forming multicentred or supplementary dative bonds.

As Moeller observed in the 1950s [106], "Although the octet rule is definitely a useful concept, its applications are limited and it should not receive the universal attention normally focussed upon it. It is much more important that attention be directed to the important phenomenon of electron pairing. The concept that the electrons seek to pair with each other is nearly universal in application and is always useful as a first approximation in predicting chemical behaviour. This "rule of two" is far more fundamental than the "rule of eight"". To contemporary sensibilities, this seems a bit harsh and generally multicentred, and dative bonds which favour adherence to the octet rule represent a convenient starting point for discussion.

3.4 *Isoelectronic Relationships*

Although the examples in the previous section have drawn attention to the possibility that molecules and ions with the same number of atoms and valence electrons may have different structures, isoelectronic relationship has proved an important way of connecting molecules with similar groups of atoms. The EAN rule or the equivalent rule based on the number of electrons in the outer shells of the inert gases emphasised the following isoelectronic relationships which proved to be particularly useful for interrelating the stoichiometries and structures of inorganic salts [42–45, 106]:

(a) No electrons H^+, D^+, T^+
(b) Two electrons: $H^- \rightarrow \mathbf{He} \leftarrow Li^+ \leftarrow Be^{2+}$
(c) Eight electrons: $N^{3-} \rightarrow O^{2-} \rightarrow F^- \rightarrow \mathbf{Ne} \leftarrow Na^+ \leftarrow Mg^{2+} \leftarrow Al^{3+}$

1. $P^{3-} \rightarrow S^{2-} \rightarrow Cl^- \rightarrow \mathbf{Ar} \leftarrow K^+ \leftarrow Ca^{2+} \leftarrow Sc^{3+}$
2. $Se^{2-} \rightarrow Br^- \rightarrow \mathbf{Kr} \leftarrow Rb^+ \leftarrow Sr^{2+} \leftarrow Y^{3+} \leftarrow Zr^{4+}$

3. $Te^{2-} \rightarrow I^- \rightarrow$ **Xe** $\leftarrow Cs^+ \leftarrow Ba^{2+} \leftarrow La^{3+} \leftarrow Ce^{4+}$
4. $At^- \rightarrow$ **Rn** $\leftarrow Fr^+ \leftarrow Ra^{2+} \leftarrow Ac^{3+} \leftarrow Th^{4+}$

It also highlighted relationships between ions which did not confirm to the EAN rule – specifically the following which had a complete d shell, but vacant s and p shells (d) or a complete s shell, but an empty p shell (e). The ions shown in (e) were described by Sidgwick as inert pair compounds [54].

(d) Filled d shell:

1. **Ni** $\leftarrow Cu^+ \leftarrow Zn^{2+} \leftarrow Ga^{3+}$
2. **Pd** $\leftarrow Ag^+ \leftarrow Cd^{2+} \leftarrow In^{3+}$
3. **Pt** $\leftarrow Au^+ \leftarrow Hg^{2+} \leftarrow Tl^{3+}$

(e) Filled d and s shell:

1. **Zn** $\leftarrow Ga^+ \leftarrow Ge^{2+} \leftarrow As^{3+}$
2. **Cd** $\leftarrow In^+ \leftarrow Sn^{2+} \leftarrow Sb^{3+}$
3. **Hg** $\leftarrow Tl^+ \leftarrow Pb^{2+} \leftarrow Bi^{3+}$

These isoelectronic relationships have resulted in recent years to compounds containing alkali metal cations, e.g. Na^-, K^-, etc., and the isolation of salts of the auride anion Au^-. They are analogues of the hydride anion [107]. As the study of transition metal and lanthanide compounds progressed, these ideas were extended to metal ions which had half-filled shells, i.e. Mn^{2+}, Eu^{2+}, Tb^{4+}, with each of the relevant d or f orbitals containing a single electron and all the electrons having parallel spins [108].

In 1919 Langmuir [42, 45] noted that molecules and ions containing the same number of atoms and the same total number of electrons invariably had identical structures. He described such series as isosteric groups and Table 2 below provides specific examples.

Isosteric and the closely related isoelectronic relationships are still widely used by inorganic chemists as an effective predictor of new molecules [109]. These isoelectronic relationships provide a good guide to the occurrence and structures of the predicted molecules, although the variation in the charges of the ions can influence their Lewis acid/base properties and their redox properties. Table 3

Table 2 Isosteric and isoelectronic molecules and ions

NO^+ N_2 CO CN^-
BH_4^- CH_4 NH_4^+
NO_2^+ N_2O CO_2 CNO^- OCN^-
NO_2^- O_3
NO_3 CO_3^{2-}
HF OH^- NH^{2-}
SO_3^{2-} PO_3^-
$S_2O_6^{2-}$ $P_2O_6^{4-}$

Table 3 Examples of isoelectronic and iso-structural main group molecules E_mXY_n ($n = 2$–6, $m = 0$–3 and represents the number of lone pairs)

Linear XY_2 molecules with 16 valence electrons			
NO_2^+	CO_2	N_2O	N_3^-
CH_2N_2	$H_2C{=}C{=}CH_2$	$H_2C{=}C{=}O$	HNCO
FCN	H_3BCN^-	H_3CCN	NCN^{2-}
	H_3BCO	BO_2^-	$H_3CC{\equiv}CH$
Trigonal XY_3 (24 valence electrons)			
BF_3	NO_3^-	CO_3^{2-}	BO_3^{3-}
F_2CO	FNO_2		
Angular EXY_2 (18 valence electrons)			
CF_2	O_3	SO_2	
$SnCl_2$	SiF_2	ClO_2^+	

Tetrahedral XY_4 (32 valence electrons)			
	BF_4^-	CF_4	NF_4^+
SiO_4^{4-}	PO_4^{3-}	SO_4^{2-}	ClO_4^-
		IO_4^-	XeO_4
Pyramidal EXY_3 (26 valence electrons)			
SF_3^+	PF_3	$SnCl_3^-$	
TeO_3^{2-}	IO_3^-	XeO_3	
Angular E_2XY_2 (20 valence electrons)			
OF_2	ClF_2^+	$SnCl_3^-$	
SF_2			

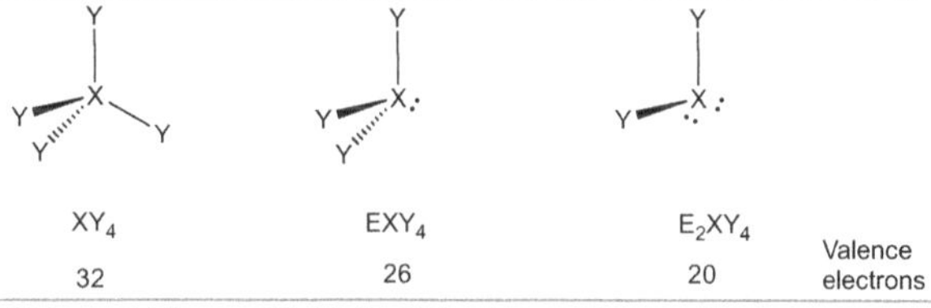

Trigonal bipyramidal XY_5 (40 valence electrons)		
AlF_5^{2-}	SiF_5^-	PF_5
GaF_5^{2-}	GeF_5^-	AsF_5
Folded square EXY_4 (34 valence electrons)		
PF_4^-	SF_4	ClF_4^+
AsF_4^-	SeF_4	BrF_4^+
$IO_2F_2^-$	XeO_2F_2	
T-shaped E_2XY_3 (28 valence electrons)		
ClF_3	XeF_3^+	
IF_3		

(continued)

Table 3 (continued)

Linear E_3XY_2 (22 valence electrons)		
IF_2^-	XeF_2	I_3^-
		Br_3^-
		Cl_3^-

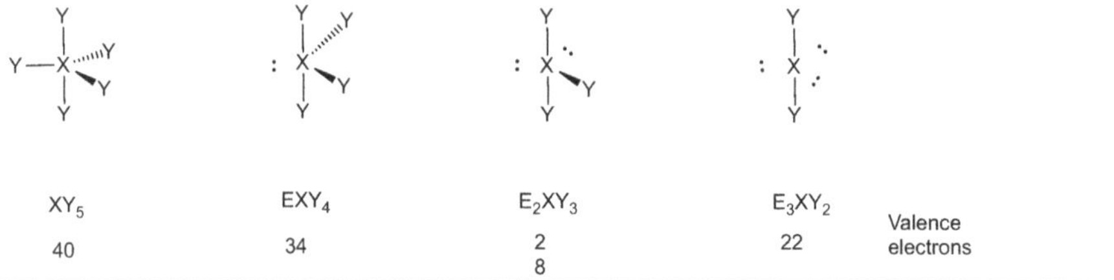

Octahedral XY_6 (48 valence electrons)			
AlF_6^{3-}	SiF_6^{2-}	PF_6^-	SF_6
	GeF_6^{3-}		
	SnF_6^{3-}		
Square-pyramidal EXY_5 (42 valence electrons)			
BrF_5			
IF_5	XeF_5^+		
Square-planar E_2XY_4 (36 valence electrons)			
ICl_4^-	XeF_4		

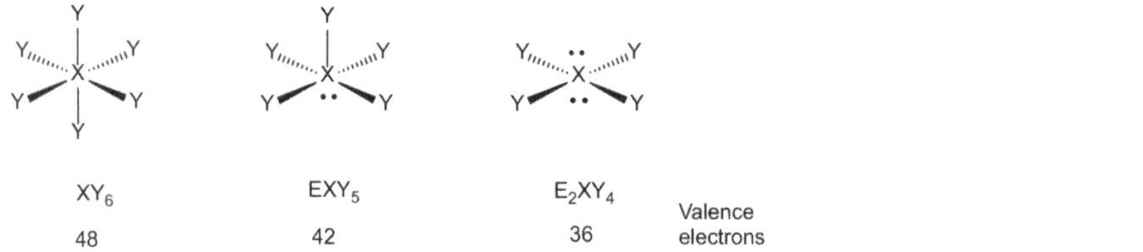

provides additional examples of inorganic main group compounds, which have the same number of valence electrons and similar geometries [109].

These isoelectronic relationships played an important role in the development of the valence shell electron-pair repulsion (VSEPR) theory and also the more general acceptance of stereochemical models based on molecular orbital models [88–92]. Within the framework of molecular orbital theory, delocalised molecular orbitals defined by symmetry considerations are calculated and then filled using the *aufbau* principle in a manner analogous to that developed for polyelectron atoms. It follows that the same basic model may be used to account for the chemical and spectroscopic properties for the iso-structural series of molecules and ions. For an isoelectronic and iso-structural series, XY_n, the same bonding molecular orbitals are occupied, but the extent of localisation on the central and peripheral atoms will reflect the electronegativity difference between X and Y. Walsh was particularly influential in introducing the molecular orbital analysis in these molecules [110–114]. Walsh diagrams were also important for highlighting differences in geometry

Table 4 Electronegativity differences [109]

Δ(Si-F) 2.4	Δ(P-F) 2.0	Δ(S-F) 1.7	Δ(Cl-F) 1.3
Δ(Si-O) 1.8	Δ(P-O) 1.4	Δ(S-O) 1.1	Δ(Cl-O) 0.7

between ground and excited states and tracing how the increase in the number of electron pairs may change as a result of increasing the number of electrons [115].

The delocalised nature of the molecular orbitals removes the strong connection inherent in Lewis structures and the valence bond method in formal bond orders. For example, the series of tetrahedral molecules SiF_4, POF_3, SO_2F_2, ClO_3F, $SiO_4{}^{4-}$, $PO_4{}^{3-}$ $SO_4{}^{2-}$ and $ClO_4{}^{-}$ are isoelectronic and have 32 valence electrons ($x = 4$ according to the formula introduced above); the replacement of fluorine by oxygen suggests increased multiple bond formation in the manner familiar to organic chemists, but within the molecular orbital framework, all the molecules have the same occupied molecular orbitals. The extent of localisation on the central atom changes according to the electronegativity difference between the central atom and the peripheral atoms (see Table 4). Also the symmetries of these molecules does not provide a clear distinction between σ and π orbitals in a way that will be familiar to organic chemists, and only precise molecular orbital calculations can provide an indication of the bond polarities and indeed the contribution made by the 3d orbitals on the central atom [116–118]. These bonding issues are discussed further in other chapters of this volume by Haaland [119] and Stalke [120].

Although isoelectronic and isosteric relationships are usually attributed to Langmuir, they were extended by Grim and Erlenmeyer. Grimm [121–123] considered all molecules with the same number of valence electrons regardless of the number of atoms involved and used isomorphism as a criterion. Erlenmeyer concluded that only the outer number of electrons should be counted in proposing iso-sterism and applied it widely to organic molecules [124].

3.5 Valence Shell Electron Pair Repulsion Theory

The observation that the lone pairs in main group molecules are stereochemically active led to the Sidgwick–Powell rules in 1940 [88], and the additional insights made by Gillespie and Nyholm [89–92] concerning the relative stereochemical roles of bond pairs and lone pairs in molecules led to its rebranding as the valence shell electron-pair repulsion (VSEPR) theory. These isoelectronic relationships may be interpreted using the valence bond model, whereby bond pairs and lone pairs occupy hybridised orbitals on the central atom [88–90]. Within the molecular orbital framework, closed shells are a consequence of the delocalised model in much the same way as found in the Schrödinger description of atoms. Furthermore, Walsh diagrams traced the evolution of molecular orbitals [110–114] as the geometries are altered and utilised the non-crossing rule to describe the mixing of MOs with the same symmetry properties. Indeed the *aufbau* filling of molecular orbitals forms an essential part of the molecular orbital methodology, and Walsh diagrams did much to educate chemists into the way in which the energies of the delocalised molecular orbitals vary as the molecular geometry was changed.

3.6 Topological Limitations of the Lewis Representations

The Lewis notation whereby a two-centre two-electron chemical bond is represented by a line was extended by organic chemists to molecules with rings, and the curly arrow notation successfully accounted for the substitution patterns of aromatic compounds [63–72]. These developments were supported theoretically by Pauling's use of the concept of resonance which developed out of the valence bond model [55–62]. For example, the resonance between the two alternative canonical forms in Fig. 13 could partially account for the stability and aromatic properties of benzene and related molecules. However, as early as 1931, Hückel established [125] that the cyclic-conjugated polyenes such as those shown in Fig. 13 were most stable if there were $4n+2$ electrons involved. The prime example is of course benzene where $n=1$ and $4n+2=6$, i.e. it has an aromatic sextet. However, cyclo-butadiene and cyclo-octatetraene are anti-aromatic because they have $4n$ electrons. This highlighted the limitations of basing arguments on resonance forms such as those shown at the top of Fig. 13. Superficially the resonance forms of C_4H_4 and C_8H_8 would appear to result in resonance stabilisation, but the resonance forms shown do not lead to cyclic delocalisation. Hückel's approach was further vindicated by the isolation of salts of the cyclopentadienyl anion and the tropylium cation which also had 6π valence electrons. The isolation of transition metal sandwich compounds of the cyclopentadienyl anion such as ferrocene, which undergoes the characteristic electrophilic substitution reactions, and the cyclo-octatetraene dianion such as uranocene underlined the limitations of the naive use of resonance arguments [126]. Pauling noted in the Third Edition of Nature of the Chemical Bond (1960) that hundreds of resonance structures were required to adequately describe the structure. These examples emphasise the dangers of necessarily linking the Lewis structures to topological features of the molecules

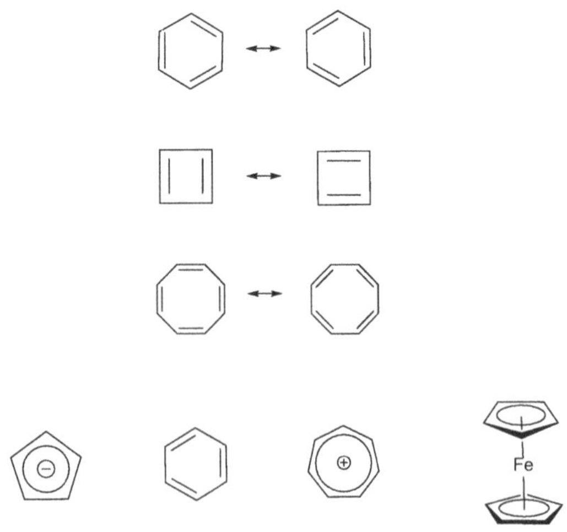

Fig. 13 Alternative resonance forms for benzene (aromatic), cyclo-butadiene and cyclo-octatetraene (anti-aromatic). Other aromatic cyclic hydrocarbons and ferrocene are illustrated at the *bottom*

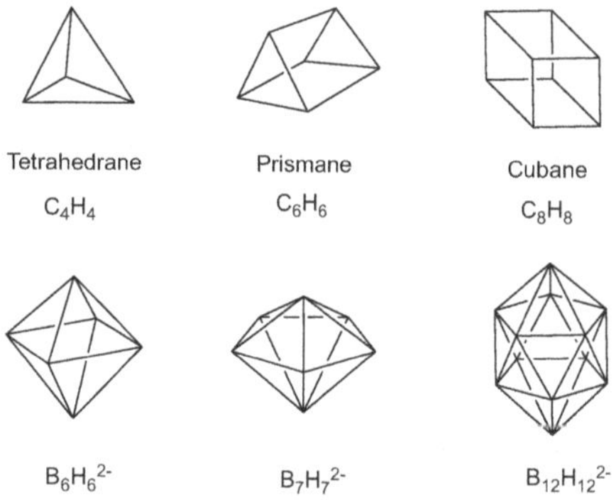

Fig. 14 Examples of polyhedral main group molecules. The hydrocarbon series shown at the *top* retains a connection between the number of C–C two-centre two-electron bonds and the edges of the polyhedron. This connection is lost for the deltahedral borane anions shown in the *bottom line*

[55, 60, 61]. When there is extensive delocalisation, it is more appropriate to use free electron-based molecular orbital models.

A similar story emerged from study of molecules which have three-dimensional structures where extensively delocalisation occurs. The Lewis description provides a good description of organic rings and three-connected polyhedral C_nH_n molecules which are illustrated in Fig. 14. The small angles in those molecules with triangular and square faces did require organic chemists to consider the chemical consequences of strained or bent C–C bonds, but essentially the Lewis picture holds. However, the topological connection between the number of edges of the polyhedron and the number of Lewis two-centre two-electron bonds is no longer valid for deltahedral boranes. Longuet–Higgins was the first to develop a three-dimensional analogue of the Hückel approximation [127] which satisfactorily accounted for the bonding in the borane anions $B_nH_n^{2-}$ and the boride anions B_6^{2-} and B_{12}^{2-}. These spherical deltahedra are characterised by $n+1$ bonding skeletal molecular orbitals, which have no direct geometric connection with either the number of faces $(2n-4)$ or edges $(3n-6)$ of deltahedra. This approach was subsequently extended to deltahedral borane anions and formed the basis of the polyhedral skeletal electron pair theory (PSEPT), which provided a three-dimensional analogue of the VSEPR theory and an extension to three dimensions of Hückel's two-dimensional cyclic delocalisation theory. Wade and Mingos in the 1970s were the primary contributors to these developments [128–133]. The connection with the Hückel approach was underpinned by Stone who developed a free-electron spherical harmonic model for deltahedral main group and transition metal clusters in the 1980s [134, 135].

3.7 *Isolobal Analogies*

The occurrence of iso-structural main group and transition metal organometallic compounds which both conform to the octet or 18-electron rules suggests possible electronic connections between the two classes of compound. The main group compounds illustrated in Fig. 14 are characterised by the following total valence electron counts: three-connected $5n$ electrons (e.g. C_nH_n) and deltahedral $4n+2$

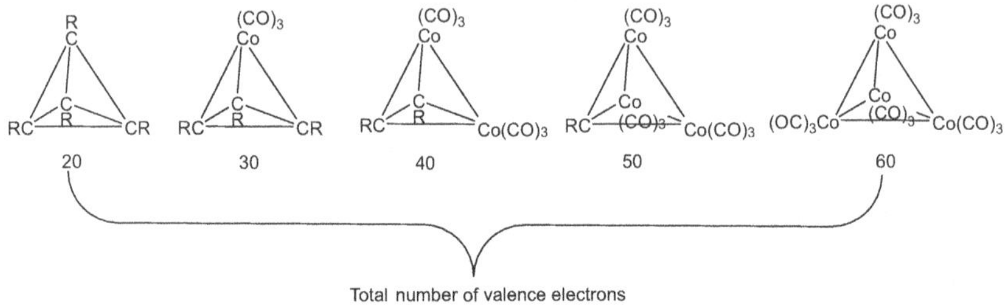

Fig. 15 Examples of iso-structural clusters, where the total number of valence electrons increments by 10 electrons for each transition metal atom introduced

electrons (e.g. $B_nH_n^{2-}$). Transition metal carbonyls form analogous compounds with $15n$ valence electrons for three-connected polyhedral molecules, e.g. tetrahedral-$[Rh_4(CO)_{12}]$ and $14n+2$ for deltahedral polyhedral molecules, e.g. octahedral $[Co_6(CO)_{14}]^{4-}$. These relationships suggest a series of iso-structural compounds whose total valence electron count increments by 10 electrons for each transition metal fragment introduced into the main group polyhedral skeleton. Figure 15 provides specific examples of tetrahedral clusters which conform to this generalisation.

This connection between main group and transition metal clusters has been underpinned by the *isolobal* analogy, which relates the properties of the constituent fragments to the symmetries and energies of the frontier orbitals [136, 137]. Interestingly the location of the terminal and bridging carbonyl ligands does not have to be defined for the relationship to work. Indeed it was shown at an early stage that the isolobal analogy depends primarily on the commonality of the antibonding molecular orbitals, which remain unoccupied in both series of molecules [131].

4 Core and Valence Electrons

The quantum mechanical description of the atom which resulted from Schrödinger's wave mechanical description of the hydrogen atom led to a much deeper understanding of electronic factors which were responsible for Mendeleev's periodic classification and Lewis and Kossel's description of the chemical bond and valency [138]. Since the Lewis approach makes a clear distinction between core and valence electrons, it was important to establish whether the quantum mechanical description and the associated spectroscopic measurements supported this assumption. When the Schrödinger model was extended to the polyelectron atom, the angular parts of the solution were assumed to a first approximation to be the same as those developed for the hydrogen atom, but the radial part were altered to take into account the differences in nuclear charge and electron repulsion effects. These effects remove the degeneracies of the hydrogen atom orbitals with the same principal quantum number n but different l quantum numbers, i.e. the energies of ns–np–nd, etc., are no longer equal. Spectral studies on electronic transitions utilised the symmetry consequences of the model to confirm the ordering of energy levels in atoms and to provide empirical evidence regarding the definition of core and valence electrons in atoms [115]. These experimental results were supported by quantum mechanically based calculations. Theoretical studies provided information concerning the probabilities of the electron distribution in atomic orbitals and led to the definition of r_{rmax}, the most probable orbital radius. The development of reliable X-ray crystallography led to data on interatomic distances in metals, salts and molecular compounds. For example, in a crystalline metal, the metallic radius of a metal atom may be defined as ½ the internuclear distance between neighbouring metal atoms in the solid state. This internuclear distance correlates well with the calculated r_{max} of the valence orbitals of metals as shown in Fig. 16. Therefore, it is not unreasonable that maximum overlap between orbitals of adjacent metal atoms occurs when the metal atoms are separated by a distance close to $2r_{\mathrm{rmax}}$ for the ns valence orbitals. The core electrons do not contribute significantly to the metal–metal bonding because their orbital radii are much smaller as shown in Table 5. The cores have volumes between one and two orders of magnitude smaller than those of the valence electrons. The empty np valence orbitals of the alkali metals are capable of overlapping with the ns and np orbitals of adjacent atoms, but their contribution does not manifest itself in the observed bond lengths.

For elements which lie towards the centre of a row of the periodic table, the ns and np valence orbitals both contribute significantly to the bonding, and plots of ns and np r_{max} against the observed covalent radii suggest that the bonding is increasingly dominated by the np orbitals (see Fig. 17) for the heavier elements of the group. For carbon both 2s and 2p contribute significantly to covalent bonding, but

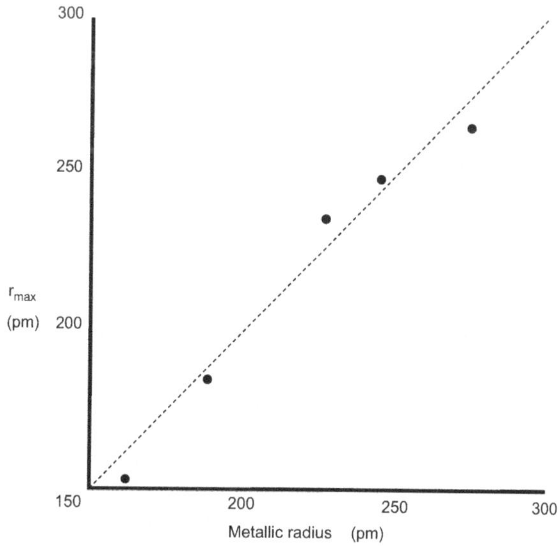

Fig. 16 Plot of r_{max} and metallic radius for alkali metals

Table 5 Values of r_{max} (relative to that for H $r_{max} = 52.918$ pm) for the alkali metal atoms. The bold numbers refer to the valence orbitals

Orbitals / Atom	1s	2s	3s	4s	5s	6s
H	**1**					
Li	0.364	**3.101**				
Na	0.093	0.607	**3.387**			
K	0.053	0.317	1.078	**4.330**		
Rb	0.026	0.149	0.450	1.270	**4.650**	
Cs	0.017	0.095	0.272	0.643	1.562	**5.138**

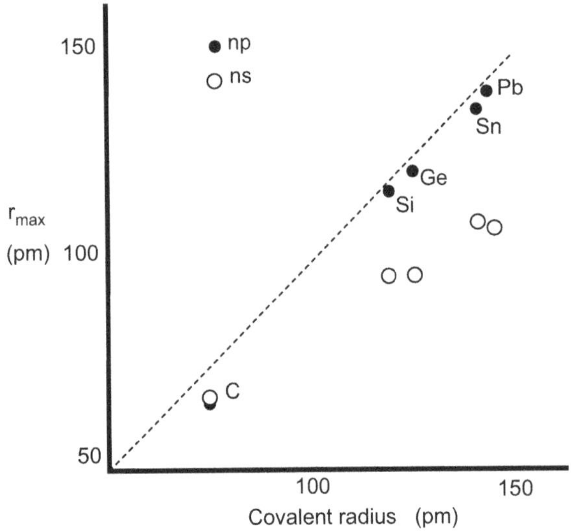

Fig. 17 Plots of r_{max} and structural determined covalent radii for the group 14 elements C, Si, Ge, Sn and Pb

the ns character in the bonds decreases for the heavier elements. This reflected in the bond angles in XY_n molecules when $n = 2$ or 3 [138].

Recent theoretical studies and accurate experimental determinations of electron densities in molecules have confirmed that the majority of electrons do indeed form a concentrated core near the nucleus which appears very atomic like [120]. The electron density is very monotonic as the radial distance from the nucleus increases. Chemists sometimes wrongly consider that the electron density in the cores of atoms has outer maxima corresponding to the shell structures. The inner electron cores are almost transferable entities and consequently endorse the valence-care partition proposed by Lewis and Kossel, and this property is utilised in frozen core approximations [138]

A pseudo-potential is an effective potential which effectively replaces the atomic all-electron potential such that core states are eliminated and the valence electrons are described by pseudo-wave functions with significantly fewer nodes. Only the chemically significant valence electrons are dealt with explicitly, while the

Table 6 Influence of sp^x hybridisation on C–C bond lengths

Bond type	C–C bond length (pm)
sp^3-sp^3	154
sp^3-sp^2	150
sp^3-sp	146
sp^2-sp^2	147
sp^2-sp	143
sp-sp	137

core electrons are "frozen". They and the nuclei are treated as rigid non-polarisable ion cores. This frozen core constraint may be relaxed by self-consistently updating the pseudo-potential with the chemical environment that it is embedded in. Pseudo-potentials derived from first principles are based on an atomic reference state, which requires that the pseudo- and all-electron valence eigenstates have the same energies and density outside a chosen core cut-off radius.

Although Lewis assumed that all the valence electrons in the outer shell of an atom had an equal bonding capability, this view has had to be modified in the light of subsequent theoretical and empirical knowledge. Specifically the bonding capabilities of the valence electrons depend not only on the principal quantum number but also the specific valence orbitals. Table 6 summarises the way in which the hybridisation of orbitals in organic compounds which have C–C single bonds is influenced by the hybridisation of the carbon atoms. The bond lengths decrease as the amount of s character is increased [139].

The differences resulting from variations in the radial distribution functions of different valence orbitals become more pronounced for atoms which have nd and nf valence orbitals, because they are significantly more contracted than the (n+1)s and (n+1)p orbitals and therefore behave in a more core like manner [140].

The transition metals have nd, (n+1)s and (n+1)p valence orbitals, but their radial distribution functions are very different [141]. For chromium the maximum of the radial distribution function of the 3d orbital is 46pm, whereas for the 4s orbital, it is 161, i.e. more than three times larger. The much more contracted nature of nd relative to (n+1)s and (n+1)p means that they are much more core like. Consequently the compounds of the transition metals show a duality of properties. For larger ligands which do not overlap strongly with the nd orbitals, they behave as if the nd orbitals are core like and form series of compounds which have similar structures. These compounds have been described as iso-stoichiometric compounds [101]. Examples of metal halides which are iso-stoichiometric are summarised in Tables 7, 8, 9, and 10. In these tables, ticks represent known compounds and crosses unknown compounds. Ligands which have smaller radii or form strong multiple bonds with the metal are capable of overlapping strongly with the nd orbitals and form strong covalent bonds. Such compounds obey the EAN rule, and some specific examples are summarised in Table 10.

Consequently transition metal carbonyls behave like main group compounds where the stoichiometry changes to satisfy the inert gas rule, and this similarity underpins the *isolobal* relationships described in Sects. 3.6 and 3.7. The f orbitals of

Table 7 Iso-stoichiometric divalent metal dihalides

	Sc	Ti	V	Cr	Mn	Fe	Co	Ni	Cu	Zn
MF_2	x	x	√	√	√	√	√	√	√	√
MCl_2	x	√	√	√	√	√	√	√	√	√
MBr_2	x	√	√	√	√	√	√	√	√	√
MI_2	x	√	√	√	√	√	√	√	√	√

Table 8 Iso-stoichiometric trivalent metal halides

	Sc	Ti	V	Cr	Mn	Fe	Co	Ni	Cu	Zn
MF_3	√	√	√	√	√	√	√	√	x	x
MCl_3	√	√	√	√	√	√	x	x	x	x
MBr_3	√	√	√	√	x	√	x	x	x	x
$MI_{3\surd}$	x	√	√	√	x	√	x	x	x	x

Table 9 Iso-stoichiometric metal complexes

	Sc	Ti	V	Cr	Mn	Fe	Co	Ni	Cu	Zn
$M(OH_2)_6^{2+}$	x	x	√	√	√	√	√	√	√	√
$M(CN)_6^{2-}$	√	√	√	√	√	√	√	√	√	√
$M(CN)_6^{3-}$	x	√	√	√	√	√	√	√	√	√
$M(acac)_3$	√	√	√	√	√	√	√			
M(edta)	x	√	√	√	√	√	√	√	√	√
$M(bipy)_3^{3+}$	√	√	√	√	√	√	√			
$M(phen)_3^{3+}$	√	√	√	√	√	√	√	√		

Table 10 Examples of metal complexes which conform to the EAN rule

	V	Cr	Mn	Fe	Co	Ni
$M(CO)_n$		$Cr(CO)_6$	$Mn_2(CO)_{10}$	$Fe(CO)_5$	$Co_2(CO)_8$	$Ni(CO)_4$
$M(CO)_n^{x-}$	$V(CO)_6^-$	$Cr(CO)_5^{2-}$	$Mn(CO)_5^-$	$Fe(CO)_4^{2-}$	$Co(CO)_4^-$	
$MCp(CO)_nH$		$CrHCp(CO)_3$	$MnCp(CO)_3$	$FeHCp(CO)_2$	$CoCp(CO)_2$	
$M(CO)_n(NO)$		$Cr(NO)_4$	$Mn(NO)_3(CO)$	$Fe(CO)_2(NO)_2$	$Co(NO)(CO)_3$	$Ni(CO)_4$

the lanthanides are even more contracted and are consequently more core like. In summary the distinction between valence and core electrons initially made by Lewis and Kossel can be justified and is generally applicable for the representative elements, but has its limitations when applied to elements with partially filled d and f orbitals, and the transition metals show Jekkyl and Hyde characteristics which depend on the abilities of the ligands to overlap with the d orbitals.

5 Odd Electron Molecules

Lewis noted in his 1913 paper [2] that the great majority of molecules have an even number of electrons, and this formed a pivotal observation in developing the concept of the electron-pair bond. He noted that NO and NO_2 provided rare examples of exceptions and that steric effects were important in influencing whether odd electron molecules could be isolated. In recent years, the incorporation of sterically demanding ligands has resulted in a wide range of molecules which either do not confirm to the Lewis description or are paramagnetic [142]. Furthermore, the study of molecules in the gas phase and under matrix isolation conditions has resulted in the structural characterisation of molecules which are not necessarily the most stable molecules under standard conditions. Also spectroscopic methods have resulted in the study of molecules in their excited states, and the geometries established by these studied have provided additional challenges for the Lewis–Kossel model, because the molecules have different metrics and at times geometries in their excited states.

The transition metals, lanthanides and actinides provide many examples of compounds with unpaired electrons because of partial core character of the nd and nf subshells for these groups of elements, as discussed above.

6 Quantum Mechanical Description of the Chemical Bond

A detailed historical account of the early development of quantum chemistry has been written by Gavroglu and Simoes [142]. Burrau published the first mathematically complete quantum description of the simplest chemically bond species, i.e. the hydrogen molecular ion, H_2^+, in 1927 [143, 144]. This demonstrated that a quantum mechanical description of the chemical bond was viable, but the mathematical methods used could not be easily extended to molecules containing two or more electrons. A more flexible, but less exact, approach was proposed in the same year by Heitler and London, and this formed the basis of what came to be described as the *valence bond theory* [145, 146]. It recognised the important contributions made by Pauli [147, 148]. The *Pauli exclusion principle* states that two particles with half-integer spin cannot occupy the same quantum state simultaneously. For electrons in atoms and molecules, it is impossible for two electrons to have the same values of the four quantum numbers (n, ℓ, m_ℓ and m_s). For two electrons residing in the same orbital, n, ℓ, and m_ℓ are the same, so m_s must be different and the electrons have opposite spins. The tendency of molecules to have electron pairs is due to the Pauli principle which allows a maximum of two electrons in the same region of space. A chemical bond can be formed with only one electron such as in H_2^+. Thus, chemical bonding is not due to the formation of an electron pair as suggested by Lewis but a result of the overlap of wave functions. Concurrently the linear combination of atomic orbitals *molecular orbital method*

(LCAO) approximation was developed by Hund, Urey and Lennard–Jones [149–151], who also suggested methods to derive the electronic structures of F_2 (fluorine) and O_2 (oxygen) molecules, from basic quantum principles. All of these approximate descriptions of chemical bonding were based on the Schrödinger quantum mechanical description of atomic orbitals which had been developed exactly for the hydrogen atom and subsequently extended in an approximate fashion to polyelectron atoms [152, 153]. The equations for bonding electrons in multielectron atoms could not be solved analytically, but approximations gave many qualitative predictions and results, which could be used by chemists.

In 1933 James and Coolidge [154] convinced the scientific community that quantum theory could give results in agreement with experiment for simple molecules such as H_2. Unlike all previous calculations, which used functions which depended only on the distance of the electron from the atomic nuclei, they used functions which also explicitly added the distance between the two electrons. With up to 13 adjustable parameters, they obtained a result very close to the experimental result for the dissociation energy. Later extensions have used up to 54 parameters and gave excellent agreement with experiments. However, this approach quantitatively accounted for the energetics and metrics of chemical bonds at the expense of diminishing the characteristic features of the valence bond and molecular orbital models.

6.1 Valence Bond Model

According to the Heitler–London model [145, 146], a covalent bond is formed by the overlap of half-filled valence atomic orbitals from two atoms, each bearing one unpaired electron. Their calculations for two neutral hydrogen atoms, which took into account the interchangeability of the two electrons as the atoms approached each other, showed that a more stable (bonding) situation resulted only when the spins of the electrons were antiparallel. Lewis' electron-pair model therefore required not only a pair of electrons but also that they had opposite spins. Thus, covalent bonding was represented as a purely quantum mechanical effect since electron spin had no classical analogue. Their calculations gave a reasonable account of the dissociation energy and internuclear distance in the hydrogen molecule. The overlapping of wave functions from the two atoms was associated with an increase in electron density in the bond region when the electrons are antiparallel. For more complex compounds, the theory focuses on pairs of electrons within molecules (the perfect pairing approximation) and suggests that these are dominant in the formation of bonds. When several perfectly paired structures have similar energies, the molecule is allowed to resonate among them and the total energy is lowered. Mathematically the resonance is represented by adding in the additional wave functions. A valence bond structure therefore incorporated the electron pairing central to the Lewis covalent bond, but if a Lewis structure cannot be written, because it does not adequately represent the symmetry or polarity of the molecule, then several valence bond wave functions are used.

Pauling developed the Heitler–London model stressing the importance of the shared electron-chemical bond and emphasised that chemical valence was the result of the Pauli exclusion principle and the quantum mechanical resonance phenomenon. He noted "*I consider myself not a stranger bringing something from outside, but rather a member of a group carrying on the work begun by Professor Lewis in 1916, for ever since I first learned of the electron pair bond, in 1920, I have devoted my efforts to attempting to understand the properties of substances from this viewpoint*". Pauling and Slater [152, 153] almost simultaneously published papers which had many common features, but Pauling was clearer in his desire to propose a universal solution. "The quantum mechanical equations permit the formulation of an extensive and powerful set of rules for the electron pair bond supplementing those of Lewis. These rules provide information regarding the relative strengths of bonds formed by different atoms, the angles between bonds, free rotation or lack of free rotation about bond axes, the relation between the quantum numbers of bonding electrons and the number and spatial arrangement of bonds". Pauling utilised the angular parts of the Schrödinger wave function of an atom to provide a simplification which yielded many important new insights into the properties and strengths of bonds. Pauling's clarity of presentation and his encyclopaedic knowledge of chemical facts provided an important role in bridging the gap between chemistry and the new quantum theory developed by physicists.

Pauling showed that the bond strength of a covalent bond was greater if two atomic orbitals are directed so that maximum electron density is located between the nuclei. He utilised the angular parts of the solutions to the Schrödinger equation to develop sp^3, sp^2 and sp hybridised orbitals which accounted for the geometries of main group and organic molecules and d^2sp^3 and dsp^2 hybrid orbitals for transition metal complexes. These hybrid orbitals are constructed using the orthogonality relationships, which result from the wave equations for the angular parts of the Schrödinger solutions. They arise naturally out of the quantum mechanical wave description, which are expressed as spherical harmonic functions. Pauling also extended the concept of resonance to account for the delocalisation of electrons in conjugated π-systems and the polarity of bonds. This required the introduction of an electronegativity scale to describe the relative abilities of atoms to attract electrons [55–62].

Pauling's *Nature of the Chemical Bond* published in 1938 [55, 60, 61, 119] had an immense impact on chemical thinking. It introduced chemists to the importance of the quantum culture by developing the concepts of resonance and underscoring the importance of atomic orbital overlaps and hybridisation in influencing the strengths and geometries of molecules. It also developed empirical principles to underline the transferability of bond lengths and energies and the importance of electronegativity in defining the partial ionic character of bonds. He extended Kossel's ionic bonding model and articulated a set of principles for understanding the structures of infinite solids and their important metrics. He underscored the importance of hydrogen bonding not only in these compounds but also in the structural principles, underlying the folding of proteins and enzymes.

The early popularity of valence bond methods declined, because the molecular orbital calculations were more amenable to writing efficient computer algorithms. Recently the programming of valence bond methods has improved because it also benefitted from the rapid development of computer technology and programs have been developed which are competitive in accuracy and economy with programs for the Hartree–Fock method and other molecular orbital-based methods. These developments are due to and have been described by Gerratt, Cooper, Karadakov and Raimondi and summarised by Li and McWeeny in 2002, van Lenthe and co-workers in 2005 and Shaik and Hiberty's reviews in 2008 [155–158].

In its simplest form, the overlapping atomic orbitals are replaced by orbitals which are expanded as linear combinations of the atom-based basis functions, forming linear combinations of atomic orbitals (LCAO). This expansion is optimised to give the lowest energy. Modern valence bond theory is thus an extension of the Coulson–Fischer method [159]. This procedure gives good energies without including ionic structures. Modern valence bond theory replaces the simple linear combination of the two atomic orbitals with a linear combination of all orbitals in a larger basis set. The two resulting valence bond orbitals look like an atomic orbital on one hydrogen atom slightly distorted towards the other hydrogen atom.

The *generalised valence bond* (GVB) method, developed by Goddard in 1970, is one of the simplest and oldest valence bond methods that use flexible orbitals in a general way. The generalised Coulson–Fischer theory for the hydrogen molecule mentioned above is used to describe every electron pair in a molecule. The orbitals for each electron pair are expanded in terms of the full basis set and are non-orthogonal. Orbitals from different pairs are forced to be orthogonal. This condition simplifies the calculations but may lead to some difficulties [160, 161].

6.2 Molecular Orbital Theory

Mulliken's interest in the electronic levels in molecules in the 1920s was stimulated by suggestions that their molecular spectra bore similarities to atomic spectra and definite relationships could be discerned for isosteric molecules. He found that the spectroscopic analogy between isosteric molecules could be extended to atoms with the same number of electrons, and this relationship was to lead subsequently to the united atom approach. He and Birge classified the electronic states in diatomic molecules using the same Russell–Saunders classification used previously for atomic states. Hund's theoretical analysis [149, 162–166] of the nature of electronic states in molecules therefore proved to be timely for Mulliken and led him to publish [167–171] a summary of the theory and provide extra experimental evidence supporting it. In the *molecular orbital theory* Hund showed how the concept of atomic orbitals and the mathematical procedures developed to define them could be extended to molecules. However, in a molecule, the electrons experience the fields exerted by all the nuclei and the other electrons. The resulting molecular orbitals, which in the simplest approximation are described as linear combinations of the atomic orbitals from each centre, are delocalised over the whole molecule. Each molecular orbital is defined by its energy, and its symmetry properties are defined by the point group of the molecule. The symmetries could be related to those of spherical atoms using the descent of symmetry procedures developed from group theory. In a one-electron approximation, the energies of the molecular orbitals are determined by their nodal characteristics and their distribution over the molecules. In more sophisticated manifestations, interelectron repulsion and correlation effects are introduced. The overlap between the orbitals and the electronegativity difference between the atoms therefore influence the distribution of the electron density. The resultant spectrum of molecular orbitals is occupied by the total number of electrons in the molecule using the *aufbau* procedure, and the wave functions obey the Pauli exclusion principle. The most stable molecular orbitals have few radial and angular nodes and are described as bonding molecular orbitals, and as more nodes are introduced, the orbitals become progressively more antibonding. In simple diatomic molecules, the energy separation between bonding and antibonding molecular orbitals is influenced primarily by the overlap between the constituent atomic orbitals and the electronegativities of the atoms. In between these two classes of molecular orbitals, non-bonding molecular orbitals are observed which generally have nodes lying in positions which nullify next-neighbour orbital overlaps. An approximate connection is made with the Lewis bond description by taking the difference in the number of doubly occupied bonding and antibonding molecular orbitals. This often coincides with the number of electron-pair bonds in the Lewis description. Occupation of non-bonding molecular orbitals has a neutral effect, and therefore occupation of these non-bonding MOs may lead to a series of related molecules and ions with the same formal bond order, e.g. the allyl cation, radical and anion have the same formal C–C bond order because the addition electrons occupy a non-bonding π-molecular orbital localised on the outer carbon atoms. Mulliken acknowledged Lewis' contribution as follows:

> The best chemical theory of valence covering all types of compounds is generally agreed to be that developed by G.N. Lewis. To a rather large extent, the essential features of this theory still stand, although their meaning has been made clearer and more specific by interpreting them in the light of the quantum theory.

He, somewhat mischievously, made the following comment on the relationship between his molecular orbital analysis and the Lewis electron-pair model: "*Now I have a favourite argument that Lewis' electron pair bonding is better described by a pair of electrons in a molecular orbital than by the Heitler-London bond. If the chemical bond has any polarity, it is necessary to add an ionic term, that is a Heitler-London plus an ionic term, to represent the bond. That is rather a messy description whereas the molecular orbital- this is not the spectroscopic but the chemical molecular orbital, the delocalized molecular orbital* fits very nicely to the

Lewis concept". He reserved his major criticisms for the Heitler–London, Pauling–Slater (HLPS) approach :

> The theories of HLPS might be called electron-pairing theories if Lewis' is called an electron-pair theory. It should also be pointed out that the HLPS electron pair differs considerably from Lewis' conception of the electron-pair bond, in that the electrons are much less closely associated; in this respect it approaches the truth much more closely than does Lewis' conception – Pauling and Slater consider a double bond to be merely two ordinary single bonds sticking out from each atom in different directions, and treat the triple bond in a similar way. In this way they do not agree very well with Lewis, nor do they agree with the results obtained from molecular orbital theory.

Given its close origins with spectral analysis, the molecular orbital description was more widely used in the early days to describe the electronic structures of molecules in their excited states, i.e. molecules where one or more electrons have been promoted from the ground state bonding and non-bonding molecular orbitals to antibonding molecular orbitals. In these excited states, the bond lengths change to reflect the different occupations of bonding and antibonding molecular orbitals, and also for polyatomic molecules, the geometry of the excited state may differ from that of the ground state [115].

The initial molecular orbital diagrams for polyatomic molecules were confirmed by Mulliken using molecular spectral data and were subsequently verified by self-consistent field molecular orbital calculations. Mulliken also developed the correlation diagram approach which used symmetry arguments and the non-crossing rule. Such diagrams show how the molecular orbital energies change as a function of the internuclear distances, R. When R is very large, the molecular orbitals must be the same as those of the isolated atoms and progress through the molecular orbitals for the polyatomic molecule with the appropriate internuclear distances R_{equil}, and as R tends to zero, the molecular orbitals must morph into the atomic orbitals of the united atom. The correlation of those molecular orbitals with the same symmetry characteristics leads to the correlation diagram. These correlation diagrams must conform to the non-crossing rule which states that atomic and molecular orbitals with the same symmetries are not able to cross. This type of analysis provides a theoretical framework for understanding why the inert gas rule is applicable to molecules as well as atoms [172, 173]. The following series of molecules illustrate the isoelectronic relationships:

8 valence electrons CH_4 NH_3 OH_2 FH Ne

These orbital correlation diagrams subsequently played a very important role in the development of Walsh diagrams and most importantly in the elucidation of the orbital symmetry rules developed by Woodward and Hoffmann which accounted for the stereochemistries of pericyclic reactions of organic molecules [174–184].

The development of molecular orbital theory in the 1950s followed two distinct paths. At that time, digital computers (and indeed electronic calculators) were not widely available and quantitative calculations were very time consuming, and the conceptual developments in organic chemistry were largely based on the Hückel approximation. Nevertheless Longuet–Higgins and Dewar [185–189] were able to provide a strong conceptual molecular orbital-based model for dealing with the bonding in organic π-systems, which utilised the alternant properties of these molecules. Their application of perturbation theory ideas resulted in an approximate and largely pictorial methodology. Dewar (not known for his understatement) noted in his book "*that the general theory of organic chemistry (based on perturbation theory) was no harder to apply than resonance theory, but infinitely more powerful and versatile*" [185]. Similar perturbation theory arguments were developed by Fukui, and his studies emphasised the important role played in the reactions of organic molecules by their frontier molecular orbitals [177–180, 183, 184]. The energies and nodal characteristics of the highest occupied molecular orbital (HOMO) and the lowest unoccupied molecular orbital (LUMO) were to be particularly important in analysing organic reactions, and these acronyms have become part of the language of the modern chemist. Hoffmann simultaneously developed the extended Hückel methodology to the σ-bonds of organic molecules and provided a platform for confirming the orbital symmetry ideas developed with Woodward [190, 191]. Hoffmann and Fukui went on to share the Nobel Prize for their contributions. Pearson [192] subsequently showed that the Lewis bond notation could be incorporated into the Woodward Hoffmann rules by taking into account the symmetries of the linear combinations of the frontier orbitals. Since that time, Hoffmann has used the extended Hückel methodology very successfully to analyse a wide range of valence problems in coordination, organometallic and solid state chemistry [193, 194].

6.3 Synergic Bonding Models

The proposal that the bonding in Lewis acid–base pairs may be described as a polarised covalent bond had been recognised soon after the publication of the Lewis papers in 1916, and Sidgwick played an important role in providing a suitable notation and applying it to Werner's coordination compounds. The transition metal carbonyls, e.g. $Ni(CO)_4$, $Fe(CO)_5$ and $Cr(CO)_6$, did conform to the EAN rule, but did provide some difficulties for the electroneutrality principle. Electron diffraction studies showed that the M–C bonds were shorter than those suggested from covalent radii considerations and the C–O bond lengths only slightly longer than that in free CO. Pauling in the Second Edition of the *Nature of the Chemical Bond* [55, 60, 61] proposed that the bonding "involves *a double bond from nickel to carbon*" (the italics are his) and proposed that the canonical form shown in Fig. 18 made a significant contribution. He added "This structure is most satisfying than the single bonded structure for the following reason: it makes the nickel atom and the other atoms neutral, whereas the single-bonded structures place a fourfold negative charge on the nickel atom, is in its general behaviour electropositive rather than electronegative". Pauling's suggestion held sway throughout the 1940s, and Hieber

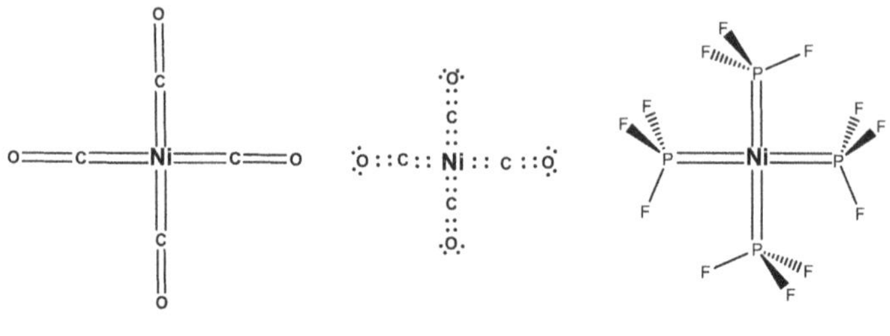

Fig. 18 Pauling's proposed Lewis structure for $Ni(CO)_4$ and its adaptation to the related PF_3 complexes

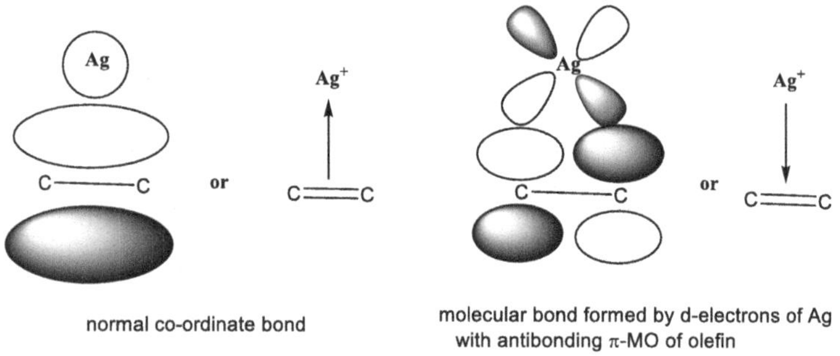

Fig. 19 Dewar's illustration of the synergic bonding in $[Ag(C_2H_4)]^+$ (his +/– signs for designating the phase relationships have been replaced by *shading*)

[195, 196], the leader in the field, accepted the possibility of double bonding and felt they were consistent with dipole moment data. Chatt [197–199] and Wilkinson [200] adapted the Pauling model to PF_3 analogues of carbonyls, by invoking the participation of the vacant 3d orbitals on phosphorus (see Fig. 18). Therefore, the possibility that certain ligands could function as both Lewis bases and Lewis acids gained currency. However, the majority of inorganic chemists did not have the theoretical background to reformulate the model within a molecular orbital framework, which required an understanding of symmetry constraints governing orbital interactions. In April 1950 at a conference held in Montpelier, Dewar presented the paper "A review of π-complex theory" and in response to a question developed the synergic bonding model illustrated in Fig. 19 [201, 202]. This led to the detailed and complete development of the synergic bonding model for metal–alkene complexes, which has become an important cornerstone of organometallic chemistry. Dewar expresses the basics very completely:

> The d electrons in heavy metals, bromine, etc. have the correct symmetry to interact with the antibonding π-molecular of an olefin, in π-complexes from the olefin and the heavy atom. If the latter carries d-electrons, it can therefore form a second dative molecular bond with the vacant antibonding π-molecular orbital, opposite in direction to the normal molecular bond. This is illustrated diagrammatically below, the phases of the lobes of the

> orbitals being identified to show the symmetry properties. The s orbital of Ag^+ has the wrong symmetry to interact with antibonding π-molecular orbital. The two molecular orbitals are therefore distinct. The combination of these two oppositely directed dative bonds would leave the olefin much less charged than it would have been in a normal π-complex; this would account for the low reactivity π-complexes from olefins with metals, where the binding energy of the d electrons is low, and also for the differences in reactivity of different metals since the stabilities of the two bonds will be affected differently by changes in overall structure.

As a result of Ingold's [63–65] contributions, organic chemists were very familiar with nucleophilic and electrophilic reagents, but Dewar established that ligands such as CO could simultaneously act as nucleophiles and electrophiles (i.e. they are ambiphilic) and the resultant synergic interaction would not only enhance the strength of the metal–ligand bond but also encourage electroneutrality. By emphasising the symmetries of the frontier orbitals of the metal and the olefin, he directed chemists towards a way of simplifying the complex interactions between fragments, which occur within the molecular orbital framework, by classifying their symmetry properties and adding the important quantum mechanical restriction that only orbitals with the same symmetries interact. The importance of frontier orbital interactions between organic molecules was also recognised by Fukui [177–180, 183, 184], who went on to widely apply the principles to a wide range of reactions.

Dewar's proposal encouraged others to recast Pauling's canonical forms (see Fig. 20) for the ligands CO, NO, CN^-, CNR, PF_3, etc., within a molecular orbital framework. The first example of a Dewar-type pictorial representation for these ligands appeared in a paper describing the bonding in $[Ni(C_5H_5)(NO)]$ in 1956 by Orgel [203], and since 1960, they have been widely used in inorganic and organometallic textbooks [108, 204]. Orgel was also influential in supplementing Dewar's symmetry arguments by providing tables of overlap integrals for the 3d orbitals of transition metals with the valence orbitals of common ligands and other metals [205, 206]. This required the classification of the primary orbital σ, π and δ

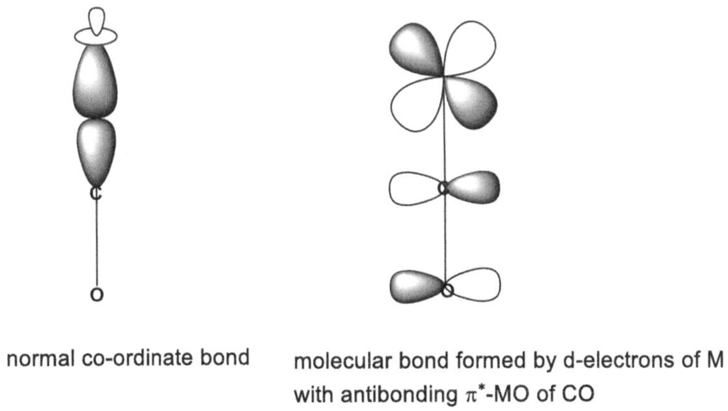

Fig. 20 Synergic bonding model for π-acceptor ligands such as CO, NO and CN^-

interactions, which could be used to calculate the necessary overlaps which incorporated the angular relationships in specific molecules. These studies clearly established that the overlap between 3d orbitals and the 2p orbitals of B, C, N and NO was significant. With the discovery of ferrocene in 1952 [207] and the structural determination, which established that it had a sandwich structure [208], Orgel [209] and Moffitt [210] independently were able to use the symmetry arguments which had been proposed by Dewar and the overlap data to establish a molecular orbital model for sandwich compounds in general. Orgel [211, 212] was also very important in reframing the crystal field concepts developed earlier for transition metal complexes into a molecular orbital-based ligand field theory. This analysis accounted for the spectral properties of transition metal complexes and established the spectrochemical series. Even at an early stage, it was recognised that CN^- was high in the spectrochemical series and Orgel proposed that the ambiphilic properties of this and related CO, CNR, NO ligands contributed to their larger d orbital splittings [213].

Analogous symmetry arguments were developed when it was established that σ-bonds could also coordinate to metals in low oxidation states by using orbitals analogous to those proposed for olefin complexes by Dewar. The dihydrogen complexes discovered by Kubas [214] were particularly important in establishing this type of coordination mode, and C–H "agostic interactions" [215] which had also been observed in a number of hypo-valent complexes were rationalised using this model. The relevant orbital interactions are illustrated in Fig. 21 and involve donation from the H_2 or C–H σ-bonds and back donation into the antibonding σ* molecular orbitals. More recently it has been shown that transition metal complexes where the metal has a low formal oxidation state are capable of behaving as Lewis bases and forming complexes with classic Lewis acids such as $AlCl_3$ [216–218]. In these complexes, the metal functions as a Lewis base by utilising a filled metal d_{z^2} as shown in Fig. 22. From the 1970s, the study of transition metal complexes also established that ligands such as N, NR, O, are able to supplement their σ coordinate

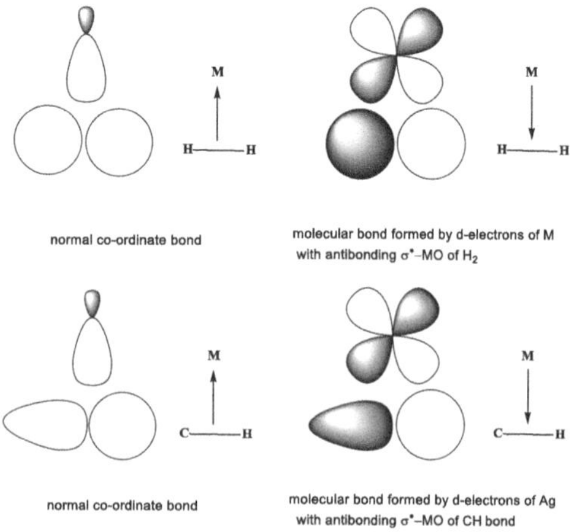

Fig. 21 Synergic bonding in complexes involving σ-bonds, e.g. H_2 and C–H bonds

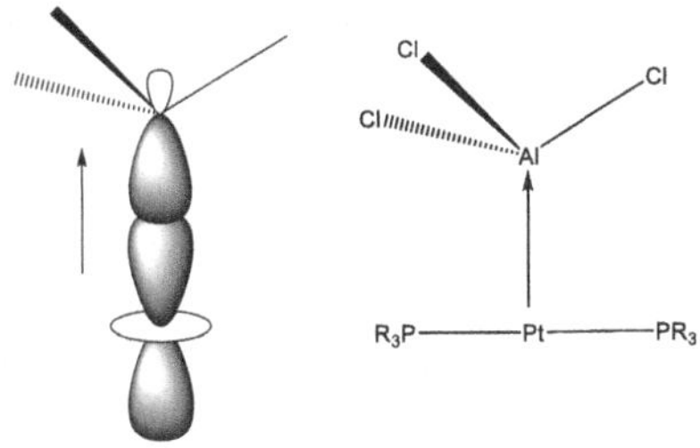

Fig. 22 An example of a complex acting as a Lewis base towards the classic Lewis acid $AlCl_3$

bonds by π-donation [219]. During the last century, the simple donor bond has evolved into a much more complex entity which involves donation not only from lone pairs but from π and σ bonds and back donation from metal d orbitals to π* and σ* antibonding orbitals on the ligand and even from metal lone pairs to a Lewis acid. The bonding permutations which result from σ and π electron-pair interactions are summarised in Fig. 23.

6.4 Ab Initio Calculations

The classic papers of Hartree, Fock, Roothan, Hall, Pople and Ahlrichs [220, 221] on the self-consistent field method allowed for the first time chemists to dream about ab initio calculations which achieved accuracies in the determination of bond lengths and angles and the thermodynamic properties of bonds which reproduced the experimental data to an accuracy of 1 kcal/mol. This resulted in a group of theoretical chemists who were primarily concerned with the a priori calculation for the electronic properties of small molecules to this degree of accuracy. Initially these calculations addressed small molecules of little interest to the average experimental chemists, but the exponential rise of the power of digital computers followed Moore's law, and within a generation, it became possible to achieve the accuracy noted above for complex organic and inorganic molecules.

Modern valence bond theory now complements molecular orbital theory, which does not adhere to the valence bond idea that electron pairs are localised between two specific atoms in a molecule but that they are distributed in sets of molecular orbitals which can extend over the entire molecule. Molecular orbital theory can predict magnetic and ionisation properties in a straightforward manner, while valence bond theory gives similar results but is more complicated. Valence bond treatments are restricted to relatively small molecules, largely due to the lack of orthogonality between valence bond orbitals and between valence bond structures, while molecular orbitals are orthogonal. On the other hand, valence bond theory provides a much more accurate picture of the reorganisation of electronic charge that takes place when bonds are broken and formed during the course of a chemical reaction. In particular, valence bond theory correctly predicts the dissociation of homonuclear diatomic molecules into separate atoms, while simple molecular

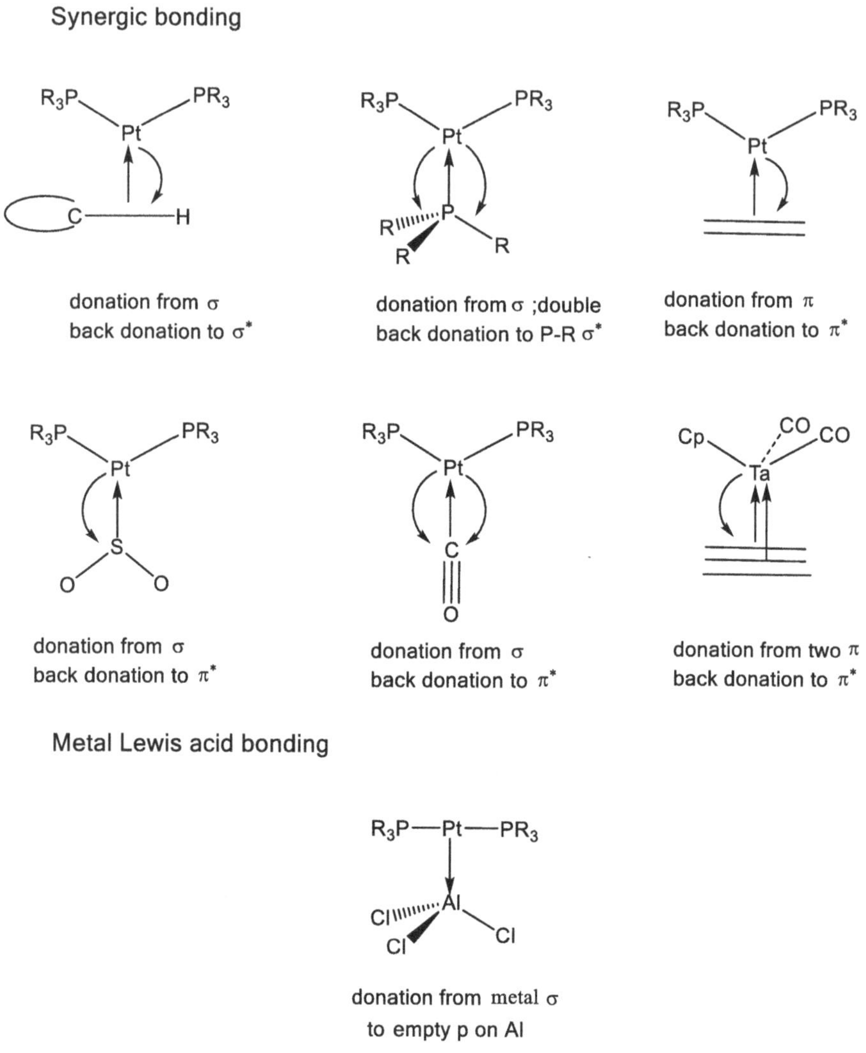

Fig. 23 Summary of metal–ligand interactions

orbital theory predicts dissociation into a mixture of atoms and ions. For example, the molecular orbital function for dihydrogen is an equal mixture of the covalent and ionic valence bond structures and so predicts incorrectly that the molecule would dissociate into an equal mixture of hydrogen atoms and hydrogen positive and negative ions.

Most quantitative calculations in modern quantum chemistry use either valence bond or molecular orbital theory as a starting point, although a third approach, density functional theory, has become increasingly popular in recent years. The impact of density functional theory over the last three decades has been amazing. Its impact on inorganic chemistry has been particularly dramatic because it has yielded reliable results for complex molecules with d and f valence orbitals. Contemporary DFT calculations combine the reasonable description of the electronic structures of transition metal compounds with excellent geometric descriptions, and it has largely displaced the "traditional" Hartree–Fock method. These developments have been covered in previous volumes of *Structure and Bonding* and the reader is referred to these developments [222–225].

In the early 1970s, a new electronic structure approach emerged from the physics community and described as density functional theory (DFT). The total energy of the molecule was expressed as a functional of the total electron density. Hohenburg and Kohn proved the unique relationship between electron density and energy, and Kohn and Sham put forward a practical variational DFT approach. Although calculations on infinite solids had been reported since the 1970s, DFT was not considered accurate enough for calculations on molecules until the 1990s when the approximations used in the theory were refined to mare exactly describe the exchange and correlation interactions. Computational costs for ab initio DFT calculations are relatively low when compared to the valence bond and molecular orbital methods. DFT began to approach the goals of computational thermochemistry to calculate the energetic properties of chemical processes to an accuracy of 1 kcal mol^{-1}. The widespread acceptance of these algorithms by the chemical community led to the award of the Nobel Prize in 1998 to Kohn and Pople.

6.5 *Natural Bond Orbitals*

The analysis of results of delocalised molecular orbital calculations for more complex molecules is problematic because there is no longer a simple relationship with the Lewis-localised bond representations. The use of fragment analyses and overlap populations has been helpful, but requires some knowledge of perturbation theory. Several methods have been developed to attempt to bridge the gap between the molecular orbital calculations and Lewis structures – one that is widely used is natural bonding orbitals (NBOs). Each bonding NBO σ_{AB} (the donor) can be written in terms of two directed valence hybrids (NHOs) h_A, h_B on atoms A and B, with corresponding polarisation coefficients c_A, c_B:

$$\sigma_{AB} = c_A h_A + c_B h_B$$

The bonds vary smoothly from covalent ($c_A = c_B$) to ionic ($c_A >> c_B$) limit. Each valence bonding NBO σ must be paired with a corresponding valence antibonding NBO σ* (the acceptor) to complete the span of the valence space:

$$\sigma^*_{AB} = c_A h_A - c_B h_B$$

Bonding NBOs of the "Lewis orbital"-type have occupation numbers near 2, and antibonding NBOs are of the "non-Lewis orbital"-type (occupation numbers near

0). In an idealised Lewis structure, full Lewis orbitals (two electrons) are complemented by formally empty non-Lewis orbitals.

Computer programs have been developed which can transform the results of molecular orbital calculations into NBOs. An optimal Lewis structure can be defined as that one with the maximum amount of electronic charge in Lewis orbitals (Lewis charge). A low amount of electronic charge in Lewis orbitals indicates strong effects of electron delocalisation [226–228]. In resonance structures, major and minor contributing structures may exist. These analyses provide results which are similar to modern valence bond theory methods.

7 Summary

The papers of Lewis and Kossel and their subsequent formulation in the mathematics and language of quantum mechanics had a major impact on chemistry. Their thinking has provided a conceptual framework, a notation and a language for discussing and evolving modern chemistry. The writing of algorithms incorporating the physics underlying quantum mechanics and the development of modern computers has led to the possibility of calculating from first principles the properties of interest to chemists (geometries, spectroscopic properties and their reaction profiles) accurately. Lewis and Kossel's generalisations arose in the absence of a quantum mechanical model, but nonetheless have proved to be enduring. They brought together a good empirical knowledge of the chemical literature and the ability to construct conceptual models which accounted for experimental observations using concepts which are well founded in modern quantum mechanics. A particularly important feature has been the representation of these ideas in a pictorial manner which chemists find easy to appreciate and apply. This also requires a notation and means of representation which allows chemists to interchange ideas in a creative manner. As McWeenie has commented [229], "If our concepts and "patterns of understanding" are sound they will be supported by rigorous calculation; if not they must be rejected". He drew attention to the following paragraph from Coulson's book [230]:

> Chemistry is an experimental subject whose results can be built into a pattern around quite elementary concepts. The role of quantum chemistry is to understand these concepts and to show what are the essential features of chemical behaviour. To say that the electronic computer shows that D(H-F) >> D(F-F) is not an explanation at all, but merely a confirmation of experiment. Any acceptable "explanation" must be in terms of repulsions between non-bonding electrons, dispersion forces between "atomic cores" and the like.

Progress has been made in a complementary way by those who have tried to answer the fundamental questions concerning the exact definition of a chemical bond, those who have translated the basic physics into efficient and computer friendly algorithms and those who have developed conceptual models and presented them in a way which could be used and adapted by experimental chemists. The subsequent chapters in this series of volumes of *Structure and*

Bonding provide up-to-date reviews by leading practitioners in the field which amplify these aims and hopefully adequately play tribute to Lewis and Kossel's major contributions 100 years ago. I hope that their insights will encourage similar progress in the coming century.

References

1. Lewis GN (1916) The atom and the molecule. J Am Chem Soc 38:762–785
2. Lewis GN (1916) Steric hindrance and the existence of odd molecules. Proc Natl Acad Sci U S A 2:588–592
3. Lewis GN (1923) Valence and the structures of atoms and molecules. The Chemical Catalog, New York
4. Kossel W (1916) Formation of molecules and its dependence on atomic structure. Ann Phys 49:229–362
5. Scerri ER (2007) The periodic table. Oxford University Press, Oxford
6. Ihde AI (1970) The development of modern chemistry. Harper Row, New York
7. Newlands JAR (1863–1866) Chem News 7:70; 10:59,94,240; 12:83;94; 13:113;130
8. Taylor W (1949) JAR Newlands: a pioneer in atomic numbers. J Chem Educ 26:491–495
9. Meyer JL (1870) Die natur der chemischen Elemente als Function ihrer Atomgewichte. Ann Supplementband, VII:354–365
10. Scerri ER (1998) The evolution of the periodic system. Sci Am 9:78–83
11. Mendeleev D (1869–1871) Osnovy Khimii (Foundations of chemistry) (in Russian). St. Petersburg: Tipogr. tovarishchestva "Obshchestvannaja Polza," 5 parts in 2 volumes, 1st Edn. English transl (Principles of chemistry). Longmans, London: 1891, 1897, 1905
12. Mendeleev D (1869) On the relationship of the properties of the elements to their atomic weights. Zhurnal 1:60–77 (in Russian). doi: 10.1002/prac.18691060141 (Abstracted in Z Chem, 1869, 12:405–406)
13. Mendeleev D (1889) The periodic law of the chemical elements (Faraday lecture, June 4th, 1889.). J Chem Soc 55:634–656. doi:10.1039/CT8895500634
14. Mendeleev D (1895) Report on argon (Report of the Russian Chemical Society, Meeting, March 14th 1895) Nature 51:543–543
15. Russell CA (1971) The history of valency. Leicester University Press, Leicester
16. Ramsey W, Raleigh L (1895) Argon a new constituent of the atmosphere. Chem News 71:51–63
17. Berzelius JJ (1832) Traite de Chemie. Jahresberichte 11:210
18. Partington JR (1964) History of chemistry, vol 4. Macmillan, London, pp 142–177
19. Frankland E (1866) Lecture notes for chemical students, embracing mineral and organic chemistry. London
20. Kekulé FA (1850) Ueber die Constitution und die Metamorpheser der Chemischen Verbindingen und über die Chemische Natur der Kohlenstoffs. Annalen 106:129–159
21. Couper AS (1859) Ueber die Constitution und Basicital der Salicylsaure. Annalen 110:46–50
22. Butlerov AM (1859) Bemerkungen über AS Couper's neue chemische Theorie. Annalen 110:51–66
23. Kolbe H (1860) Bermorkungen über A.S. Couper's neu chemische theorie. Annalen 115:157–206
24. Frankland E (1850) On the isolation of organic radicals. J Chem Soc 2:263
25. Kekulé FA (1850) Annalen 104:129–30
26. Kekulé FA (1861) Lehrbuch der organische chemie. Erlagen
27. Crum Brown A (1865) On the theory of isomeric compounds. J Chem Soc 18:230–245

28. McNaught AD, Wilkinson A (1997) IUPAC compendium of chemical terminology, 2nd edn. Blackwell Scientific, Oxford
29. Abegg R, Bodlander G (1899) Die Valenz und das Periodische System; Versuch einer Theorie der Molekular Verbindingen. Z Anorg Chem 20:453–496
30. Abegg R (1904) Z Anorg Chem 39:335–380
31. Drude P (1904) Optische Eigenschaften und Electronentheorie. Ann Phys 14:677–725
32. Thomson JJ (1897) Discovery of the electron. Philos Mag 44:293–303
33. Thomson JJ (1907) The corpuscular theory of matter. Archibald and Constable, London
34. Rutherford E (1911) The scattering of particles by matter and the structure of the atom. Philos Mag 21:669–689
35. Moseley HGJ (1913) The high frequency spectra of the elements. Philos Mag 26:1024–1044
36. Moseley HGJ (1914) Philos Mag 27:703–723
37. Bohr N (1913) The constitution of atoms and molecules. Philos Mag 26:1–25
38. Bohr N (1922) Atomic structure and the chemical and physical properties of the elements. Z Phys 9:1–67
39. Sommerfeld A (1916) The Drude dispersion theory from the standpoint of Bohr's model, and the constitution of hydrogen, oxygen, and nitrogen. Ann Phys 51:1
40. Sommerfeld A (1916) Ann Phys 53:497–550
41. Sommerfeld A (1918) Phys Z 17:497–550
42. Langmuir I (1919) The arrangement of electrons in atoms and molecules. J Am Chem Soc 41:868–934
43. Langmuir I (1919) Isomorphism, isosterism and covalence. J Am Chem Soc 41:1543–1559
44. Langmuir I (1919) The structure of atoms and octet theory of valence. Proc Natl Acad Sci U S A 5:252–259
45. Langmuir I (1921) Types of valence. Science 54:59–67
46. Dirac PAM (1925) The fundamental equations of quantum mechanics. Proc Roy Soc 109:642–653
47. Dirac PAM (1926) Proc Roy Soc 110:561–572
48. Pauli W (1927) Ueber Garsentartung der Paramagnetismum. Z Phys 31:765–785
49. Lennard-Jones JK (1952) The spatial correlation of electrons in molecules. J Chem Phys 20:1024–1030
50. Linnett JW (1961) A modification of the Lewis-Langmuir octet rule. J Am Chem Soc 83:2643–2653
51. Van't Hoff JH (1874) A Suggestion looking to the extension into space of the structural formulae at present used in chemistry. And a note upon the relation between the optical activity and the chemical constitution of organic compounds. Arch Neerl Sci Exactes Nat 9:445–454
52. leBel JA (1874) On the relations which exist between the atomic formulas of organic compounds and the rotatory power of their solutions. Bull Soc Chem Fr 22:337–367
53. Sidgwick NV (1923) Co-ordination compounds and the Bohr atom. J Chem Soc 123:725–730
54. Sidgwick NV (1927) The electronic theory of valency. Clarendon, Oxford
55. Pauling L (1938) The nature of the chemical bond, 1st edn. Cornell University Press, Ithaca
56. Pauling L, Huggins ML (1934) Covalent radii of atoms and interatomic distances in crystals containing electron-pair bond. Z Krist 87:205–238
57. Pauling L (1938) The nature of inter-atomic forces in metals. Phys Rev 54:899
58. Pauling L (1947) Atomic radii and intra-atomic distances in metals. J Am Chem Soc 69:542–553
59. Pauling L (1932) The nature of the chemical bond III, the transition from one bond extreme to another. J Am Chem Soc 54:988–1003
60. Pauling L (1940) The nature of the chemical bond, 2nd edn. Cornell University Press, Ithaca
61. Pauling L (1960) The nature of the chemical bond, 3rd edn. Cornell University Press, Ithaca
62. Pauling L (1932) The nature of the chemical bond IV. The energy of single bonds and the relative electronegativity of atoms. J Am Chem Soc 54:3570–3582

63. Ingold CK (1922) The structure of the benzene nucleus. Part 1. Intra-nuclear tautomerism. J Chem Soc Trans 121:1133–1143
64. Ingold CK (1925) The nature of the alternating effect in carbon chains. Part I. The directive influence of the nitroso-group. J Chem Soc 127:513–518
65. Holmes EL, Ingold CK (1925) The nature of the alternating effect in carbon chains. Part III. A comparative study of the directive efficiencies of oxygen nitrogen atoms in aromatic substitution. J Chem Soc 127:1800–1821
66. Kermach WO, Robinson R (1922) Explanation of the property of induced polarity of atoms and an interpretation of the theory of partial valency on an electronic basis. J Chem Soc Trans 121:427–440
67. Lapworth A (1922) Theoretical derivation of the principle of induced alternate polarities. J Chem Soc Trans 121:416–427
68. Robinson Sir Robert (1976) The memoirs of a minor prophet – 70 years of organic chemistry. Elsevier, Oxford
69. Todd L, Cornforth JW (1976) Sir Robert Robinson. Biographical Memoirs of the Royal Society 22:415–527
70. Pauling L, Wheland GW (1933) Nature of the chemical bond V. The quantum mechanical calculation of the resonance energies in benzene and naphthalene and the hydrocarbon free radicals. J Chem Phys 1:362–374
71. Wheland GW (1942) A quantum mechanical investigation of the orientation of substituents in aromatic molecules. J Am Chem Soc 64:900–908
72. Pauling L, Wheland GW (1935) The quantum mechanical discussion of the orientation of substituents in organic molecules. J Am Chem Soc 57:2086–2095
73. Sutton LE (1940) Electron diffraction by gases and vapours and electric dipole moments. Ann Rep Chem Soc 37:36–80
74. Sidgwick NV, Sutton LE, Thomas W (1933) Dipole moments and the structures of organic azides and aliphatic diazo compounds. J Chem Soc 406–412
75. Sidgwick NV, Sutton LE, Hammick DL, New RCA (1930) Structures of the isocyanides and other compounds of bivalent compounds. J Chem Soc 1876–1887
76. Haaland A (2008) Molecules and models. Oxford University Press, Oxford
77. Power PP (1998) Homonuclear multiple bonding in heavier main group elements. J Chem Soc Dalton Trans 2939
78. Weidenbruch M (1999) Some recent advances in the chemistry of silicon and its homologues in low co-ordination states. J Organomet Chem 646:39–52
79. Bury CR (1921) Langmuir's theory of the arrangement of electrons in molecules. J Am Chem Soc 43:1602–1609
80. Blanchard AA, Gulliland WL (1926) The constitution of nickel carbonyl and the nature of secondary valency. J Am Chem Soc 48:872–882
81. Sidgwick NV (1923) The nature of non-polar link. Trans Faraday Soc 19:469–475
82. Reiff F (1931) Konstitution und Eigenschaften der $Co(NO)(CO)_3$. Z Anorg Chem 202:375–381
83. Sidgwick NV, Bailey RM (1934) Structures of metallic carbonyls and nitrosyls. Proc Roy Soc A144:521–537
84. Mitchell PR, Parish RV (1969) The eighteen-electron rule. J Chem Educ 46:811–814
85. Jensen WB (2005) The origin of the 18 electron rule. J Chem Educ 82:28–29
86. Green JC, Green MLH, Parkin G (2012) The occurrence and representation of three-centre two-electron bonds in covalent inorganic compounds. J Chem Soc Chem Commun 48:11481–11503
87. Green MLH (1995) A new approach to the classification of covalent compounds of the elements. J Organomet Chem 500:127–148
88. Sidgwick NV, Powell HW (1940) Stereochemical types valency groups. Proc R Soc London Ser A 176:153–180

89. Gillespie RJ, Nyholm RS (1958) Stereochemistry of inorganic molecules and complex ions inorganic stereochemistry. Progr Stereochem (Academic, New York) 2:261–305
90. Gillespie RJ, Nyholm R (1957) Inorganic stereochemistry. Q Rev Chem Soc 17:339–380
91. Gillespie RJ, Popelier PLA (2001) Chemical bonding and molecular geometry. Oxford University Press, New York, pp 154–155
92. Gillespie RJ, Robinson EA (2006) Gilbert N Lewis and the chemical bond: the electron pair and the octet rule from 1916 to the present day. J Comput Chem 28:87–97
93. Mingos DMP, Hawes JC (1985) Complementary spherical electron density model. Struct Bond 63:1–63
94. Mingos DMP, Zhenyang L (1989) Non-bonding orbitals in co-ordination, hydrocarbon and cluster compounds. Struct Bond 71:1–56
95. Mingos DMP (2004) Complementary spherical electron density model and its implications to 18 electron rule. J Organomet Chem 689:4420–4436
96. Mingos DMP (2006) The relevance of the complementary spherical electron density model to organometallic intermediates in homogeneous catalysis. J Organomet Chem 691:3165–3175
97. Lin Z (2016) Lewis description of bonding in transition metal complexes. Struct Bond. doi:10.1007/430_2015_182
98. Hoffman RV (2004) Organic chemistry an intermediate text. Wiley, New York
99. Mingos DMP, Crabtree R (2007) Comprehensive organometallic chemistry 3. Elsevier, Oxford
100. Ghosh A, Berg S (2014) Arrow pushing in inorganic chemistry: a logical approach to the chemistry of main group elements. Wiley, New York
101. Mingos DMP (1998) Essential trends in inorganic chemistry. Oxford University Press, Oxford
102. Longuet-Higgins HC, Bell RP (1943) The structure of boron hydrides. J Chem Soc 250–255
103. Lipscomb WN (1959) Recent studies on boron hydrides. Adv Inorg Radiochem 1:118–157
104. Bartlett N (1962) Xenonhexafluoroplatinate. Proc Chem Soc, pp 218
105. Bartlett N, Lohmann DH (1964) Fluorides of the noble metals III fluorides of platinum. J Chem Soc 619–626
106. Moeller T (1952) Inorganic chemistry. Wiley, New York
107. Dye JL, Andrews CW, Mathews SE (1975) Strategies for the preparation of compounds of alkali metal anions. J Phys Chem 79:3065–3070
108. Orgel LE (1960) Introduction to transition metal chemistry. Methuen, London
109. Kotz JC, Treichel P (1999) Chemistry and chemical reactivity, 4th edn. Saunders College, New York
110. Walsh AD (1953) The electronic orbitals, shapes and spectra of polyatomic molecules Part I. J Chem Soc 2260–2265
111. Walsh AD (1953) The electronic orbitals, shapes and spectra of polyatomic molecules Part II. J Chem Soc 2266–2295
112. Walsh AD (1953), The electronic orbitals, shapes and spectra of polyatomic molecules Part III. J Chem Soc 2296–2300
113. Walsh AD (1953) The electronic orbitals, shapes and spectra of polyatomic molecules Part IV. J Chem Soc 2301–2305
114. Walsh AD (1953) The lectronic electronic orbitals, shapes and spectra of polyatomic molecules Part V. J Chem Soc 2306–2310
115. Herzberg G (1967) Molecular structure and molecular spectra. Van Nostrand, New York
116. Molina JM, Dobado JA (2001) The three-centre four –electron model revisited. An atoms in molecules approach (AIM) and ELF study. Theor Chem Acta 105:328–337
117. Noury S, Silvi B, Gillespie RJ (2002) Chemical bonding in hypervalent molecules: is the octet rule relevant? Inorg Chem 41:2164–2172
118. See RF (2009) Which method of assigning bond orders in Lewis structures best reflects experimental data? J Chem Educ 86:1241–1247

119. Haaland A, Tilset M (2016) Lewis and Kossel's legacy – structure and bonding in main group compounds. Struct Bond. doi:10.1007/430_2015_192
120. Stalke D (2016) Charge density and chemical bonding. Struct Bond. doi:10.1007/430_2015_199
121. Grimm HG, Gunther M, Tittus H (1931) Zur Kenntnis der isomorphen Vetretbarkeit nichtpolar gebundener Atome und Atomgruppen. Z Phys Chem 14B:169
122. Grimm HG (1925) Z Elektrochem 31:474
123. Grimm HG (1929) Naturzisseenchaften 17:535
124. Erlenmeyer H, Leo M (1932) Uber pseudoatome. Helv Chim Acta 15:1171–1186
125. Hückel E (1931) Quantum theoretical contributions to the benzene problem I, II. The electron configuration of benzene and related compounds. Z Phys 70(204–309):310–349
126. Dewar MJS (1969) The molecular orbital theory of organic chemistry. McGraw-Hill, New York
127. Longuet-Higgins HC, de V Roberts M (1955) The electronic structure of an icosahedron of boron atoms. Proc Roy Soc 230:110–119
128. Mingos DMP, Johnston RL (1987) Theoretical models of cluster bonding. Struct Bond 68:31–82
129. Williams RE (1971) Carboranes and boranes; polyhedra and polyhedral fragments. Inorg Chem 10:210–214
130. Wade K (1971) Structural significance of the number of skeletal bonding electron-pairs in carboranes, the higher boranes, and borane anions, and various transition metal carbonyl cluster compounds. J Chem Soc Chem Commun 792–793
131. Mingos DMP (1971) A general theory for cluster and ring compounds of the main group and transition elements. Nat Phys Sci 236:99–102
132. Rudolph RW (1976) Boranes and heteroboranes: a paradigm for the electron requirements of clusters? Acc Chem Res 9:446–452
133. Mingos DMP, Johnston RL (1987) Group theoretical paradigms for describing the skeletal molecular orbitals of cluster compounds I and II. J Chem Soc Dalton Trans 647–656; 1445–1456
134. Stone AJ (1981) New approach to bonding in transition-metal clusters and related compounds. Inorg Chem 20:563–571
135. Stone AJ (1981) The bonding in boron and transition-metal cluster compounds. Polyhedron 3:1299–1306
136. Elian M, Chen MML, Mingos DMP, Hoffmann R (1976) A comparative study of conical fragments. Inorg Chem 15:1148–1155
137. Hoffmann R (1982) Building bridges between inorganic and organic chemistry, Nobel Prize Lecture. Angew Chem Int Ed 21:711–724
138. Weinhold F, Landis C (2005) Valency and bonding. Cambridge University Press, Cambridge, pp 96–100
139. Allen HF, Kennard O, Taylor R (1983) Systematic analysis of structural data as a research technique in organic chemistry. Acc Chem Res 16:146–153
140. Cotton FA, Walton RA (1993) Multiple bonds between metal atoms, 2nd edn. Clarendon University Press, Oxford
141. Desclaux JP (1973) Relativistic Dirac-Fock expectation values for atoms with Z =1 to Z = 120. At Data Nucl Data Tables 12:311–406
142. Gavroglu K, Simoes A (1994) Early ideas in the history of quantum chemistry. Hist Stud Phys Sci 25:47–110
143. Burrau Ø (1927) Berechnung des Energiewertes des Wasserstoffmolekel-Ions (H_2^+) im Normalzustand. Danske Vidensk. Selskab. Math.-fys. Meddel (in German) M 7:14: 1–18
144. Burrau Ø (1927) The calculation of the energy value of hydrogen molecule ions (H2+) in their normal position. Naturwissen 15:16–17
145. Heitler W, London F (1927) Interaction of neutral atoms according and homoplar binding according to the quantum mechanics. Z Phys 44:455–472

146. Hettema H (2000) English translation - Quantum chemistry: classic scientific papers. World Scientific, London, pp 140–161
147. Pauli W (1922) Über das Modell des Wasserstoffmolekülions. Ann Phys 373:177–240
148. Pauli W (1922) Extended doctoral dissertation; received 4 Mar 1922, published in issue no 11 of 3 Aug 1922
149. Hund F (1926) Zur Deutung einiger Erscheinungen in den Molekelspektren (On the interpretation of some phenomena in molecular spectra). Z Phys 36:657–674
150. Urey HC (1925) The structure of the hydrogen molecule ion. Proc Natl Acad Sci U S A 11:618–621
151. Lennard-Jones JK (1929) The electronic structure of some diatomic molecules. Trans Faraday Soc 25:668–686
152. Pauling L (1928) The application of the quantum mechanics to the structure of the hydrogen molecule and hydrogen molecule-ion and to related problems. Chem Rev 5:173–213
153. Slater JC (1931) Directed valence in polyatomic molecules. Phys Rev 37:481–489
154. James HM, Coolidge AS (1933) The ground state of the hydrogen molecule. J Chem Phys 1:825–835
155. Li J, McWeeny R (2002) Pushing valence bond to new limits. Int J Quantum Chem 89:208–216
156. Engelberts JJ, Havenith RWA, van Lenthe JH, Jenneskens LW, Fowler PW (2005) The electronic structure of inorganic benzenes: valence bond and ring current descriptions. Inorg Chem 44:5266–5274
157. Shaik S, Hiberty PC (2004) Valence bond theory, its history, fundamentals and applications: a primer. Rev Comput Chem 20:1–100
158. Shaik S, Hiberty PC (1995) Valence bond mixing and curve crossing diagrams in chemical reactivity and bonding. Adv Quantum Chem 26:99–163
159. Coulson CA, Fisher I (1949) Notes on the molecular orbital theory of the hydrogen molecule. The London, Edinburgh, and Dublin Philosophical Magazine and Journal of Science: Series 7 40:386–393
160. Hunt WJ, Hay PJ, Goddard WA (1972) Self-consistent procedures for generalised valence bond wavefunctions. Applications H_3, BH, H_2O, C_2H_6 and O_3. J Chem Phys 57:738–748
161. Goodgame MM, Goddard WA (1985) Modified generalized valence-bond method: a simple correction for the electron correlation missing in generalized valence-bond wave functions; prediction of double-well states for Cr_2 and Mo_2. Phys Rev Lett 54:661–664
162. Hund F (1927) Zur Deutung der Molekelspektren. I. Z Phys 40:742–764
163. Hund F (1928) Zur Deutung der Molekelspektren. II. Z Phys 42:93–120
164. Hund F (1928) Zur Deutung der Molekelspektren. III. Z Phys 43:805–826
165. Hund F (1928) Zur Deutung der Molekelspektren. IV. Z Phys 51:759–795
166. Hund F (1930) Zur Deutung der Molekelspektren. V. Z Phys 63:719–751
167. Mulliken RS (1927) Electronic states. IV. Hund's theory; second positive nitrogen and Swan bands; alternate intensities. Phys Rev 29:637–649
168. Mulliken RS (1928) The assignment of quantum numbers for electrons in molecules. Phys Rev 32:186–222
169. Kutzelnigg W (1966) Hund and chemistry -the occasion of Hund's 100th birthday. Angew Chem Int Ed 35:573–586
170. Mulliken RS (1967) Spectroscopy, molecular orbitals and chemical bonding, Nobel Lecture. Science 157:13–24
171. Mulliken RS (1932) The interpretation of band spectra part III. Electron quantum numbers and states of molecules and their atoms. Rev Mod Phys 4:1
172. Fehlner TP, Bowser JR (1988) Proton power: an intuitive approach to the electronic structures of molecule hydrides. J Chem Educ 65:976–980
173. Lowe JP (1978) Quantum chemistry. Academic, New York, pp 168–192
174. Woodward RB, Hoffmann R (1965) Stereochemistry of electrocyclic reactions. J Am Chem Soc 87:395–397

175. Woodward RB, Hoffmann R (1968) Conservation of orbital symmetry. Acc Chem Res 1:17–22
176. Longuet-Higgins HC, Abrahamson EW (1965) The electronic mechanism of electrocyclic reactions. J Am Chem Soc 87:2045–2046
177. Fukui K (1992) Nobel Prize Lecture, Chemistry, 1981–1990. The role of frontier orbitals in chemical reactions, vol 3. World Scientific, New York, p 27
178. Fukui K (1981) The path of chemical reactions – the IRC approach. Acc Chem Res 14:363–368
179. Fukui K (1965) The stereoselectivity associated with noncycloaddition to unsaturated bonds. Tetrahedron Lett 28:2427–2432
180. Fukui K, Fujimoto H (1966) Bull Chem Soc Jpn 39:498
181. Zimmerman HE (1966) Molecular orbital correlation diagrams, Möbius systems and factors controlling ground and excited state reactions I and II. J Am Chem Soc 88 (1564–1566):1566–1567
182. Dewar MJS (1966) A molecular orbital theory of organic chemistry VII – aromaticity and electrocyclic reactions. Tetrahedron 22(Suppl 8):75–92
183. Fukui K (1965) Stereospecificity with respect to cyclic reactions. Tetrahedron Lett 6:2009–2015
184. Fukui K, Fujimoto H (1966) Bull Chem Soc Jpn 39:2116
185. Dewar MJS (1969) The molecular orbital theory of organic chemistry. McGraw-Hill & Co, New York
186. Dewar MJS (1949) The electronic theory of organic chemistry. Oxford University Press, Oxford
187. Dewar MJS (1952) A molecular orbital theory of organic chemistry I-VI. J Am Chem Soc 74:3341; 3345; 3350; 3353; 3355; 3357
188. Dewar MJS, Longuet-Higgins HC (1952) The correspondence between resonance and molecular orbital theories. Proc Roy Soc Ser A 21:482–493
189. Longuet-Higgins HC (1950) Molecular orbital theory 1-III. J Chem Phys 18:265–274; 275–282; 283–291
190. Hoffmann R (1963) Extended Hückel theory I – hydrocarbons. J Chem Phys 39:1397–1412
191. Albright TA, Burdett JK, Whangbo M-H (1985) Orbital interactions in chemistry. Wiley, New York
192. Pearson RG (1976) The symmetry rules for chemical reactions. Wiley, New York
193. Gimarc BM (1979) Molecular structure and bonding. Academic, New York
194. Hoffmann R (1988) Solids and surfaces: a chemist's view of bonding in extended structures. VCH, Weinheim
195. Hieber W (1942) The present status of the chemistry of metal-carbonyls. Die Chem 55:24–28
196. Hieber W (1970) Metal carbonyls, forty years of research. Adv Organomet Res 8:1–28
197. Chatt J (1950) The co-ordinate link in chemistry. Nature (London) 165:637
198. Chatt J, Duncanson LA (1953) Olefin co-ordination compounds. Part III. Infra-red spectra and structure: attempted preparation of acetylene complexes. J Chem Soc 2939–2953
199. Chatt J, Willians AA (1951) The nature of the co-ordinate link IV. Complex formation by phosphorus trifluoride. J Chem Soc 3061–3078
200. Wilkinson G (1951) The preparation and properties of tetrakistribromophosphine-nickel and tetrakistrifluorophosphinenickel. J Am Chem Soc 73:5501–5502
201. Dewar MJS (1951) A review of π complex theory. Bull Soc Chim Fr 18:C79
202. Mingos DMP (2002) A historical perspective on Dewar's landmark contribution to organometallic chemistry. J Organomet Chem 635:1–8
203. Orgel LE (1956) Electronic structures of some mixed compounds of cyclopentadienyl and carbon monoxide or nitric oxide with the transition metals. J Inorg Nucl Chem 2:315–322
204. Cotton FA, Wilkinson G (1962) Advanced inorganic chemistry: a comprehensive text. Intersciene/Wiley, New York

205. Craig DP, Macoll A, Nyholm RS, Orgel LE (1954) Chemical bonds involving d orbitals I. J Chem Soc 332–353
206. Craig DP, Macoll A, Nyholm RS, Orgel LE (1954) Chemical bonds involving d orbitals II. J Chem Soc 354–357
207. Wilkinson G, Rosenblum M, Whiting MC, Woodward RB (1952) The structure of iron bis (cyclopentadienyl). J Am Chem Soc 74:2125–2156
208. Dunitz JD, Orgel LE (1953) Dicyclopentadienyliron: a molecular sandwich. Nature 171:121–122
209. Orgel LE (1955) Electronic structure of metal bis-cyclopentadienyls. J Chem Phys 23:954–958
210. Moffitt W (1954) The electronic structure of bis-cyclopentadienyl compounds. J Am Chem Soc 76:3386–3392
211. Orgel LE (1952) The effects of crystal field theory on the properties of transition metal ions. J Chem Soc 4756–4761
212. Orgel LE (1955) Electronic structure of transition metal complexes. J Chem Phys 23:1819–1823
213. Orgel LE (1955) Spectra of transition metal complexes. J Chem Phys 23:1004–1014
214. Kubas G (2014) Activation of dihydrogen and co-ordiantion of molecular H_2 to transition metals. J Organomet Chem 751:33–49
215. Brookhart M, Green MLH (1988) Carbon-hydrogen-transition metal bonds. Prog Inorg Chem 36:1–124
216. Braunschweig H, Gruss K, Radachki K (2007) Interaction between d and p block metals. Synthesis and structure of platinum alane adducts. Angew Chem Int Ed 46:7782–7784
217. Amgoune A, Bourissou D (2011) σ-Acceptor Z-type ligands for transition metals. J Chem Soc Chem Commun 47:859–871
218. Mingos DMP (2014) A theoretical analysis of ambivalent and ambiphilic Lewis acid/bases with symmetry signatures. Coord Chem Rev 293–294:2–18
219. Winkler JR, Gray HB (2012) Electronic structures of oxo-metal ions. Struct Bond 142:17–28
220. Hinchcliffe A (2000) Modelling molecular structures, 2nd edn. Wiley, Chichester
221. Szabo A, Ostlund NS (1996) Modern quantum chemistry. Dover, Mineola
222. Kaltsoyanis N, McGrady JE (2004) Principles and applications of density functional theory in inorganic chemistry I. Struct Bond 112:1–189
223. Kaltsoyanis N, McGrady JE (2004) Principles and applications of density functional theory in inorganic chemistry II. Struct Bond 113:1–232
224. Putz MV, Mingos DMP (2012) Application of density functional theory to chemical reactivity. Struct Bond 149:1–186
225. Putz MV, Mingos DMP (2012) Application of density functional theory biological and bioinorganic chemistry. Struct Bond 150:1–231
226. Weinhold F, Landis CR (2012) Discovering chemistry with natural bond orbitals. Wiley, New Jersey
227. Weinhold F, Landis CR (2001) Natural bond orbitals and extensions of localized bonding concepts. Chem Educ Res Pract 2:91–104
228. Reed AE, Curtiss LA, Weinhold F (1988) Intermolecular interactions from a natural bond orbital, donor-acceptor viewpoint. Chem Rev 88:899–926
229. McWeeny R (1979) Coulson's valence, 3rd edn. Oxford University Press, Oxford
230. Coulson CA (1961) Valence, 2nd edn. Oxford University Press, Oxford

Struct Bond (2016) 169: 57–88
DOI: 10.1007/430_2015_199

Published online: 19 March 2016

Charge Density and Chemical Bonding

Dietmar Stalke

Abstract In the past 100 years, the Lewis diagram has frequently been challenged, modified, extended and rejected as being too simplistic. Those who teach chemistry to freshman, however, appreciate the diagram as one of the didactical rocks in the wild sea of ever developing science, because it is simple, easy to understand and long ranged in mediate basic chemistry. This article is aimed at the evaluation of the Lewis diagram in the light of modern charge density investigations and the topological analysis based on the quantum theory of atoms in molecules. Some old molecules like boranes, sulfate, and high-coordinate silicon will be revisited as well as some recent low-valent silicon species that were regarded impossible to make only some years ago. Can the Lewis diagram cope with new results from experiment and theory and be extended to "impossible" molecules? The answer is yes and that makes a model a good model: easy to adapt by and by and not suggesting any scientific dead ends, because the model might eventually be mistaken to be real from the inexperienced applicant.

Keywords Bond theory • Charge density • Silylene • Silylone

Contents

D. Stalke (✉)
Georg-August-Universität Göttingen, Institut für Anorganische Chemie, Tammannstraße 4, 37077 Göttingen, Germany
e-mail: dstalke@chemie.uni-goettingen.de

1 Introduction

The diagram introduced 100 years ago by Gilbert N. Lewis has emerged to be one of the most resilient concepts of all disciplines of chemistry [1]. From the freshman to the highly advanced theoretician, all chemists understand the electron pairing and the dash between two element symbols as a 2-centre 2-electron bond. Two or three dashes consequently represent a 2-centre 4-electron and a 2-centre 6-electron bond, referred to as double or triple bond, respectively. Hence, the simple concept of the Lewis diagram not only contains information about the atomic connectivities in molecules but also a judgment on the quality of the bond. This is even more remarkable as it was developed 10 years before Erwin Schrödinger applied his equation to the hydrogen atom and founded modern quantum chemistry [2], nowadays able to predict and rationalize with physically meaningful numbers the bonding in molecules and materials [3]. Still the Lewis diagram frequently is the first model for chemists to approach unprecedented, newly synthesized molecules, because it is the least common denominator to get their ideas across. It is a sort of a chemical world language that differently to Esperanto is actively used and advanced. Despite all trials and tribulations from modern chemistry, it remains to be very flexible and open to interpretation, hence a perfect basis to start a discussion (Table 1).

Lewis established a theory of cubical atoms consisting of a kernel and a shell, referring to Richard Abegg's extension of the coordination number to valency [4]. For his part this goes back to Alfred Werner's observation that only carbon is the lucky case where the maximum coordination number is identical to the valency,

Table 1 Lewis diagrams of some key compounds

		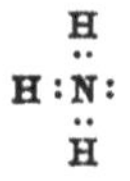	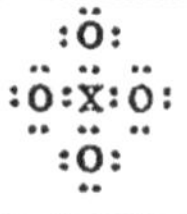	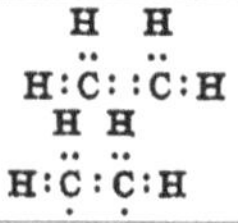	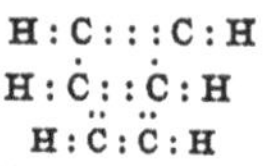
I_2	H_2O	NH_3	XO_4^{n-}	C_2H_4	C_2H_2
Iodine	Water	Ammonia	Silicate	Ethene	Ethyne
			Phosphate		
			Sulfate		
			Perchlorate		

If X is Cl the ion has one negative charge; if S it has two negative charges, and so on (taken from [1])

but for all other elements there is a discrepancy [5]. Lewis's observation that the number of electrons in the shell might vary during chemical changes between zero and eight established the still operational eight-electron rule or octet rule, mirrored by Irving Langmuir's 18-electron rule for stable transition metal complexes [6]. Well aware of high-coordinated very polar substances, Lewis emphasized that the degree of bonding between the two extremes polar and non-polar is "not a sudden and discontinuous change, but by imperceptible gradations" [1]. This already suggests that the bonding in high-coordinated molecules should conceptionally not be different from that in low-coordinated species.

Characteristic to any rule there are also various exceptions to the octet rule. Even more prominent than radicals and compounds with a Group 13 central atom (BF_3 from Fig. 1), where the number of eight electrons in the valance shell is underrun, are the examples where formally the central atom accommodates more than eight electrons. Elements heavier than the second period can apparently engage up to twelve valence electrons. This can be exemplified in a series of perfluorinated main group compounds. From boron one to the right in the periodic table, CF_4 is the paradigmatic Lewis-precise eight-electron molecule where coordination number and valency are the same. This is also valid for the heavier congener SiF_4, but already silicon can adopt the hexafluorosilicate dianion with six fluorine atoms coordinated to the central atom. If every dash was a 2-centre 2-electron bond than the silicon would formally convey a double negative charge, which is tremendously counterintuitive for a metal. Obviously a formal charge deduced from Lewis diagrams containing strict 2-centre 2-electron bonds does not at all hint towards the

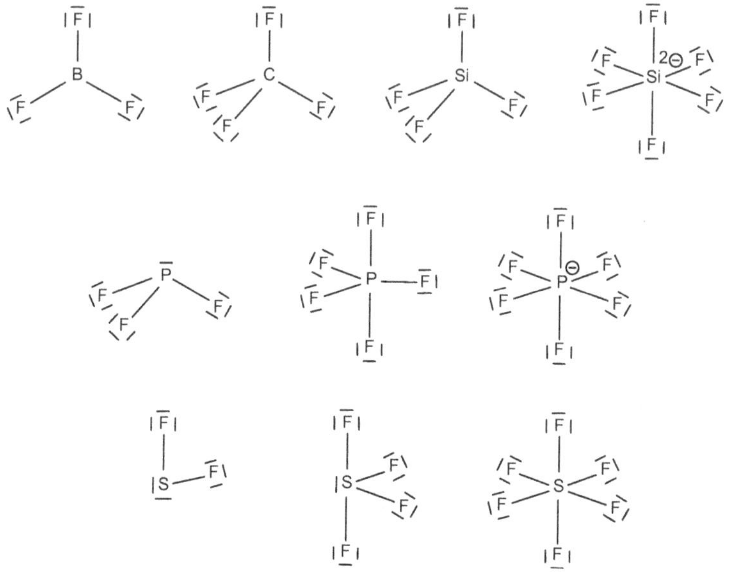

Fig. 1 Lewis diagrams of some textbook main group molecules, apparently underrunning, obeying, and exceeding the eight-electron rule

real charge distribution in the molecule, but that is difficult to understand to the beginner. The polarity of the bond due to electronegativity differences obviously needs to be taken account for as well. But clearly it would be most instructive if the formal charge would be in tune with the real charge. The shape of PF_3 is pyramidal because the central phosphorus atom accommodates at the apical position a lone pair of electrons in addition to the three 2-centre 2-electron bonds towards the fluorine atoms. This Lewis diagram is expected from the eight-electron rule. However, PF_5 apparently breaks that rule because the central phosphorus atom takes up ten valence electrons from five bonding pairs. Even worth, there seem to be twelve electrons from six pairs in PF_6^-. Counterintuitively the central phosphorus atom has to carry the negative formal charge, while the fluorine atoms are much more electronegative (4.0 for F vs 2.2 for P). In the series of sulfur fluorides, SF_2 embraces the bent structure caused by two lone pairs and two bonding pairs. Interestingly the FSF angle of 98° [7] is more acute than anticipated from OF_2 (103° [8]) and H_2O (107° [9], all from the gas phase by vibrational spectroscopy). Obviously the heavier elements are much more reluctant to hybridize and bond predominately with their orthogonal p-orbitals, while one lone pair resides in the remaining p-orbital and the second in the stereochemically inactive s-orbital.

To rationalize the high coordination numbers, the sp^3d hybridization for trigonal bipyramidal molecules like PF_5 and the sp^3d^2 hybridization for octahedral molecules like PF_6^- were employed by Linus Pauling [10]. However, this formulation turned out to be in contrast to theoretical investigations from the mid-1980s, which verified that d-orbitals cannot participate in this bonding due to large energy differences between the main group element p- and d-orbitals [11–14]. Furthermore, these MO calculations on second-row atoms in “hypervalent” molecules (a comprehensive introduction into that topic is provided in [15, 16]) showed that the d-orbitals are mainly needed as polarization functions rather than as bonding orbitals, because they are only occupied of 0.3e at most. They showed that the central atoms carry a considerable amount of positive charge, i.e., +2.9e in SF_6 [13]. This illustrates the dilemma of the Lewis diagram. How can we draw canonical forms obeying the eight-electron rule with high-coordinated species and employ the resulting formal charges to mirror at least qualitatively the real charge distributions in the molecules? This can easily be achieved by charge separated species depicted in Fig. 2. The boron atom in BF_3 is only three coordinated and accommodates only six electrons in the valence shell. If there was one B=F double bond contribution, this would of course reflect the high Lewis acidity of the empty boron p-orbital and the Lewis basicity of the fluorine atom from the doubly occupied p-orbital at the same time. Unfortunately this octet-rule conform connotation would cause formal charges against the polarity of the bond with a negative formal charge at the electropositive boron atom. In contrast the diagram with the charge separated species would reflect the real charge distribution and would as well explain the short B–F distance. The electrostatic contribution would obviously further shorten the covalent bond similar to the double bond contribution. The charge separated species of SO_2 is in tune with ozone O_3 and explains the S–O bond shortening with some π- and electrostatic contribution at the same time,

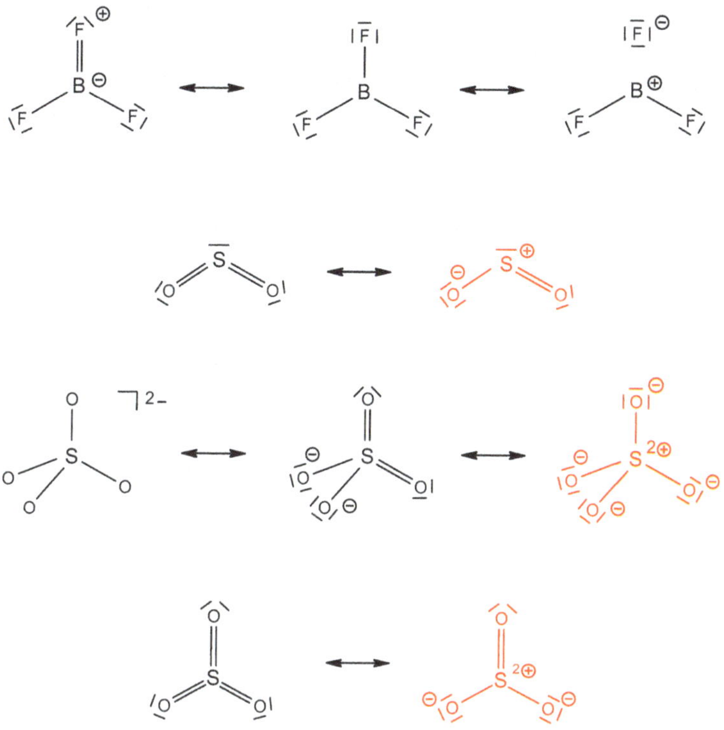

Fig. 2 Lewis diagrams in favour for the charge separated species (coloured in *red*)

obeying the eight-electron rule. Obviously there is no need for valence expansion and recruiting high-energy 3d-orbitals. The same is valid for sulfate SO_4^{2-} and SO_3. In the first, the doubly positive-charged sulfur atom and the four negatively charged oxygen atoms all obey the eight-electron rule. SO_3 can be written with one double bond without violating the eight-electron rule, indicating some π-contribution to the bonding on top of the electrostatics. Beneficially the formal positive charges always accumulate at the electropositive sulfur atom and qualitatively reproduce the real charge distribution.

A different bonding model that would not need to invoke d-orbitals was already suggested by George Pimentel and Robert Rundle [17, 18]. It involved a 3-centre 4-electron bond, and the planarity of the SO_x units in the sulfur oxides allows the formation of a delocalized π-electron system, leading to m-centre–n-electron bonding. Nowadays this simple model is challenged to provide more sophisticated explanations for "hypervalent" molecules (e.g., [19]).

The issue was investigated in detail by the whole cornucopia of contemporary theoretical methods, among them are the density functional theory (DFT), the electron localization function (ELF), the electron localization indicator (ELI), the quantum theory of atoms in molecules (QTAIM, vide infra Chap. 2) and the valence electron equivalent (γ). DFT calculations and QTAIM topological analyses could only establish a 3-centre 4-electron bond contribution to linear molecules like $[FHF]^-$ or $[F_3]^-$, a minute contribution in T-shaped molecules like ClF_3 of SF_3, but none in trigonal-bipyramidal complexes with no lone pair like PF_5 or $[SiF_5]^-$. The topological analyses yielded the fluorine bonds to the central atom with a high ionic character. In the series SF_2, $[SF_3]^-$, SF_4, to SF_6, the QTAIM charge at the fluorine atoms stays almost the same (−0.54, −0.67, −0.56 and −0.56e) so that the charge at the electropositive sulfur atom has to counterbalance the increasing negative charge (+1.80, +0.91, +2.20 and +3.34e). The same is valid for SiF_4 and $[SiF_5]^-$ (−0.80, −0.85 and +3.20, +3.23e, respectively) [20]. This suggests that the bond quality from the eight-electron-precise (SF_2) to the eight-electron-exceeding molecules (SF_6) is the same, but that the central electropositive atom is increasingly electronically depleted, proposing that the sum of valence electrons at the central atom is getting less rather than more than eight. Obviously the electron density gets increasingly accumulated at the fluorine atoms. This is visualized by the charge separated Lewis diagrams in red in Fig. 3 (vide infra). The ELF method, however, finds the effective valence density at the central sulfur atom from SF_2 and SF_4 to SF_6 to increase from 4.97 to 5.80 to finally 6.18e, but still not exceeding eight. It is only in the molecules with a small difference between the electronegativities of the substituents and the central atom that the valence density surpasses the number of eight. In SeF_6 it is only 2.18, while it is 10.98e in $SeMe_6$ (ΔEN Se/F: 1.43; Se/C: 0). In $TeMe_6$ it even reaches 11.10e [21]. It remains very interesting to see what numbers the ELI might give, because different from the ELF, this is not to use an arbitrary reference to the uniform electron gas [22]. Currently ELI-D seems to be a very promising method [23]. The most recent examination with the topic is most elementary and convincing at the same time. Marcus Durrant suggests the valence electron equivalent $\gamma(X)$ which stems from the formal charge and the electron count of each contributing resonance form. When $\gamma(X)$ equals eight, the atom obeys the original Lewis octet rule. If $\gamma(X) < 8$, the atom obeys the "modified octet rule" that takes care of the polarization of the bond. Only if $\gamma(X) > 8$, then neither form of the octet rule is obeyed and the atom is hypervalent [24]. For example, the central sulfur atom in SO_2 and SO_4^{2-} shows $\gamma(S)$ values of 5.25 and 4.34, respectively, hence no sign of hypervalency. Interestingly for the central oxygen atom in ozone, a traditional paradigmatic textbook example where the eight-electron rule must be obeyed, $\gamma(O)$ is 9.52, hence indicates hypervalency. The same is valid for the central sulfur atom in S_3 because here $\gamma(S)$ adopts 9.60. Again, the vanishing ΔEN promotes hypervalency; thus $\gamma(Cl)$ in perchlorate ClO_4^- is 9.11. Remarkably in this approach, QTAIM and NBO charges give the more reasonable γ values compared to Mulliken and Hirshfeld charges.

With this whole discussion in mind, we found it most important for teaching reasons to separate the dash employed in the Lewis diagram with the connotation of a 2-centre 2-electron bond (red diagrams in Fig. 3) from that of a simple topological line guiding the way to the arrangement of atoms in molecules tied up to structure their overall shape (black diagrams in Fig. 3) and coordination. In the visualization of the rock salt structure NaCl nobody would interpret the six dashes from each cation to the next neighbouring anions and vice versa as 2-centre 2-electron bonds

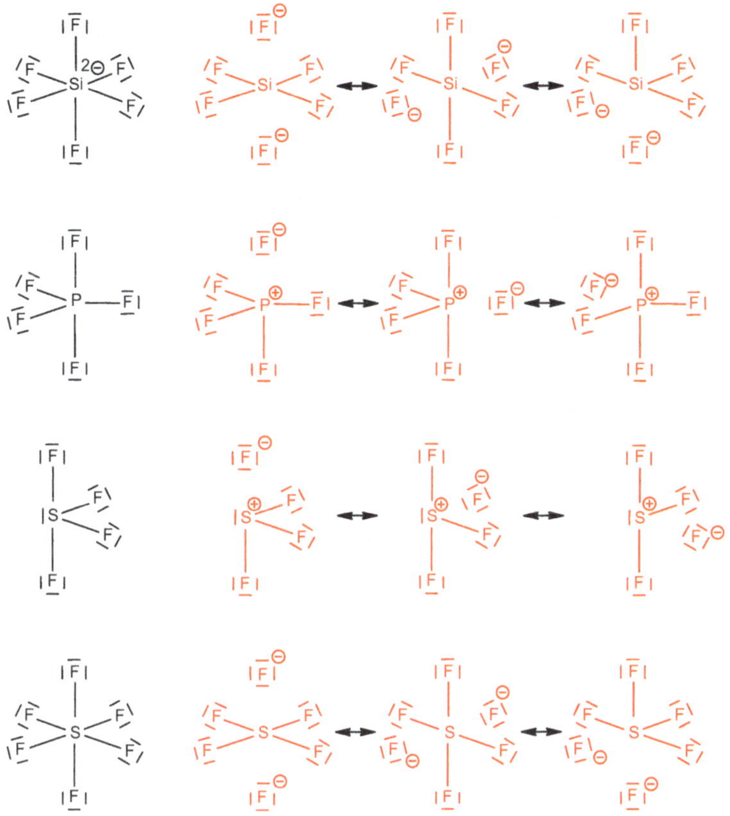

Fig. 3 Topological diagrams in *black* and Lewis diagrams in *red*

but only as topological lines. The same modus operandi is helpful in the high-coordinated species. The up to six dashes in the black diagrams in Fig. 3 are topological lines featuring the coordination, while the charge separated species denote the Lewis diagrams with the formal charges, qualitatively representing the real charge distribution. Obviously there is no need to draw the octet-compliant structures of, e.g., SiF_4 charge separated ($[SiF_2]^{2+}$ $2F^-$), although the Si–F bonding obviously is the same in SiF_4 and SiF_6^{2-}. On the other hand, the charge separated structures would neither prevent nor contradict the coordination pattern deduced from VSEPR, because the mere number of coordination items (central atom, ligands and lone pairs) is not affected by the charge separated Lewis diagram. It is only emphasizing the bond polarity.

Anyhow, the question remains whether there are any physically meaningful experimental measures to the validity of the Lewis model, a very resilient model over the last 100 years though, but nevertheless only a model.

2 Bond Properties from Atoms in Molecules

The electron density (ED) in molecules and materials is the overarching observable in natural science, and hence investigation of the ED provides an ideal tool to understand bonding [25–28]. It can be approached either way from the theoretical wave function and experimentally from the X-ray diffraction experiment. The latter still needs a model because in the experiment the phase information is lost and the electron density can only be recombined from the diffraction data by modelled phases. If the arrangement of the atomic nuclei in the crystal lattice is known, the structure factors can be calculated from a parameterized model. These calculated structure factors F_{calc} are then compared to the measured structure factors F_{obs}. By optimizing the parameters of the calculated model, F_{calc} is adjusted to F_{obs} in a least-square refinement; hence, the model is adjusted to the observation. The atom-centred multipole model (MM) by Niels Hansen and Philip Coppens [29] still provides the best method to describe the aspherical density. Within this approach the atomic density $\rho(\mathbf{r})$ is partitioned into three components: the spherical core density $\rho_{\mathrm{c}}(\mathbf{r})$, the spherical valence density $\rho_{\mathrm{v}}(\kappa\ \mathbf{r})$ and the aspherical valence density $\rho_{\mathrm{d}}(\kappa'\mathbf{r})$. κ and κ' represent radial scaling parameters that allow for an expansion or a contraction of the spherical valence density and the aspherical valence density, respectively. The aspherical valence shell density $\rho_{\mathrm{d}}(\kappa'\mathbf{r})$ is modelled by spherical harmonics, the multipoles, which provide a most flexible way to describe the measured density:

$$\rho_{\mathrm{atom}}(r) = P_{\mathrm{c}}\rho_{\mathrm{c}}(\mathbf{r}) + P_{\mathrm{v}}\kappa^3\rho_{\mathrm{v}}(\kappa\mathbf{r}) + \rho_{\mathrm{d}}(\kappa'\mathbf{r})$$

In addition to the Hansen and Coppens MM, Richard Bader's quantum theory of atoms in molecules (QTAIM) provided a most important analytical tool for the experimental charge density investigation [30]. QTAIM is based on the assumption that properties of a molecule are composed by the properties the single atoms contribute. Bader showed that the electron density distribution (EDD) can be partitioned uniquely into subsystems, the atoms in molecules. Mathematically the ED $\rho(\mathbf{r})$ is a scalar field. Thus, an atom can be defined as a basin that is limited by a surface where the gradient $\nabla\rho(\mathbf{r})$ vanishes, the so-called zero flux surface. Since the gradient always follows the highest increase, the gradient path must link a minimum or saddle point in the ED with a maximum or saddle point. Each of these basins thus includes only one core (maximum in $\rho(\mathbf{r})$). Integration of the ED over such a basin gives the Bader charge, reminiscent to the chemical charge of an atom or the oxidation state. If two cores are linked by a gradient path, this is a privileged exchange channel [31]. Within QTAIM such a path is called a bond path, a sufficient and necessary condition for a chemical bond. The beginning of such a path as well as its minimum is a so-called critical point in the electron density. At these extrema in the ED, the gradient $\nabla\rho(\mathbf{r})$ vanishes. For a further classification of the bonds, the Laplacian $\nabla^2\rho(\mathbf{r})$ is examined. The sign of $\nabla^2\rho(\mathbf{r})$ at the bond critical point (BCP) determines the kinetic energy $G(\mathbf{r})$ or the potential energy $V(\mathbf{r})$

to be dominant [30, 32]. A negative Laplacian indicates a local concentration of the ED, whereas charge depletion is shown by a positive Laplacian. This facilitates the identification of valence shell charge concentrations (VSCCs). They can either be located in interatomic bonding regions or in non-bonding areas. VSCCs not involved in any bonding are potential lone pairs [33–35]. Naturally VSCCs also provide information about the bonding type. If the VSCCs along a bond path overlap, this interaction is characterized as a covalent interaction. The Laplacian at the BCP is negative and the ED is relatively high. A closed shell interaction is characterized by a positive Laplacian at the BCP (the VSCCs do not overlap) and a low ED. Furthermore, a closed shell interaction goes along with a ratio of the eigenvalues $\eta = |\lambda_l|/\lambda_3$ much smaller than unity. However, especially for weak or very polar bonds where the Laplacian is close to zero, the characterization just on the basis of the values of $\rho(\mathbf{r})$ and $\nabla^2\rho(\mathbf{r})$ at the BCP is not sufficient. Therefore, the distribution along the whole bond path should be taken into account. Cremer and Kraka introduced the total electronic energy density $H(\mathbf{r})$, which is the sum of the potential $G(\mathbf{r})$ and the kinetic energy $V(\mathbf{r})$ [36, 37]:

$$H(\mathbf{r}) = G(\mathbf{r}) + V(\mathbf{r})$$

These energies can be calculated from the experimental EDD according to the approximation by Abramov [38] A negative value of the total energy density at the BCP indicates shared interaction, while values bigger than zero are indicative for closed shell interactions. Moreover the ratio G/ρ at the bond critical point should be less than unity for open shell and greater for closed shell interactions. For a comprehensive discussion and for further classification schemes, we refer to the reviews by Gatti [39–41]. Additional information about the bond can be gained from the Eigenvalues themselves, i.e., the ellipticity of a bond ε ($\varepsilon = |\lambda_1/\lambda_2| - 1$) can be used to analyze its shape. A perfect cylindric homonuclear non-polar σ-single bond is characterized by $\varepsilon = 0$ at the BCP. Larger values show deviation from such a symmetry and thus can indicate, for example, double bond character (Table 2).

Obviously the properties at the bond path are quite indicative to the nature of the bond, but it is much more advisable to monitor them all along the bond path because the position of the BCP sometimes is quite unsettled and the values change massively even by only slight changes in the BCP position.

In the following some benchmark molecules will be discussed from the QTAIM perspective to shine some light on the Lewis diagram and to evaluate whether this

Table 2 Some quantum theory of atoms in molecules properties at the bond critical point to characterize the nature of a bond

Properties at the BCP	Covalent	Closed shell
Electron density $\rho(\mathbf{r})$	High	Low
Laplacian $\nabla^2\rho(\mathbf{r})$	Negative	Positive
$\eta = |\lambda_1|/\lambda_3$	Bigger than unity	Smaller than unity
Total energy density $H(\mathbf{r})$	Negative	Positive

simple model endures in the light of physical meaningful bond properties. It seems clear from the beginning that such a simple model cannot transport all the detailed features provided from this sophisticated approach, but it should be persistent enough to serve as a didactical vehicle to get the basic ideas across. Although a model has to be reduced, it should be useful to get the main ideas. It will be seen that the Lewis diagram in its original flavor still is valid and explains a lot.

3 Valence: To be Underrun and Expanded?

Obviously, the two territories separated by the eight-electron rule are those of hypovalent, i.e., electron deficient and hypervalent, i.e., electron rich, compounds. The first contain preferentially a Group 13 central atom with only three bonding partners. Hence, the vacant p-orbital has to be filled either by back-bonding or aggregation. The first alternative is already exemplified by partial B=F double bonding in BF_3 (Fig. 2, top); the second is the domain of, e.g., diborane and polyboranes, containing a bridging Y-shape 3-centre 2-electron BHB bond. The other side is populated with hypervalent species, containing as central atom a 3p-block element or heavier accommodating more partners in the coordination sphere than tolerated by the amount of valence electrons and lone pairs, seemingly exceeding the eight-electron rule. In those cases it seems particularly interesting to look at the bonding by experimental and theoretical charge density investigations (Fig. 4).

3.1 *Boron as the Central Atom*

Group 13 element compounds play a leading role in materials chemistry, i.e., as Ziegler-Natta co-catalysts [42, 43] as well as in frustrated Lewis pair (FLP) hydrogen activation and transfer [44–46]. In such reactions it is inevitably important to fine-tune the Lewis acidity of borane compounds. In bis(pentafluorophenyl)(*N*-pyrrolyl) borane (**1**) (Fig. 5, top), only one C_6F_5 substituent in the omnipresent $B(C_6F_5)_3$ is substituted by a pyrrolyl heterocycle and changes the reactivity only slightly.

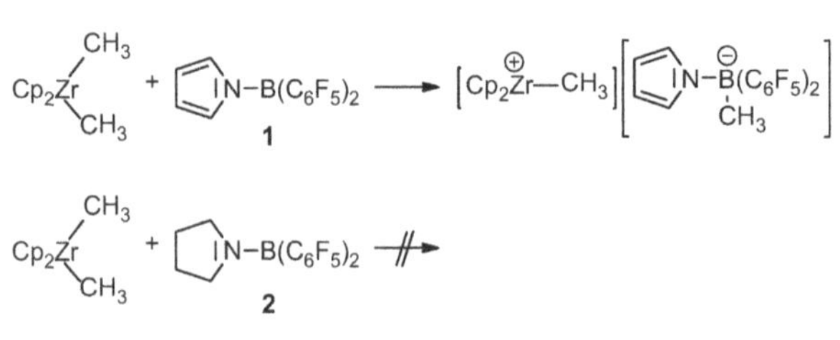

Fig. 4 Different reactivity of the boranes **1** and **2** just slightly different in Lewis basicity

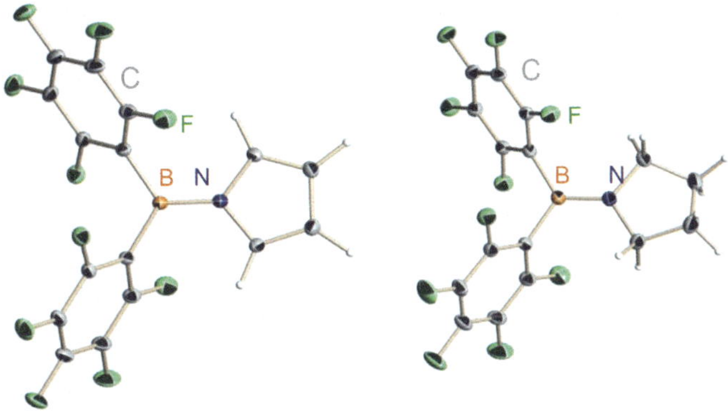

Fig. 5 Molecular structures of **1** and **2** (anisotropic displacement parameters are depicted at the 50% probability level)

Alkyl anion equivalents are still abstracted from zirconocene complexes and related compounds. In contrast the non-aromatic bis(pentafluorophenyl)(*N*-pyrrolidinyl) borane (**2**) (Fig. 5, right) does not show this ability. Obviously the less Lewis acidic boron atom in **2** is less attractive to the methanide than that one in **1**.

1 and **2** were determined by low-temperature high-resolution X-ray diffraction experiments, and their structures were subsequently refined by the multipole model. In addition, quantum chemical computations of the structures and electron densities, using density functional theory (DFT), were carried out. As anticipated from the different electronegativities, the BCPs of the B–N bonds are located closer to the boron than to the nitrogen atom. In **1** the distance of the BCP from the boron atom is 0.47 Å and 0.46 in **2**. The values of the electron densities, $\rho(\mathbf{r}_{BCP})$, the Laplacians, $\nabla^2\rho(\mathbf{r}_{BCP})$, and the ellipticities, $\varepsilon(\mathbf{r}_{BCP})$, at the BCPs of the B–N bonds in **1** and **2** are presented in Table 3. At the BCP of the shorter bond path in **2**, $\rho(\mathbf{r}_{BCP})$ is 0.11 eÅ^{-3}, $\nabla^2\rho(\mathbf{r}_{BCP})$ 5.56 eÅ^{-5} and $\varepsilon(\mathbf{r}_{BCP})$ 0.08 higher than in **1**. Comparison of the experimental and theoretical data in Table 3 reveals a satisfactory qualitative agreement for all properties, i.e., in particular all trends between **1** and **2** are the same. Especially the values of the Laplacian at the BCP deviate strongly which is well known from other comparative studies. One has to interpret those values for polar bonds with special caution, because the BCP lies at the rampant edge of the Laplacian and a small local change causes a tremendous change in $\nabla^2\rho(\mathbf{r}_{BCP})$. This is quite clear from Fig. 6 where the Laplacian for the B–N bonds of **1** and **2** is depicted along the bond path. It is slightly compressed for **2** relative to **1**, which leads to the observed increase at the BCP. The charge concentrations are almost exclusively located in the nitrogen basins, while they are depleted in the boron basins. Nevertheless, two minima in the interatomic regions, originating from the valence shell charge concentrations, are observed in both molecules, which indicate shared, even though severely polarized, interactions of a covalent bond. In homoatomic bonds a noticeable ellipticity at the BCP is regarded to be a sign of

Table 3 Topological properties at the B–N bond critical point of **1** and **2** (exp. in plain, theo. values (DFT-PBEh/TZVPP) in *italics*)

	B–N bond length [Å]	Bond path length [Å]	$\rho(\mathbf{r}_{\mathrm{BCP}})$ [$e\text{Å}^{-3}$]	$\nabla^2\rho(\mathbf{r}_{\mathrm{BCP}})$ [$e\text{Å}^{-5}$]	$\varepsilon(\mathbf{r}_{\mathrm{BCP}})$
 N–B(C_6F_5)$_2$ **1**	1.409/*1.415*	1.41/*1.42*	1.47/*1.40*	0.50/*13.48*	0.12/*0.02*
 N–B(C_6F_5)$_2$ **2**	1.374/*1.379*	1.38/*1.38*	1.58/*1.50*	6.06/*17.64*	0.20/*0.15*

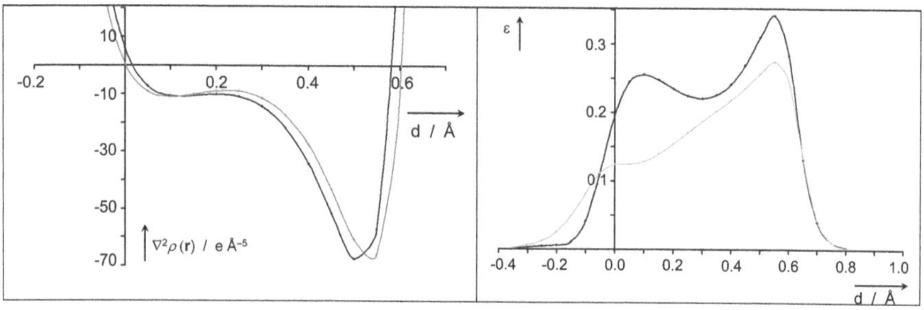

Fig. 6 Laplacian $\nabla^2\rho(\mathbf{r})$ (*left*) and ellipticity ε ($\varepsilon = (\lambda_1/\lambda_2) - 1$) (*right*) along the B–N bond for **1** (*light grey*) and **2** (*dark grey*) with d being the distance from the BCP (at 0.0 Å) and the boron basins spanning the negative values, while the nitrogen basins span the positive ones

at least partial π-bonding or a double bond, while a low ellipticity is typical for single or triple bonds. In both compounds, ε is higher than zero along the whole B–N bond path (see Fig. 5, right) Therefore, π-contributions to the B–N bonds in both molecules can be assumed. For **2** this π-contribution at any point is more pronounced than for **1**. Together with the more pronounced valence shell charge concentration (VSCC) at the nitrogen atom towards the boron atom in **2** (−70.2 vs −67.7 eÅ^{-5} in **1**), this gives a shorter B–N bond in **2**. While in **1** the density is more delocalized in the aromatic NC_4 five-membered aromatic ring, it is more pushed forward to the vacant p-orbital of the boron atom in **2**, lowering the positive electrostatic potential and hence the Lewis acidity.

In both molecules, in addition to the B–N σ-bonds, electron density from the nitrogen atoms couples into the B–N bond at two different levels. In **1** the charge density is restrained to a higher extent in the aromatic system, nevertheless, permitting delocalization over the nitrogen atom into the vacant p-orbitals of the boron atom. In **2** we find more electron density shifted from the nitrogen atom into the B–N bond (Fig. 7), leading to an increased boron-directed VSCC, very distinct π-contribution to the shortened B–N bond, and a less pronounced charge at the boron atom with some spatial shielding from nucleophilic attack. Hence, the topological analyses of the electron density distributions obtained from experiment and quantum chemical calculations can consistently explain marginal changes in bonding and rationalize different catalytic abilities [47]. Translated back to the

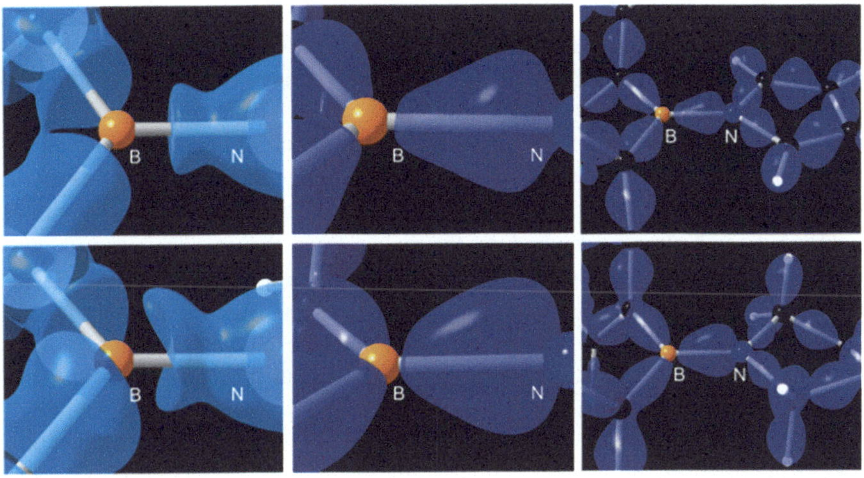

Fig. 7 Reactive surface ($\nabla^2\rho(\mathbf{r}) = 0$ eÅ^{-5}) around the boron atoms and the static deformation density $\Delta\rho_{\text{static}}(\mathbf{r})$ at the level of 0.2 eÅ^{-3} for **1** (*upper row*) and **2** (*bottom row*)

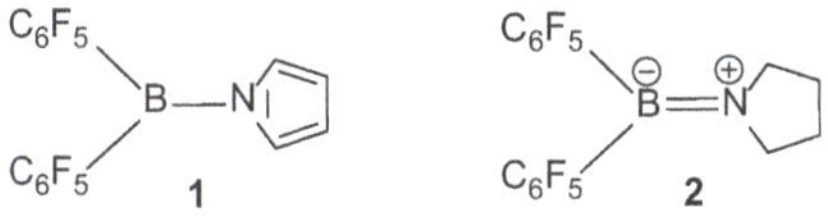

Fig. 8 Various weighted Lewis diagrams for **1** and **2**

Lewis diagram that would imply more B=N double bond character in **2** than in **1** in Fig. 8 contributes more to describe the bonding in **2** adequately than in **1**. The negative formal charge at the boron in **2** would not indicate the real charge density distribution but the lowered Lewis acidity compared to **1**.

3.2 *Sulfur as the Central Atom*

Ever since the introduction of the Lewis bonding concept, polyoxoanions such as SO_4^{2-}, PO_4^{3-} and ClO_4^- have been paradigmatic examples for the concept of hypervalency. Since Pauling's extended eight-electron rule the double bonds are envisaged to resonate among the oxygen atoms. Hence, we examined the S–O bonding by investigating the charge density of the sulfate group, SO_4^{2-}, within a crystalline K_2SO_4 environment based on both, experimental and theoretical methods. High-quality, very low-temperature (10 K) single crystal X-ray diffraction data were collected using a small crystal (~30 μm) and a high-energy (30 keV) synchrotron beam, because bigger crystals suffer from extinction and absorption. The experimental charge density was determined by multipole modelling, whereas a theoretical density was obtained from periodic ab initio DFT calculations [48]. Both approaches simultaneously characterize the S–O interactions as highly polarized, covalent bonds, with the single-bond charge separated description discharging the double bond picture totally and utterly. The average bond distance of 1.46 Å from the three crystallographically independent S–O bonds is much closer to the value normally quoted for an S=O double bond (1.43 Å) than an S–O single bond (1.57 Å) [49]. However, the ED at the S–O bond critical point is only 2.03 eÅ^{-3} and almost identical to the value from theory. The experimental Laplacian at the BCP is only slightly positive (0.75 eÅ^{-5}), while the theoretical (11 eÅ^{-5}) clearly indicates substantial closed shell contributions. This is further substantiated by the atom-like spherical shape of the oxygen and sulfur atoms (Fig. 9). The net charge at sulfur from the experiment is +4.3 but −1.4 at any oxygen atom, resulting in a charge of −1.36 for the sulfate anion. In theory, to all oxygen atoms three VSCCs in the non-bonding region could be resolved, pointing away from the sulfur atom (Fig. 10). They are indicative to the lone pairs. The only difference in the experiment is that for O2 only two VSCC regions pointing away from the sulfur atom are found. At O2 the non-bonded VSCCs are strong but very broad, and consequently different maxima could not be resolved.

In addition, the Source Function analysis provides additional insights. It quantifies the different extent to which the S- and O-bonded atoms determine the density at the BCPs, hence giving a measure of bond polarity and deviation from covalency. It also revealed a significant contribution (6%) from the neighbouring oxygen atoms to each S–O bond. When the S–O distance is decreased, the S–O bond unexpectedly becomes less covalent and even more polar. At the same time the neighbouring oxygen atom involvement increases, and electronic charge is transferred from the oxygen lone pairs to the S–O bonding region. From the Source

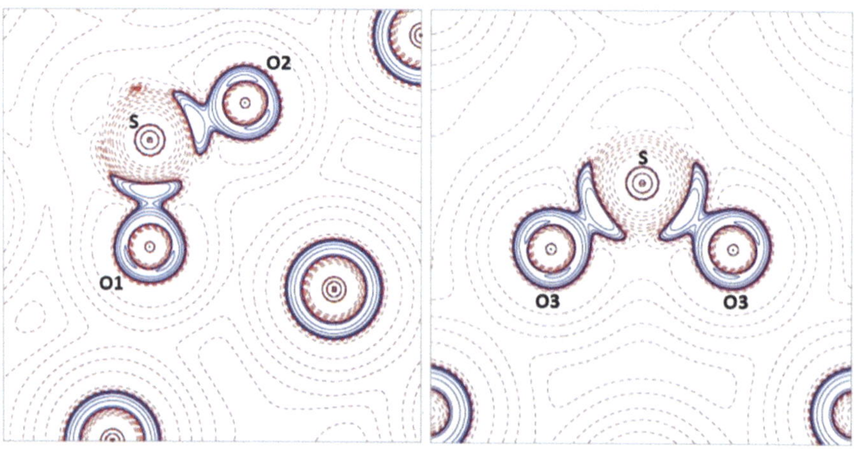

Fig. 9 Experimental $\nabla^2\rho$ maps cut in the O1–S–O2 (*left*) and O3–S–O3′ (*right*) plane of crystalline K_2SO_4. *Solid blue lines* mark regions of charge accumulation (negative $\nabla^2\rho$), while *dotted red lines* mark regions of charge depletion (positive $\nabla^2\rho$)

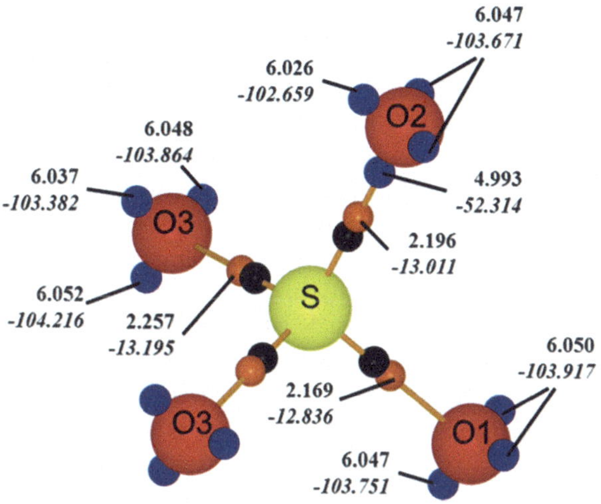

Fig. 10 Bonding and non-bonding maxima for the S–O interactions from theory; the *blue spheres* represent the oxygen VSCCs, while the *orange spheres* are the ones lying along the S–O bond paths, closer to the sulfur atom (~0.74 Å from the S nucleus) but still located in the oxygen basins. The values of ρ and the Laplacian (*italic*) are given for each maximum. The *small black spheres* mark the position of the BCPs

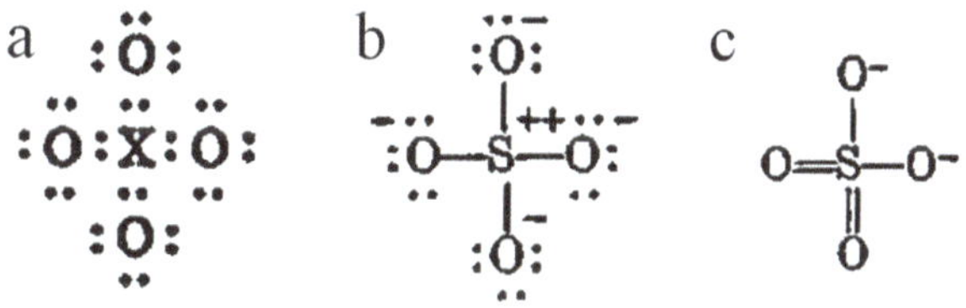

Fig. 11 Polyoxoanion formula introduced by Lewis (**a**) [1], the extreme sulfate structure, challenged by Pauling (**b**), and the representation with resonating double bonds, which he regarded as more satisfactorily (**c**) [10]

Function approach a picture emerges in which there is a partial contribution to bonding in the sulfate ion from resonant forms, enabling a significant electron exchange among all oxygen atoms. However, the S–O bond has such a polarized character that the actual S–O bond order does not exceed that of a standard covalent single bond. The present bonding analysis therefore rules out the hypervalent description of the S atom (formula **c** in Fig. 11).

To trace back those findings to the Lewis concept modified over the years, it is clear that the original approach was right:

> On the other hand, we may now write formulae in which an atom of oxygen is tied by only one pair of electrons to another atom and yet have every element in the compound completely saturated. To illustrate this important point we may write the formula of perchlorate, sulfate, orthophosphate and orthosilicate ions, in which each atom has a complete shell of eight electrons. Thus [formula **a** in Fig. 11, substituted by this author]

represents all of these ions. If **X** is **Cl** the ion has one negative charge; if **S** it has two negative charges, and so on. [1]

Ahead of the sulfate ion, we already found that the same is valid when the oxygen atoms in sulfate are valence isoelectronically replaced by imide groups (NR, S–N 1.69 and S=N 1.52 Å [49]) or methylene groups (CR_2, S–C 1.83 and S=C 1.63 Å [49]). $H_2C\{S(NtBu)_2(NHtBu)\}_2$ (**3**) contains three different S–N bonds: one sulfur-amide single bond, one sulfur-imide bond as a hydrogen bond acceptor and one pending sulfur-imide formal double bond [50, 51]. All were studied by experimental charge density investigations and none of them showed S=N double bond character. The first S–N(H)*t*Bu bond clearly shows the topological features of a S–N covalent single bond: $d_{(S\text{–}N1)}$ 1.65 Å $\rho(\mathbf{r}_{BCP}) = 1.89$ eÅ^{-3} and $\nabla^2\rho(\mathbf{r}_{BCP})$ −13.4 eÅ^{-5}, but the two others are quite similar despite the distances normally quoted as S=N double bond distances, i.e., $d_{(S\text{–}N2)} = 1.53$ Å $\rho(\mathbf{r}_{BCP}) = 2.31$ eÅ^{-3} $\nabla^2\rho(\mathbf{r}_{BCP}) = -16.6$ eÅ^{-5} and $d_{(S\text{–}N3)} = 1.52$ Å $\rho(\mathbf{r}_{BCP}) = 2.37$ eÅ^{-3} $\nabla^2\rho(\mathbf{r}_{BCP}) = -16.4$ eÅ^{-5}. At both imide nitrogen atoms, we could deconvolute two VSCCs in the non-bonding regions, indicative for the two lone pairs. Hence, all topological features suggest that the S–N/S=N bonding situation is much better described as a $S^{\delta+}$–$N^{\delta-}$ bond which is shortened by electrostatic reinforcement, resulting from bending of the nitrogen lone pairs towards the electropositive sulfur atom. This is also where the appreciable ellipticities of the S–N bonds are from. Remarkably that of the S1–N1 single bond is the highest of $\varepsilon = 0.11$ (Fig. 12).

The two S–N bonds in sulfur diimide $S(NtBu)_2$ (**4**), valence isoelectronic to SO_2, show virtually the same topological properties, i.e., $d_{(S\text{–}N1)} = 1.54$ Å $\rho(\mathbf{r}_{BCP}) = 1.93$ eÅ^{-3} $\nabla^2\rho(\mathbf{r}_{BCP}) = -9.44$ eÅ^{-5} and $d_{(S\text{–}N2)} = 1.53$ Å $\rho(\mathbf{r}_{BCP}) = 2.24$ eÅ^{-3} $\nabla^2\rho(\mathbf{r}_{BCP}) = -9.38$ eÅ^{-5} [52, 53]. Furthermore, these values are almost identical to those found for sulfur triimide $S(NtBu)_3$ (**5**), valence isoelectronic to SO_3 (Fig. 13) [52, 53].

Retranslating those findings back to Lewis diagrams, it is clear that the S–N bond is much more covalent than the S–O bond, already anticipated from the electronegativities. However, even here there is no sign of valence expansion of sulfur. The charge separated species obeying the eight-electron rule describe the

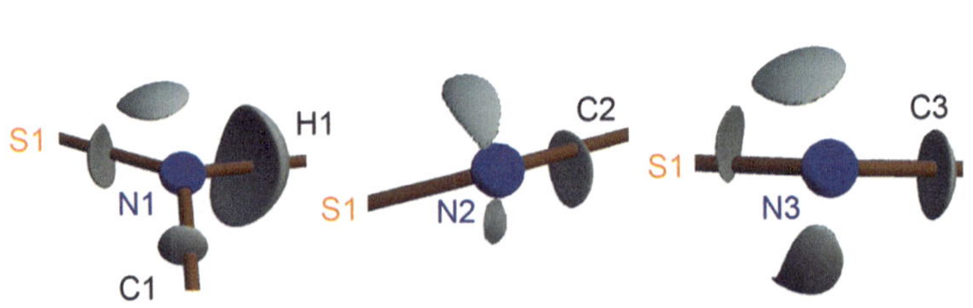

Fig. 12 Various VSCCs at the three differently coordinated nitrogen atoms in $H_2C\{S(NtBu)_2(NHtBu)\}_2$ (**3**)

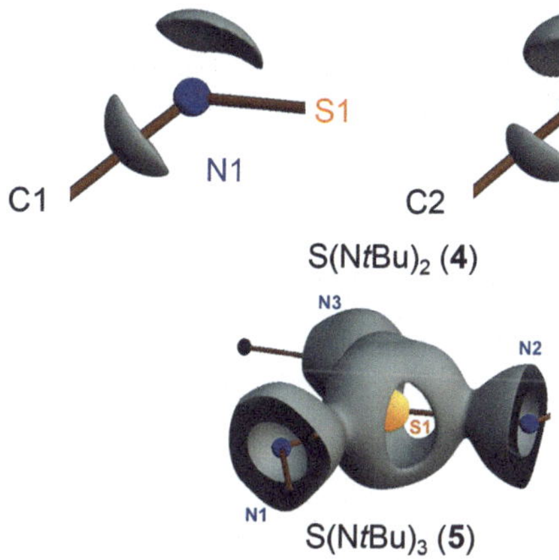

Fig. 13 VSCCs at the two nitrogen atoms in $S(NtBu)_2$ (**4**) (*top*) and the reactive surface in $S(NtBu)_3$ (**5**) (*bottom*)

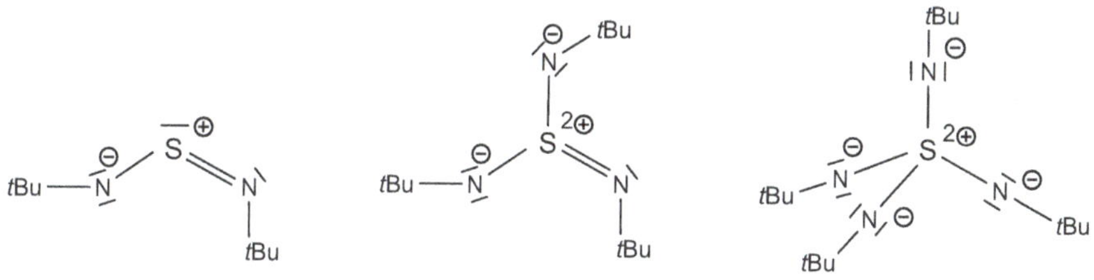

Fig. 14 Lewis diagrams for the charge separated species $S(NtBu)_2$ (**4**), $S(NtBu)_3$ (**5**), and $S(NtBu)_4^{2-}$

bonding much better than the hypervalent diagram favoured by Pauling to avoid the formal charges. The short bond distances are caused by electrostatic reinforcement rather than by double bonding. In fact here the formal charges in the Lewis diagram are beneficial because they qualitatively describe the real electron density distribution and the bond order of 1.5 in SO_2 and 1.33 in SO_3 correctly. Recently the solid-state structure of SO_2 was determined. The bond order from the X-ray wave function refinement was determined to 1.5, hence in tune with the charge separated species (Fig. 14) [54, 55].

3.3 *Silicon as the Central Atom*

The bonding in the hexacoordinated difluoro-bis-[*N*-(dimethylamino)phenyl-acetamide-*N*,*O*]silicon [$F_2Si\{OC(Ph){=}NNMe_2\}_2$] (**6**, Fig. 15a) might serve as a paradigmatic case of hypervalent central silicon because in this complex, it is coordinated by two nitrogen, two oxygen and two fluorine atoms [56]. Such high-coordinated silicon complexes are frequently referred to as hypervalent [57]. However, this cannot be confirmed by the facts. The electron density at the BCP is very low and only slightly above 1 for the Si–F bond (Table 4). Furthermore, the Laplacian at the BCP clearly is positive and only turns negative deep inside the N, O or F basine (Fig. 15c). Consequently η is less than unity, indicating only very few covalent character. This is further substantiated by the ratio G/ρ, which clearly is bigger than one and the electronic energy density H of nearly zero. The Si–F bonds can be characterized as almost purely ionic (Fig. 15b, c, Table 4). The bonds to oxygen show predominantly ionic character. All atoms in the SiO_2F_2 plane show ionic behaviour. The oxygen and fluorine atoms are almost spherical and silicon is electronically depleted. The integrated charges of +2.78e for silicon and −0.8e and −1.2e for fluorine and oxygen, respectively, encourage the description of the bonds as mainly ionic. Only the donating nitrogen

Table 4 Topological properties at the Si–E bond critical point of $[F_2Si\{OC(Ph){=}NNMe_2\}_2]$ (**6**)

Bond	$\rho(\mathbf{r}_{BCP})$ [eÅ^{-3}]	$\nabla^2\rho(\mathbf{r}_{BCP})$ [eÅ^{-5}]	H_{BCP}	G_{BCP}/ρ_{BCP}	η
Si ← N	0.501(16)	7.78(3)	−0.011	1.23	0.20
Si–O	0.766(13)	7.37(3)	−0.052	1.14	0.29
Si–F	1.015(13)	13.47(3)	−0.076	1.43	0.26

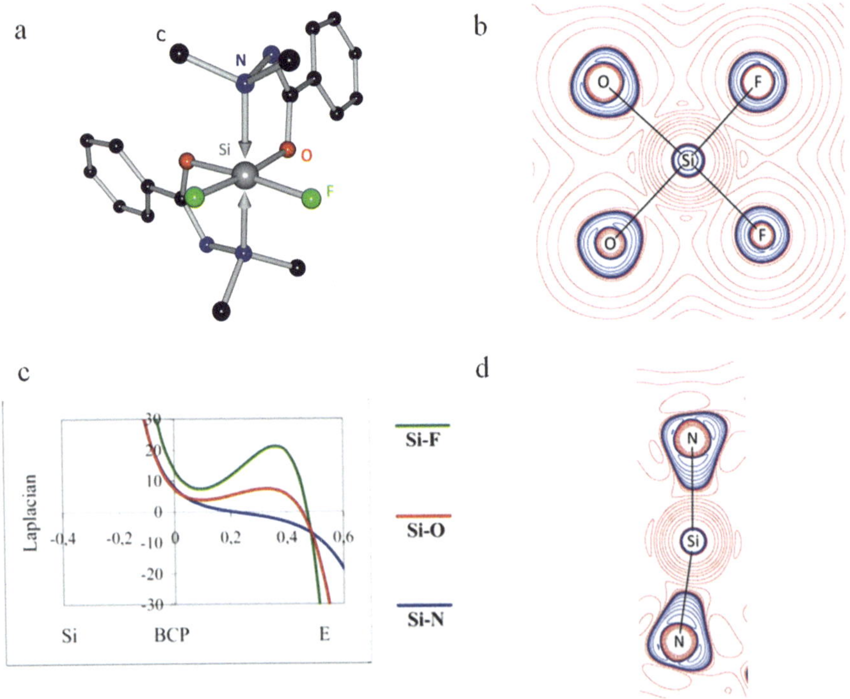

Fig. 15 (**a**) Molecular structure of $[F_2Si\{OC(Ph){=}NNMe_2\}_2]$ (**6**); (**c**) Laplacian distributions along the Si ← N, Si–O and Si–F bond paths; Contour plots of the Laplacian distribution in the O–Si–F (**b**) and N–Si–N (**d**) plane. Charge concentration depicted in *blue*, depletion in *red lines*

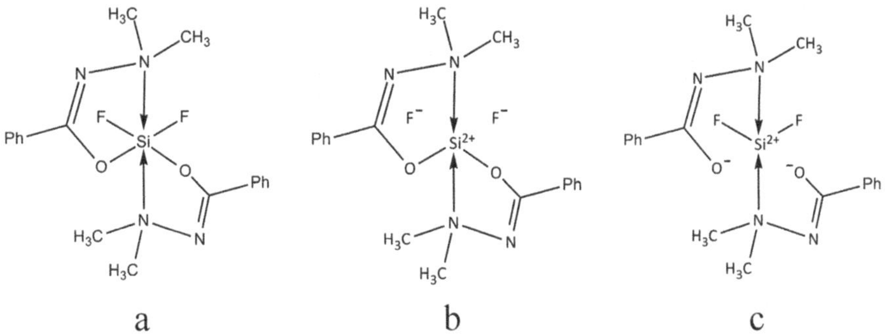

Fig. 16 Lewis diagrams for the charge separated species of $[F_2Si\{OC(Ph)=NNMe_2\}_2]$ (**6**)

atoms are polarized towards the silicon atom, because they direct their lone pair at the apical position of the amine side arm for electrostatic reasons to the centre of the depletion, hence the silicon atom (Fig. 15d). The Lewis structure of **6** should thus be described by three charge separated resonance structures of which **b** and **c** by far contribute most to the real bonding situation (Fig. 16). The arrow for the Si ← N donor bond certainly is the most useful connotation in these Lewis diagrams [58–63].

In quintessence, hypervalency and recruitment of energetically out-of-reach d-orbitals is a superfluous and misleading concept. Pauling's statement that only a Lewis diagram describes the bonding adequately if it avoids formal charges has to be withdrawn. Those charge separated species not only obey the eight-electron rule but also place the formal charges frequently with the right polarity at the right atoms. They avoid additional assumptions difficult to teach to beginners. Why should we draw ozone with a single double bond and sulfur dioxide with two double bonds when the single double bond obeys the eight-electron rule and even puts the formal charges in the places where the real ones are? Maybe we have to think about drawing ozone with two double bonds rather than SO_2 [24].

4 Aromaticity

Aromaticity is one of the most fundamental but also most fuzzy concepts in chemistry. First established to explain the unprecedented stability and unique reactivity observed in benzenoid chemistry, its validity has been expanded to countless further examples. The quantification of aromaticity, however, remains a large and vigorously discussed field of research, and different methods for the validation of the degree of aromaticity are still under debate (e.g., [64–66] and the whole thematic issue "Delocalization – Pi and Sigma" in Chem Rev 105(10)). In benzene C_6H_6, the king of aromaticity, the effect is visualized by resonating the three double bonds in the Lewis diagram between the two possible conjugated positions. Ever since aromaticity saw the light, it was interesting to replace a single or even more carbon atoms in that aromatic ring system by silicon atoms. The first stable compound containing a C_5Si six-membered ring has only been isolated and characterized recently [67]. That set the signal for the synthesis

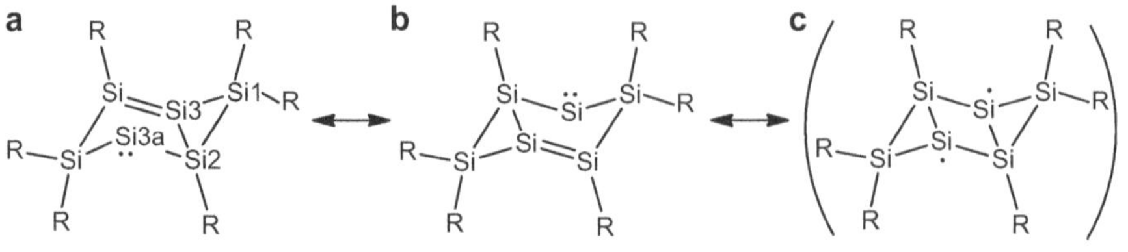

Fig. 17 Lewis diagrams for $(RSi)_6$ (**7**); (R = Tip = 2,4,6-triisopropylphenyl)

Table 5 Topological properties at the Si–Si bond critical point $(TipSi)_6$ (**7**)

Bond A–B	Si1–Si2	Si1–Si3	Si2–Si3a	Si2–Si3
$d_{A\text{–}B}$ [Å]	2.3275(1)	2.3245(1)	2.3089(1)	2.3245(1)
$\rho(\mathbf{r}_{BCP})$ [eÅ^{-3}]	0.537(5)	0.545(6)	0.595(11)	0.466(7)
$\nabla^2\rho(\mathbf{r}_{BCP})$ [eÅ^{-5}]	−1.645(8)	−1.628(8)	−2.285(12)	−1.164(9)
η	1.32	1.28	1.72	1.07

of a vast variety of aromatic perimeters containing various amounts of silicon atoms [68, 69]. We contributed an amidinato supported 1,4-disilabenzene [70]. However, the up-to-now most remarkable molecule certainly is the ring isomer of hexasilabenzene $(RSi)_6$, (R = Tip = 2,4,6-triisopropylphenyl) synthesized by David Scheschkewitz et al. [71]. The molecule forms a puckered six-membered ring that can synthesis-wise be rationalized as a head-to-tail dimer of two three-membered rings reminiscent to the structure of cyclohexane. Unlike the evenly hydrogen-substituted aromatic benzene, in $(TipSi)_6$ (**7**) hexasilabenzene the Tip substituents are not uniformly distributed among the six silicon ring atoms. Two silicon atoms (Si1 and Si1a in Fig. 17) of oxidation state +II are substituted by two Tip substituents, two (Si2 and Si2a) of oxidation state +I by only one, and two (Si3 and Si3a) of oxidation state 0 by none.

The authors came up with the new concept of aromaticity in the central planar Si_4 four-membered ring. There are six required electrons to suit the Hückel rule which can be cyclically delocalized: two from the transannular Si2–Si3 bond, two from the formal Si3=Si2a double bond and two from the lone pair at Si3a. That should give rise to dismutation and a merged oxidation state plus 6π-aromaticity, termed dismutational aromaticity. Obviously a standard structure determination is not suitable to provide any insight in such issues; hence, we performed a high-resolution structure determination $(\sin\theta/\lambda)_{max} = 1.11$) at 100 K followed by a multipole refinement and a topological analysis (Table 5) [72, 73].

Just looking at the distances in the molecule would not provide any clue. The Si–Si bond length range from 2.3089(1) to 2.3275(1) Å, and the expected double bond would only be two hundredths of an Angstrom shorter than the single bond. However, we found all bond paths and bond and ring critical points, suggesting that there is transannular Si2–Si3 bonding. At the divalent silicon atom Si3 we found a distinct VSCC of −2.08 eÅ^{-5} at the position where the lone pair is expected. Remarkably the oxidation states of +II for Si1, +I for Si2 and 0 for Si3 match quite well with their Bader charges obtained by integration of the atomic basins, i.e., +1.6e, +0.6e and −0.3e, respectively. The values of $\rho(\mathbf{r})$ at the BCP of the Si1–Si2 and Si1–Si3 bonds are similar, whereas the transannular Si2–Si3 bond accumulates 0.1 eÅ^{-3} less. The highest value of $\rho(\mathbf{r})$ is found at the shortest Si2–Si3a bond which should show some double bond character. The same is valid for the Laplacians. At the Si1–Si2 and Si1–Si3 bonds it is similar with −1.645(8) eA^{-5} and −1.628(8) eA^{-5}, respectively, while the transannular Si2–Si3 bond shows a much less pronounced value of −1.164(9) eA^{-5}. The most negative Laplacian is present at the Si2–Si3a bond with −2.285(12) eA^{-5}. The η values for the Si–Si bonds in **7** are by the factor 5 higher than for the Si–X bonds in the high-coordinated complex **6**. They all are above unity and indicate covalent Si–Si bonds. Again, the highest value is found at the Si2–Si3a bond (1.72) and the smallest at the weakest transannular Si2–Si3 bond (1.07).

In conclusion, our experimental charge density investigation proves that the assumption of dismutational aromaticity in the dark-green ring isomer of hexasilabenzene $(TipSi)_6$ **6** is valid. Clearly there is a VSCC present in the non-bonding region in the apical position of Si3, the silicon of oxidation state 0 (Fig. 18b). Furthermore, the transannular VSCCs of opposite silicon atoms

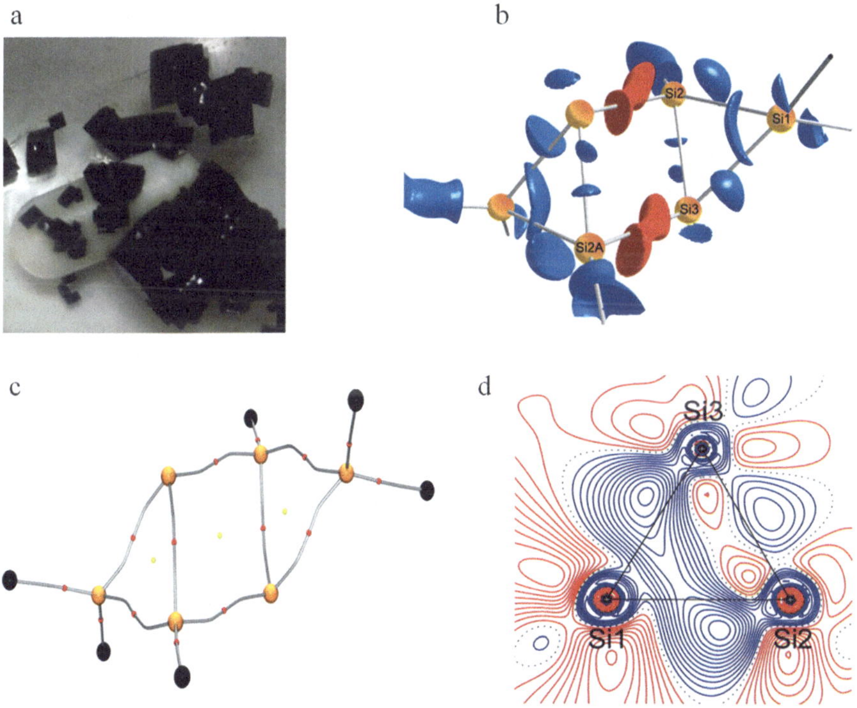

Fig. 18 Dark-green crystals of $(TipSi)_6$ (**7**, Tip=2,4,6-triisopropylphenyl) (**a**), Laplacian distribution around the silicon atoms of **7** at an isosurface level of −1.9 eÅ^{-5} (**b**), bond paths in **7** with the BCPs as *red spheres* and the RCP as *yellow spheres* (**c**), static deformation density contour plot of **7**, contour lines are drawn at ±0.015, 0.03, … eÅ^{-3} interval level; *blue*: positive; *red*: negative (**d**)

indicate the presence of two transannular bonds. This is further substantiated by the presence of bond paths and BCPs (Fig. 18c). The static deformation density shows distinctly bent bonds in the Si_3 triangles, because the maxima are clearly outside the straight lines forming that triangle (Fig. 18d). Additionally there is charge accumulated between Si2 and Si3, accounting for the presence of the transannular bond. Clearly a standard structure determination from an independent atom model would not have shown that subtle features, but the resonance forms from the Lewis diagram are remarkably good in visualizing the main features once they have physically established.

5 Donor Bonds

Most of today's silicon chemistry particularly on industrial scale still rests upon Si (IV) conversion. While the lower oxidation state +II gradually gets more stable while descending Group 14, the +IV oxidation state still is favoured by silicon. While $GeCl_2$ commercially is readily available, $SiCl_2$ was only known as a reaction intermediate and is known to polymerize at room temperature [74, 75]. Hence, it remains very challenging to generate and stabilize this reactive species without choking the reactivity by very bulky substituents. Almost all silylenes or low-valent silicon compounds reported so far are prepared by reductions of their parent halide compounds using strong reducing agents, such as potassium. We reported the synthesis of $SiCl_2$ under mild and metal-free reaction conditions by reductive elimination of HCl from trichlorosilane in the presence of an *N*-heterocyclic carbene (NHC) to give NHC → $SiCl_2$ (**8**) (Fig. 19) [76, 77]. At the same time the synthesis of NHC → $SiBr_2$ by potassium graphite reduction of NHC → $SiBr_3$ was reported [78, 79].

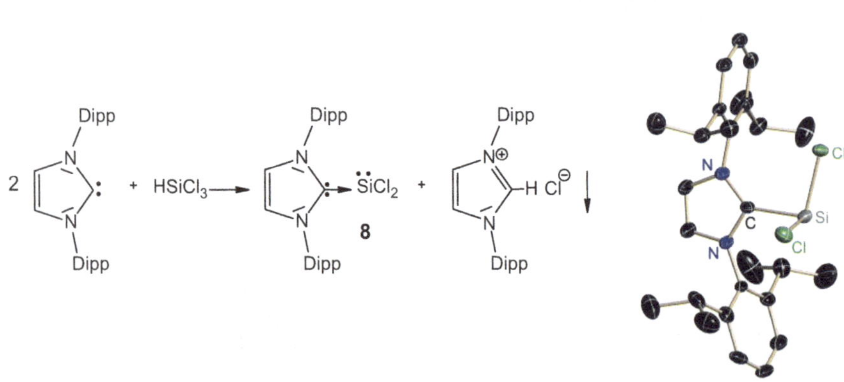

Fig. 19 Synthesis and structure of NHC stabilized dichlorosilylene NHC → $SiCl_2$ (**8**); (Dipp = 2,6-diisopropylphenyl)

5.1 Carbene Stabilized Dichlorosilylene

Since Paul von Ragué Schleyer and Gregory Robinson et al. published their silicon (0) compound with a Si=Si double bond, NHC→:Si=Si:←NHC [80], the dative and non-oxidative nature of carbene ligands in silicon chemistry is vigorously discussed [58–63]. Following the first statements than the carbene ligand would hardly deliver any electrons to the silicon and just kinetically shield the Si_2 moiety in the middle (**a** in Fig. 20). This might be reminiscent to the carbonyl ligand in transition metal chemistry where the σ-bonding is compensated by the π*-back donation. This interpretation would transfer some of the concepts known from coordination chemistry to main group chemistry. On the other hand the carbene atom in the NHC ligand might be regarded as a one-electron donor substituent like any aryl group in organometallic chemistry, yielding a Zwitterionic form with a formal negative charge at silicon and the positive delocalized in the NCN moiety of the five-membered imidazolyl ring. This of course is well established and would not need any further reasoning (**b** in Fig. 20). However, there is something special about the NHC ligand. The publications containing NHC ligands went up from the year 2000 with virtually none to 900 in 2014 with a total of 35,000 citations [81]. The ligand is superb to stabilize low oxidation states in main group as well as in transition metal chemistry, providing the first hint that it acts as a neutral ligand rather than a carbanion. So it might be worth looking at the NHC–Si bond by means of charge density determination [76, 77]. The structure of NHC → $SiCl_2$ (**8**) shows an almost orthogonal arrangement of the ligands at the silicon atom (Cl1–Si–Cl2 97.25(6), C–Si–Cl2 94.66(13), C–Si–Cl1 98.80(12)°). This suggests predominant p-orbital bonding at silicon with very little, if any, s-orbital participation, hence no hybridization at silicon. The topological analysis confirms the s-character of the lone pair at silicon (Fig. 21a). The Laplacian in the non-bonding region of silicon confirms the spherical shape of the VSCC and the sweeping envelopment of the silicon core by the density (Fig. 21b, d). This is far-off from a directed, stereochemically active lone pair expected to be accommodated in a sp^3 hybrid orbital. The Si–C bond different from the Si–Cl bond shows considerable ellipticity, and therefore we suggest the following qualitative bonding scenario (Fig. 21c). The NHC ligand accommodates the singlet lone pair in the directed sp^2 hybrid orbital, donating to the vacant p-orbital at silicon. Synergistically the silicon lone pair in the s-orbital donates back to the unoccupied p-orbital of the carbene carbon atom. Therefore, the arrangement of the NHC ligand relative to the $SiCl_2$ plane has

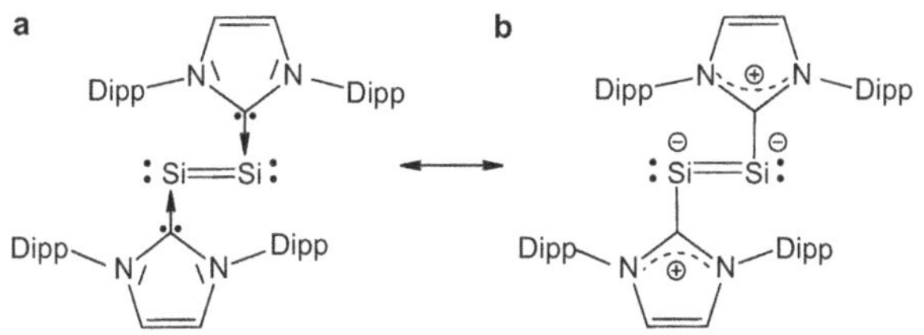

Fig. 20 Two Lewis diagrams to handle the NHC–Si bonding (from [58, 59])

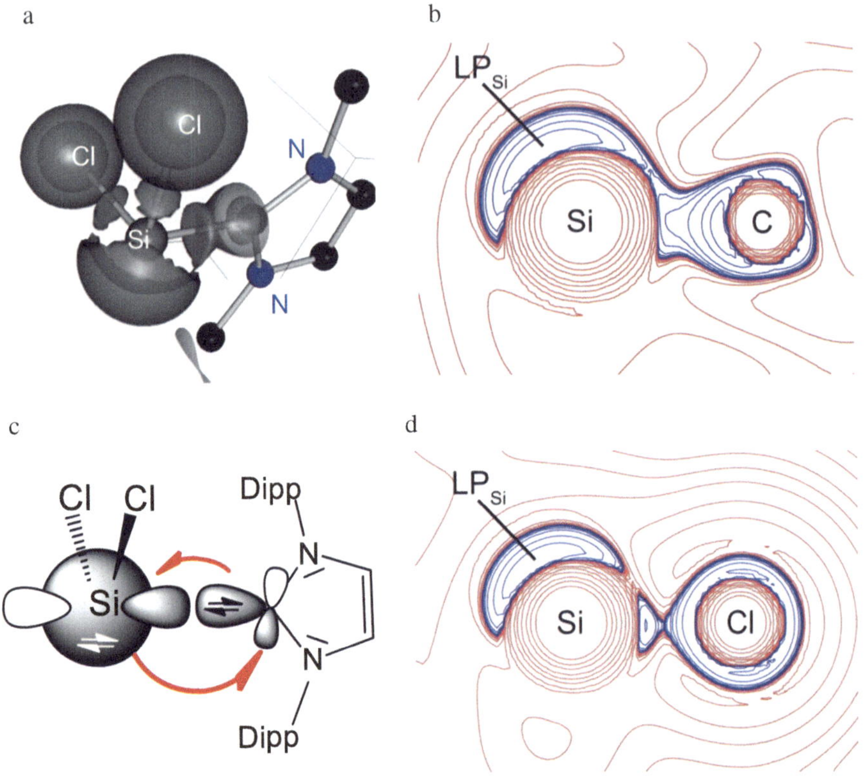

Fig. 21 Isosurface plot of $\nabla^2\rho(\mathbf{r})$ NHC$\rightarrow$$SiCl_2$ (**8**) at the -0.53 eÅ^{-5} level around Si (**a**), contour plots of $-\nabla^2\rho(\mathbf{r})$ in the C–Si–VSCC plane (**b**), orbital diagram for the dual donor-acceptor Si–NHC bond (**c**), and in the Cl(inplane)–Si–VSCC (**d**). Local charge concentrations are depicted in *blue*, charge depletions in *red*. The contour values are at 0.2×10^n, 0.4×10^n, and 0.8×10^n with $n = -3, \pm 2, \pm 1$

to be wider than 90° (96° in **8** and 97° in NHC$\rightarrow$$SiBr_2$), because otherwise the s-/p-orbital overlap integral would be zero. This of course is reminiscent to s-donation and p-back donation in coordination chemistry. Here the back donation is not facilitated with a π*-orbital but would just add to the heteroaromatic ring system of the ligand. Taking into account that the aromatic ring density is mainly accumulated at the ring nitrogen atoms [82, 83] then the depleted carbene carbon atom is attractive to that additional interaction.

From this point of view, we would prefer the Lewis diagram **a** in Fig. 20 emphasizing the donating arrow, because this transition metal perspective really adds to the problem. The average Si–C bond dissociation energy is 435(21) kJ/mol, whilst that in NHC$\rightarrow$$SiCl_2$ (**8**) is only 169 kJ/mol. Although this is still an appreciable value, it is not even half the value contributed from the one-electron donating organometallic standard substituents. Hence, a Lewis diagram with a double bond between the carbene and the silicon would suggest much stronger bonding than found. The arrow suites as well the found distances. The average Si–C_{ar} bond length from the CCDC [84] (746 bonds) is 189 pm, while the bonds in NHC silicon complexes are much longer (1.985(4) Å in **8**, 1.989(3) Å in NHC → $SiBr_2$ and 1.927(2) Å in NHC→:Si=Si:←NHC).

5.2 *Carbene Stabilized Silylone*

To further investigate the carbene → Si bond we synthesized and studied a single silicon(0) atom stabilized by two cyclic alkyl amino carbene (cAAC) ligands. As starting material serves the base stabilized dichlorosilylene NHC → $SiCl_2$ (**8**). From the three added equivalents cAAC the first is needed to replace the NHC in **8**, the second additionally coordinates the silicon atom and the third equivalent forms a C–C bond with the NHC (Fig. 22). $(cAAC)_2SiCl_2$ (**9**) forms two polymorphs (I and II) of blue-black block shaped crystals which differ only in the size of the unit cell [85, 86]. The molecule shows a tetra-coordinated central silicon atom bound to two chlorine atoms and two cAAC ligands. Both, the Si–Cl and the Si–C bond distances in **9** are significantly shorter than in **8** (av. Si1–C 1.847(2) vs 1.989(3) and av. Si–Cl 2.068(2) vs 2.166(2) Å). This already hints to a quite different bonding. **9** is a unpaired biradical and thus shows an EPR resonance of the related hyperfine structure. Both unpaired electrons are located at the carbene carbon atoms. The singlet-triplet excitation energy required for NHC (89 kcal/mol) is much higher than for cAAC (50 kcal/mol). The calculated bond dissociation energy of 227 kcal/mol in **9** is sufficiently compensating for the singlet-triplet excitation energy for $SiCl_2$ (60 kcal/mol) and two times cAAC (100 kcal/mol). The much higher singlet-triplet gap in NHC explains thermodynamically why a triplet state $(NHC)_2SiCl_2$ is not formed. Different from **8**, in **9** each cAAC forms a 2-centre 2-electron bond (see Figs. 22 and 23). Apart from the unpaired electron now the cAAC ligand seems to behave as a one-electron donating organometallic ligand and hence can be bound

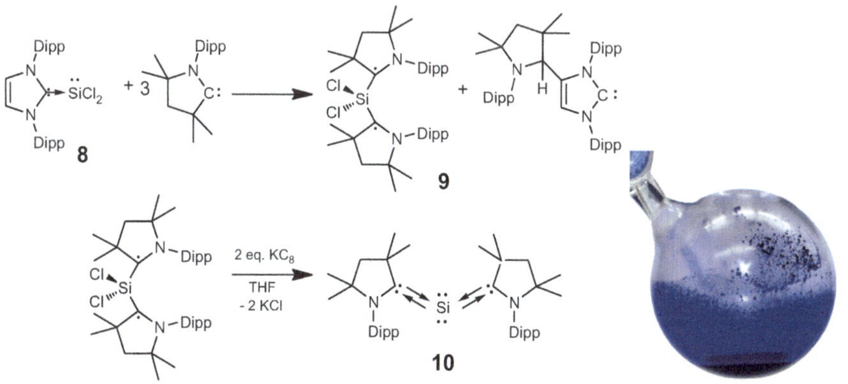

Fig. 22 Synthesis of $(cAAC)_2SiCl_2$ (**9**) and $Si(cAAC)_2$ (**10**) from NHC → $SiCl_2$ (**8**) and a reaction flask containing **9**

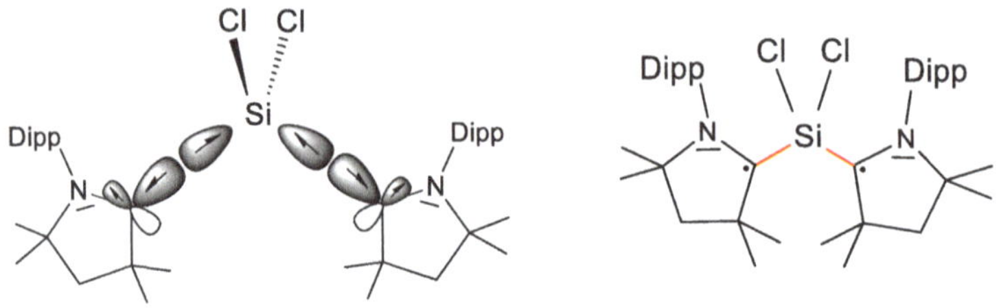

Fig. 23 Bonding in $(cAAC)_2SiCl_2$ (**9**) from the triplet cAAC and dichlorosilylene state overcompensating the required energy by forming two Si–C bonds in a deep blue biradical

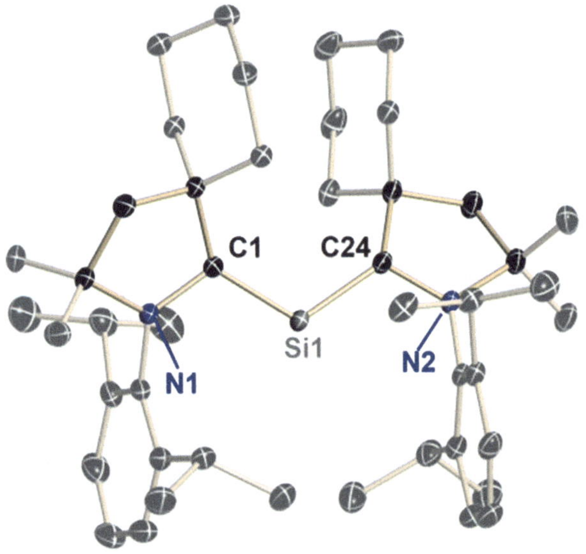

Fig. 24 Solid-state structure of $Si(cAAC)_2$ (**10**)

by a dash to the silicon atom in the Lewis diagram. Interestingly $(cAAC)_2SiCl_2$ (**9**) can dihalogenated twice by two equivalents of potassium graphite to give Si $(cAAC)_2$ (**10**) (see Figs. 22 and 24) [87, 88].

Formally this molecule now contains a silicon(0) central atom and could be interpreted as a silaallene if there were two Si=C double bonds (analogue **a** in Fig. 25). The substituents, however, make a difference. If the adjacent substituents are carbenes, then the molecule should rather be called a carbodicarbene or a carbone (**b** in Fig. 25). They have been predicted from theory by Gernot Frenking et al. to be bent (C–C–C $\approx$ 135°) and show two remarkable proton affinities (292 and 155 kcal/mol) [89, 90]. A year later those molecules were synthesized and employed in metal coordination by Guy Bertrand et al. and Alois Fürstner et al. [91–94]. Remarkably even the trisilaallene synthesized by Kira et al. shows a bent structure of 136° at the central silicon atom (**c** in Fig. 25) [95].

	L = NHC		L = cAAC
a	**b**	**c**	**d**
cumulenes	carbodicarbenes	trisilaallenes	siladicarbenes
allenes	carbones		silylones
	[89-94]	[95]	[96, 97]

Fig. 25 Various ways to rationalize the bonding in $Si(cAAC)_2$ (**10**)

Table 6 Topological properties at the Si–C bond critical points in $Si(cAAC)_2$ (**10**); theor. values in squared brackets

Bond	$\rho(\mathbf{r}_{BCP})$ [eÅ^{-3}]	$\nabla^2\rho(\mathbf{r}_{BCP})$ [eÅ^{-5}]	ε	η
Si(1)–C(1)	0.726 [0.742]	6.901 [10.838]	0.56 [0.29]	0.30 [0.21]
Si(1)–C(24)	0.741 [0.762]	5.095 [10.301]	0.13 [0.21]	0.31 [0.21]

$Si(cAAC)_2$ (**10**) shows a bond angle of 119.10(1)° at the central silicon(0) atom, so we embarked to study the bonding in more detail in an charge density investigation [96, 97]. There are small but significant differences in the bond lengths of Si1–C1 (1.8454(2) Å) and Si1–C24 (1.8615(2) Å), in the torsion angle between the five-membered ring and the silicon (C5–N1–C1–Si1 18.7(1)°, C28–N2–C24–Si1 11.5(1)°, in $\rho(\mathbf{r}_{BCP})$ of 0.726 for the Si1–C1 and 0.741 eÅ^{-3} for the Si1–C24 bond and the Bader charges of −0.51 for C1 and −0.37e for C24 (Table 6 and Fig. 24). Such differences were also reported for a similar germylone [98]. In the non-bonding region of Si1, we found two distinct VSCCs of −2.82 and −2.80 eÅ^{-5} in the position where one would expect the lone pairs of a potential silicon(0) atom (Figs. 26 and 27). Being not involved in any chemical bonding renders them potential lone pair indicators (see, e.g., Table 6). From this data both Si–C bonds have to be regarded as very polar bonds with a slight covalent contribution. However, by analyzing the ellipticity ε along the bond paths, a significant difference in the Si1–C1 and Si1–C24 bonds is recognized, possibly indicating different π-contributions.

The integrated Bader charges are 1.44 for Si1 and −0.51 for C1 and −0.37e for C24, respectively. An earlier NBO analysis [85, 86] including the option of a multicentre bonding gives one σ lone pair orbital and a three-centre C–Si–C π-orbital of which 40% is at Si and 30% at each C atom. This would lead to an expected Bader charge of 1.2e for silicon and −0.6e for each carbene carbon atom, matching reasonably the experimental values. The Laplacians along the Si1–C1 and Si1–C24 bond paths feature a very similar shape. At the BCP they have a slightly positive value and reach their minimum at about −30 eÅ^{-5} close to the carbene carbon atoms. The rather low ED at the BCP and the charge concentration and depletion along the different directions at the Si1–C1 and Si1–C24 BCPs support the very polar character of both Si–C bonds. In both cases η is <1 and even smaller than in other previously discussed cases for various analyzed Si–C, Si–O, Si–N [99], and S–N [100].

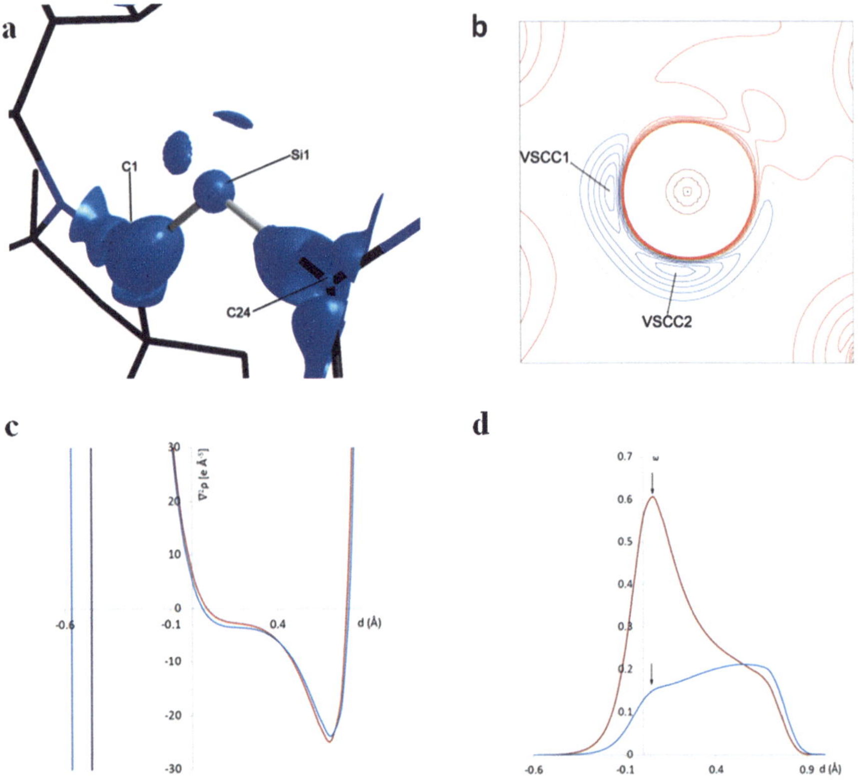

Fig. 26 Laplacian distribution at the silicon atom in **10** at an isosurface level of -2.5 eÅ^{-5} (**a**), orthogonal to the C1···C24 vector (**b**) and along the Si1–C1 (*red*) and Si1–C24 (*blue*) bond paths; ellipticity (ε) along the Si1–C1 (*red*) and Si1–C24 (*blue*) bond (**c, d**)

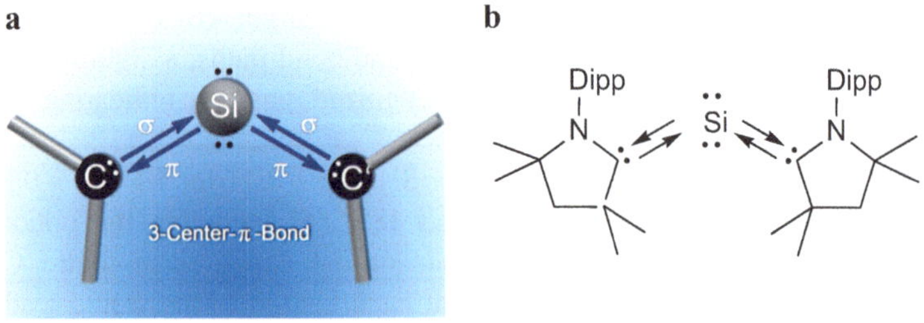

Fig. 27 Bonding in **10** from one σ lone pair orbital and a three-centre C–Si–C π-orbital of which 40% is located at Si and 30% at each C atom

In conclusion, we have shown by an experimental and theoretical charge density study that the interpretation of **10** as a silylone is valid. We were able to find two separated VSCCs present in the non-bonding region of the central silicon. The Bader charges correlate well with the values expected from NBO analysis. Furthermore, we were able to show that there are significant differences in the bonding situation of the two silicon carbon bonds, indicating a different amount of π-contribution. Hence, the donor bonds indicated by the arrows in the Lewis diagram are valid and a valuable tool to emphasize the different bonding. **10** is definitely not a silaallene where the central carbon(0) from allenes is simply replaced by a silicon atom.

6 Conclusion

It turned out that the original Lewis diagram is remarkably resilient and flexible to respond to the various new findings in traditional and established as well as in newly discovered main group molecules. Most of the reservations over the times seem to stem from the fact that the dash is equally used to indicate electron pairing in bonding in the Lewis sense as well as to indicate coordination in a topological meaning. As soon as we are aware of the two different connotations and we agree in what sense they are used, most of the difficulties disappear. It is clear that we have to discard hypervalency from very polar bonds and that we should accept formal charges in Lewis diagrams. There is nothing wrong with them. Apart from Group 13 elements, they even get the formal charges at the electropositive atoms if we accept charge separated diagrams. To obey the eight-electron rule prevents double bonding and causes formal charges. Double bonding is the privilege of low polar bonds anyway, and electrostatic contribution to covalent bonding explains bond shortening at least as good as multiple bonding. Although the Lewis diagram gets 100 years old, it is still much more vital than many other models.

Acknowledgement This work was supported by the Deutsche Forschungsgemeinschaft within the priority program 1178 "Experimental charge density as the key to understand chemical interactions," the DNRF-funded Center for Materials Crystallography, the PhD program CaSuS, Catalysis for Sustainable Synthesis, funded from the Land Niedersachsen, Chemetall, Frankfurt and the Volkswagenstiftung. The author is particularly indebted to many capable students for providing the results that form the basis of this article.

References

1. Lewis GN (1916) J Am Chem Soc 38:762–785
2. Schrödinger E (1926) Phys Rev 28:1049–1070
3. Frenking G, Shaik S (eds) (2014) The chemical bond. Wiley-VCH, Weinheim. ISBN 978-3-527-33318-9
4. Abegg R (1904) Z Anorg Chem 39:330–380
5. Werner A (1893) Z Anorg Chem 3:267–330
6. Langmuir I (1921) Science 54(1386):59–67
7. Johnson DR, Powell FX (1969) Science 164:950–951
8. Carberry JJ, Retton RH (1961) J Chem Phys 35:2240–2241
9. Shibata S, Bartell LS (1965) J Chem Phys 42:1147–1151

10. Pauling L (1960) The nature of the chemical bond, 3rd edn. Cornell University Press, Ithaca, Chapter 5
11. Kutzelnigg W (1984) Angew Chem 96:262–286
12. Kutzelnigg W (1984) Angew Chem Int Ed Engl 23:272–295
13. Reed AE, Weinhold FJ (1986) J Am Chem Soc 108:3586–3593
14. Reed AE, Schleyer PR (1990) J Am Chem Soc 112:1434–1445
15. Jensen WB (2006) J Chem Educ 83:1751–1752
16. See RF (2009) J Chem Educ 86:1241–1247
17. Pimentel GC (1951) J Chem Phys 19:446–448
18. Hach RJ, Rundle RE (1951) J Am Chem Soc 73:4321–4324
19. Braïda B, Hilberty PC (2013) Nat Chem 5:417–422
20. Molina JM, Dobado JA (2001) Theor Chem Acc 105:328–337
21. Noury S, Silvi B, Gillespie RJ (2002) Inorg Chem 41:2164–2172
22. Kohout M (2004) Int J Quantum Chem 97:651–658
23. Finzel K, Grin Y, Kohout M (2012) Theor Chem Acc 131:1106–1113
24. Durrant MC (2015) Chem Sci 6:6614–6623
25. Koritsanszky T, Coppens P (2001) Chem Rev 101:1583–1627
26. Gatti C, Macchi P (eds) (2011) Modern charge density analysis. Springer, Heidelberg
27. Stalke D (2011) Chem Eur J 17:9264–9278
28. Stalke D (ed) (2012) Electron density and chemical bonding I (experimental, vol 146) and II (theoretical, vol 147) structure and bonding. Springer, Berlin. ISBN 0081-5993
29. Hansen NK, Coppens P (1978) Acta Crystallogr A 34:909–921
30. Bader RFW (1990) Atoms in molecules – a quantum theory. Oxford University Press, New York
31. Pendás AM, Francisco E, Blanco MA, Gatti C (2007) Chem Eur J 13:9362–9371
32. Bader RFW (1998) J Phys Chem A 102:7314–7323
33. Henn J, Meindl K, Oechsner A, Schwab G, Koritsanszky T, Stalke D (2010) Angew Chem 122:2472–2476
34. Henn J, Meindl K, Oechsner A, Schwab G, Koritsanszky T, Stalke D (2010) Angew Chem Int Ed 49:2422–2426
35. Hey J, Leusser D, Kratzert D, Fliegl H, Dieterich JM, Mata RA, Stalke D (2013) Phys Chem Chem Phys 15:20600–20610
36. Cremer D, Kraka E (1984) Angew Chem 96:612–614
37. Cremer D, Kraka E (1984) Angew Chem Int Ed Engl 23:627–628
38. Abramov Y (1997) Acta Crystallogr A 53:264–272
39. Gatti C (2005) Z Kristallogr 220:399–457
40. Gatti C (2012) Struct Bond 147:193–285
41. Engels B, Schmidt TC, Gatti C, Schirmeister T, Fink RF (2012) Struct Bond 147:47–97
42. Sinn H, Kaminsky W (1980) Adv Organomet Chem 18:99–149
43. Storre J, Schnitter C, Roesky HW, Schmidt H-G, Noltemeyer M, Fleischer R, Stalke D (1997) J Am Chem Soc 119:7505–7513
44. Stephan DW (2009) Dalton Trans 3129–3136
45. Stephan DW, Erker G (2010) Angew Chem Int Ed 49:46–76
46. Stephan DW (2010) Chem Commun 46:8526–8533
47. Flierler U, Leusser D, Ott H, Kehr G, Erker G, Grimme S, Stalke D (2009) Chem Eur J 15: 4595–4601
48. Schmøkel MS, Cenedese S, Overgaard J, Jørgensen MRV, Chen Y-S, Gatti C, Stalke D, Iversen BB (2012) Inorg Chem 51:8607–8616
49. Rademacher P (1987) Strukturen organischer Moleküle. VCH, New York
50. Leusser D, Walfort B, Stalke D (2002) Angew Chem 114:2183–2186
51. Leusser D, Walfort B, Stalke D (2002) Angew Chem Int Ed Engl 41:2079–2082
52. Leusser D, Henn J, Kocher N, Engels B, Stalke D (2004) J Am Chem Soc 126:1781–1793
53. Henn J, Leusser D, Ilge D, Stalke D, Engels B (2004) J Phys Chem A 108:9442–9452

54. Grabowsky S, Luger P, Buschmann J, Schneider T, Schirmeister T, Sobolev AN, Jayatilaka D (2012) Angew Chem 124:6880–6884
55. Grabowsky S, Luger P, Buschmann J, Schneider T, Schirmeister T, Sobolev AN, Jayatilaka D (2012) Angew Chem Int Ed 51:6776–6779
56. Kocher N, Henn J, Gostevskii B, Kost D, Kalikhman I, Engels B, Stalke D (2004) J Am Chem Soc 126:5563–5568
57. Kost D, Kalikhman I (1998) In: Rappoport Z, Apeloig Y (eds) The chemistry of organic silicon compounds. Wiley, Chichester, pp 1339–1445
58. Himmel D, Krossing I, Schnepf A (2014) Angew Chem 126:378–382
59. Himmel D, Krossing I, Schnepf A (2014) Angew Chem Int Ed 53:370–374
60. Frenking G (2014) Angew Chem 126:6152–6158
61. Frenking G (2014) Angew Chem Int Ed 53:6040–6046
62. Himmel D, Krossing I, Schnepf A (2014) Angew Chem 126:6159–6160
63. Himmel D, Krossing I, Schnepf A (2014) Angew Chem Int Ed 53:6047–6048
64. Gleiter R, Herberhauer G (2012) Aromaticity and other conjugated effects. Wiley-VCH, Weinheim. ISBN 978-3-527-32934-2
65. Matito E, Poater J, Solà M, Schleyer PR (2009) In: Chattaray PK (ed) Chemical reactivity theory. Taylor and Francis, Boca Ratón, pp 419–438
66. Schleyer PR (2005) Chem Rev 105:3433
67. Tokitoh N, Wakita K, Okazaki R, Nagase S, Schleyer PR, Jiao H (1997) J Am Chem Soc 759(119):6951–6952
68. Lee VY, Sekiguchi A (2007) Angew Chem 119:6716–6740
69. Lee VY, Sekiguchi A (2007) Angew Chem Int Ed 46:6596–6620
70. Sen SS, Roesky HW, Meindl K, Stern D, Henn J, Stückl AC, Stalke D (2010) Chem Commun 46:5873–5875
71. Abersfelder K, White AJP, Rzepa HS, Scheschkewitz D (2010) Science 327:564–566
72. Kratzert D, Leusser D, Holstein JJ, Dittrich B, Abersfelder K, Scheschkewitz D, Stalke D (2013) Angew Chem 125:4574–4578
73. Kratzert D, Leusser D, Holstein JJ, Dittrich B, Abersfelder K, Scheschkewitz D, Stalke D (2013) Angew Chem Int Ed 52:4478–4482
74. Schmeisser VM, Voss P (1964) Z Anorg Allg Chem 334:50–56
75. Schenk VPW, Bloching H (1964) Z Anorg Allg Chem 334:57–65
76. Ghadwal RS, Roesky HW, Merkel S, Henn J, Stalke D (2009) Angew Chem 121:5793–5796
77. Ghadwal RS, Roesky HW, Merkel S, Henn J, Stalke D (2009) Angew Chem Int Ed 48: 5683–5686
78. Filippou AC, Chernov O, Schnakenburg G (2009) Angew Chem 121:5797–5800
79. Filippou AC, Chernov O, Schnakenburg G (2009) Angew Chem Int Ed 48:5867–5870
80. Wang Y, Xie Y, Wei P, King RB, Schaefer HF III, Schleyer PR, Robinson GH (2008) Science 321:1069–1071
81. Thomas Reuters (2015) Web of Science
82. Ott H, Pieper U, Leusser D, Flierler U, Henn J, Stalke D (2009) Angew Chem 121:3022–3026
83. Ott H, Pieper U, Leusser D, Flierler U, Henn J, Stalke D (2009) Angew Chem Int Ed 48: 2978–2982
84. Cambridge Structural Database, Version 5.36 (2014) Cambridge Crystallographic Data Centre, Cambridge
85. Mondal KC, Roesky HW, Schwarzer MC, Frenking G, Neudeck S, Tkach I, Wolf H, Kratzert D, Herbst-Irmer R, Niepötter B, Stalke D (2013) Angew Chem 125:3036–3040
86. Mondal KC, Roesky HW, Schwarzer MC, Frenking G, Neudeck S, Tkach I, Wolf H, Kratzert D, Herbst-Irmer R, Niepötter B, Stalke D (2013) Angew Chem Int Ed 52:2963–2967
87. Mondal KC, Roesky HW, Klinke F, Schwarzer MC, Frenking G, Niepötter B, Wolf H, Herbst-Irmer R, Stalke D (2013) Angew Chem 125:1845–1850
88. Mondal KC, Roesky HW, Klinke F, Schwarzer MC, Frenking G, Niepötter B, Wolf H, Herbst-Irmer R, Stalke D (2013) Angew Chem Int Ed 52:1801–1805

89. Tonner R, Frenking G (2007) Angew Chem 119:8850–8853
90. Tonner R, Frenking G (2007) Angew Chem Int Ed 46:8695–8698
91. Dyker CA, Lavallo V, Donnadieu B, Bertrand G (2008) Angew Chem 120:3250–3253
92. Dyker CA, Lavallo V, Donnadieu B, Bertrand G (2008) Angew Chem Int Ed 47:3206–3209
93. Fürstner A, Alcarazo M, Gooard R, Lehmann CW (2008) Angew Chem 120:3254–3258
94. Fürstner A, Alcarazo M, Gooard R, Lehmann CW (2008) Angew Chem Int Ed 47:3210–3214
95. Ishida S, Iwamoto T, Kabuto C, Kira M (2003) Nature 421:725–727
96. Niepötter B, Herbst-Irmer R, Kratzert D, Samuel PP, Mondal KC, Roesky HW, Jerabek P, Frenking G, Stalke D (2014) Angew Chem 126:2806–2811
97. Niepötter B, Herbst-Irmer R, Kratzert D, Samuel PP, Mondal KC, Roesky HW, Jerabek P, Frenking G, Stalke D (2014) Angew Chem Int Ed 53:2766–2770
98. Li Y, Mondal KC, Roesky HW, Zhu H, Stollberg P, Herbst-Irmer R, Stalke D, Andrada DM (2013) J Am Chem Soc 135:12422–12428
99. Kocher N, Selinka C, Leusser D, Kost D, Kalikhman I, Stalke D (2004) Z Anorg Allg Chem 630:1777–1793
100. Stalke D (2012) Chem Commun 48:9559–9573

Struct Bond (2016) 169: 89–130
DOI: 10.1007/430_2015_182

Published online: 15 October 2015

Lewis Description of Bonding in Transition Metal Complexes

Fu Kit Sheong, Wen-Jie Chen, and Zhenyang Lin

Abstract Transition metal complexes have been playing an increasingly important role in modern chemistry in the past century, and this is partly due to their distinctive structure and bonding features that allow them to play a special role in organometallic reactions. Despite their importance, the current understanding of their structure and bonding relies to a large extent on sophisticated quantum chemical treatments, which do not encourage the formulation of more generalized rules. In this review, commemorating the centennial anniversary of the seminal Lewis paper, we would like to go back to basics and start from the classical Lewis description and then combine some observations we obtain from modern molecular orbital theory to give a simple but general bonding picture for transition metal complexes. This model, albeit simple, provides a localized description to metal–ligand interactions in these complexes and allows us to easily treat various cases with atypical metal–ligand or even metal–metal interactions in a modular manner.

Keywords Electron counting • Lewis structure • Structure and bonding • Transition metal complexes

Contents

F.K. Sheong, W.-J. Chen, and Z. Lin (✉)
Department of Chemistry, The Hong Kong University of Science and Technology, Clear Water Bay, Kowloon, Hong Kong
e-mail: chzlin@ust.hk

1 Introduction

The structure and bonding in molecules have always posed the most basic questions in chemistry. They not only are the fundamental bases to the intrinsic properties of all compounds but are also highly related to the possible reactions compounds can undergo. Variations in structure and bonding behaviors with respect to variations of elements are central to the general trends we observe for compounds, and therefore a thorough understanding of structure and bonding from overall trends to detailed analyses is essential in the comprehensive understanding of chemistry.

Historically, numerous efforts have been made in this direction. Most notably, there is the Lewis description of bonding as sharing of electron pairs, as well as a "rule of eight" (which is what we called "octet rule" nowadays) formulated initially on a cubic arrangement of valence electrons [1]. Even though we now know that electrons do not arrange around a cube, this localization of electrons into bonds and the "octet rule" are still highly influential to our way of rationalizing chemical compounds and still appear in elementary chemistry textbooks which have influenced generations of students and researchers.

This very basic approach to understanding the electron arrangement in compounds has later been adopted to explain the electronic structures of a much wider variety of examples beyond simple compounds. For example, Sidgwick extended the electron-counting procedure by a concept known as "effective atomic number" (EAN) [2]. Instead of handling only the valence electrons, the core electrons are also included in the counting to give rise to an inert-gas-like configuration. Indeed his EAN approach was not only outstanding in describing simple organic and inorganic compounds but also successful when applied to many examples of transition metal coordination compounds known at that time.

Another important development in valence electron counting comes from Langmuir, who (unlike Sidgwick) only considered the valence electrons and stated the primitive version of the 18-electron rule based on the transition metal carbonyl compounds simply by counting the valencies of transition metal centers [3]. This rule was found to be applicable to a very large group of transition metal complexes. Even almost a century after its proposal, this rule is still of fundamental importance nowadays in the field of coordination and organometallic chemistry.

Indeed, even a cursory glance of the literature shows a large number of examples that conform to the predictions of these approaches. Even if we confine ourselves to consider only the transition metal compounds, which will be the main focus of this chapter, we can still see a wide range of examples that conform to the 18-electron rule (Fig. 1).

The abovementioned approaches have been quite useful in handling the electron counts and determining the formal charges of compounds based on the chemical topologies. However, as become apparent in Fig. 1, coordination compounds adopt a wide range of geometries, yet the relationship between electrons and compound geometries is not explicitly handled in these theories (not even for main group compounds). Though as time passes, the Lewis concept of electron pair localization in bonding was later exemplified by two very important models, namely, the valence shell electron pair repulsion (VSEPR) model [12, 13] and the valence bond (VB) model (or sometimes known as the "hybridization" model) [14–17]. These two models partially addressed the relationships between the bonding and the geometries of compounds.

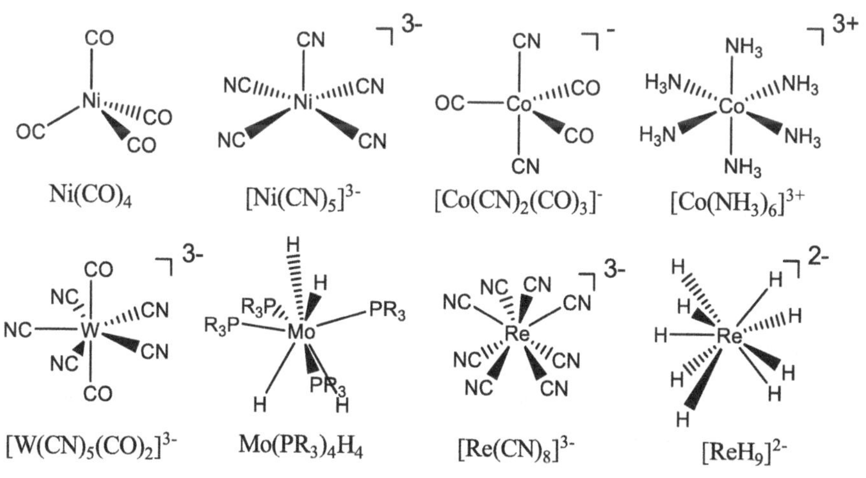

Fig. 1 Examples of various transition metal compounds conforming to the 18-electron rule [4–11]

The VSEPR model is an approach whereby bonding pairs and "lone pairs" are considered as electron domains. These domains interact with each other through an "exclusion" process (according to Gillespie, one of the original developers of the VSEPR model, the model should be more properly called as "valence shell electron pair domain," as they do not interact as a repulsive process [13]). These exclusions will then give rise to different geometries for various compounds.

On the other hand, the VB model developed by Pauling was based on a more quantum-mechanical-based model of valence electrons. Instead of simply putting electrons into spatially separated regions (domains), atomic orbitals are involved in the discussion, and all bonding interactions are considered as overlaps between orbitals of different atoms. This approach has been very useful in bridging the Lewis model to the more modern molecular orbital treatment, and similar to VSEPR, the "hybridization" scheme developed by Pauling has provided an important theoretical underpinning for chemistry over the years.

These models are successful in explaining many relationships between structure and bonding, but still if we follow their original formulation, there are actually various limitations that prevent them from being directly applied to transition metal complexes. For example, when applying VSEPR directly to a transition metal complex, it is only the bonding pairs that are usually considered in the exclusion process to give geometries like octahedron or square antiprism, whereas the effect of the presence of d electrons (which might to certain extent resemble the "lone pairs" in main group compounds) is seldom discussed, even though they do play a crucial role in forming geometries like square planar complexes.

At the same time, even in the field of main group chemistry, these models are still proven to be inadequate in modern applications like spectroscopy, and therefore researchers have moved on to more elaborated models for describing various compounds. Accompanied by many important developments in quantum mechanics, molecular orbital theory caught researchers' attention and has since become commonly used in understanding various aspects of chemistry [18]. Not only can it give powerful predictions (except those that are dominated by correlative effects or relativistic effects), it is also an important foundation of modern quantum chemistry. Still, despite its extreme usefulness, it only gives a highly delocalized picture, meaning that a modular description of complex molecules becomes inconvenient or even impossible (in other words, it is not easy for us to analyze a compound based on their fragments or "functional groups" under the molecular orbital paradigm). For example, in the description of a simple methane molecule (CH_4), the bondings between the central C and all the H atoms are partitioned into the interactions of the ligand group orbitals (LGOs) with one 2s and three 2p orbitals on carbon. Although this description is spectroscopically accurate [19], it by no means reflects our "intuitive" understanding of the four equivalent C–H bonds.

Emphasis on increased accuracy often leads to sacrifice of simple pictures/concepts, though this seems to become a trend for modern theoretical chemistry. Yet, on the other hand, most of our understanding and a priori predictions on reactions are actually still based on simple models/concepts such as Lewis structures, functional groups, lone pairs, etc. These models, albeit simple, usually give us very good approximations on structure and bonding and serve as good bases for discussion in various reactions. The fact that these models are used from blackboards to top-notch journals should not be a mere historical coincidence, but instead this should reflect their important role in chemistry.

However, despite the widespread use of functional group concepts and Lewis pictures in main group compounds (especially organic chemistry), in the discussion of transition metal complexes, these simple models are not used as thoroughly as in their main group counterparts. Most descriptions stop at the number of "bonds" formed with the metal center, and the actual metal–ligand interactions are usually handled as the interaction between the atomic orbitals of a metal center and the LGOs as in the methane case. The d electron count as well as the geometry of a metal complex is also often handled purely by molecular orbital description. People seldom go back to build a simple and modular picture in order to give more insightful descriptions (even though people seldom use molecular orbitals to describe methane or other simple organic compounds except in spectroscopy). In a more general setting, crystal structures or computational results are usually accepted as a fact, instead of considering them to follow an underlying trend that can be derived from more general understanding. This seems to be an unnecessary downfall for the simple models, as we believe they can be adopted to handle more complicated cases in transition metal complexes.

In view of this clear research gap, some efforts have been previously proposed. One notable effort is Hoffmann's fragment orbital approach that localizes orbitals into fragments and makes extensive analogies with main group compounds [20]. Another example is the Complementary Spherical Electron Density (CSED) model by Mingos [21, 22], which provides a very interesting view on the relationship between bonding and nonbonding orbitals in transition metal complexes and has significantly influenced the localization picture we put forward in this work.

In this work, we will take into account both CSED and Pauling's hybridization scheme [14, 15], together with a revised view on the concept of "lone pair," and construct a modular picture for the structure and bonding of transition metal complexes. In the process, we will make use of the orbitals resulting from the CSED analysis, relate them to the "lone pair" concept in the Lewis description, and then infer their impact on the overall molecular geometry. We hope that through our analysis we can provide chemists with a simple localized picture that could serve as an extensible building block for understanding transition metal complexes of different complexities.

2 Octahedral Transition Metal Complexes

We commence our discussion of various examples of transition metal complexes, by comparing and contrasting the molecular orbital (delocalized) approach and the Lewis (localized) approach based on the most well-known geometry, i.e., octahedral transition metal complex.

2.1 *Molecular Orbital Approach*

As mentioned in the Introduction (Sect. 1), an approach commonly used in recent years to handle transition metal complexes is the molecular orbital approach. To set the scene for development of a localized model, we will begin our discussion by first giving a general account of this molecular orbital approach (which represents the "delocalized model") based on the octahedral complexes.

For all transition metal complexes, the valence orbitals provided by the transition metal center are always the nd, (n+1)s, and (n+1)p orbitals. On the ligand side, the ligand orbitals are mixed into ligand group orbitals (LGOs) governed by both the number of ligands present and also the symmetry of their configuration [23]. In the case of an octahedral complex, where the six ligands reside on all six vertices of an octahedron, the ligand orbitals will be mixed into six ligand group orbitals ($a_{1g}+t_{1u}+e_g$), as seen in Fig. 2.

In an octahedral complex, the interactions between the metal center and the six σ-bonded ligands are usually described by considering the overlap between the nine valence atomic orbitals of the metal center and the six ligand group orbitals (Fig. 3). Among all the orbitals formed via their overlap, the lowest one is formed by the

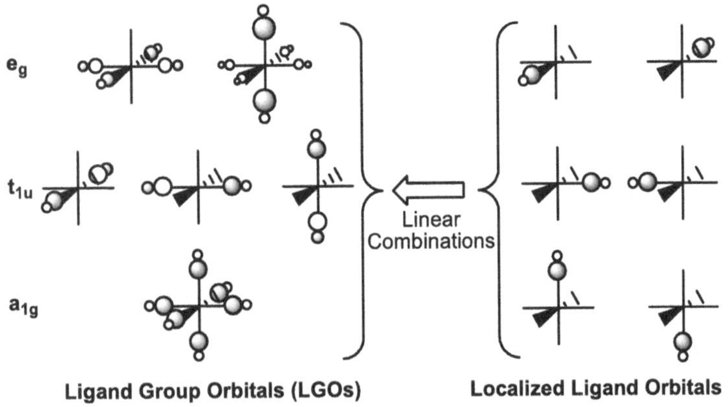

Fig. 2 Mixing of localized ligand orbitals into ligand group orbitals (LGOs). Due to the symmetry constraint of the coordination complex, the ligand orbitals are usually not considered one by one but mixed into ligand group orbitals for interaction with the transition metal center

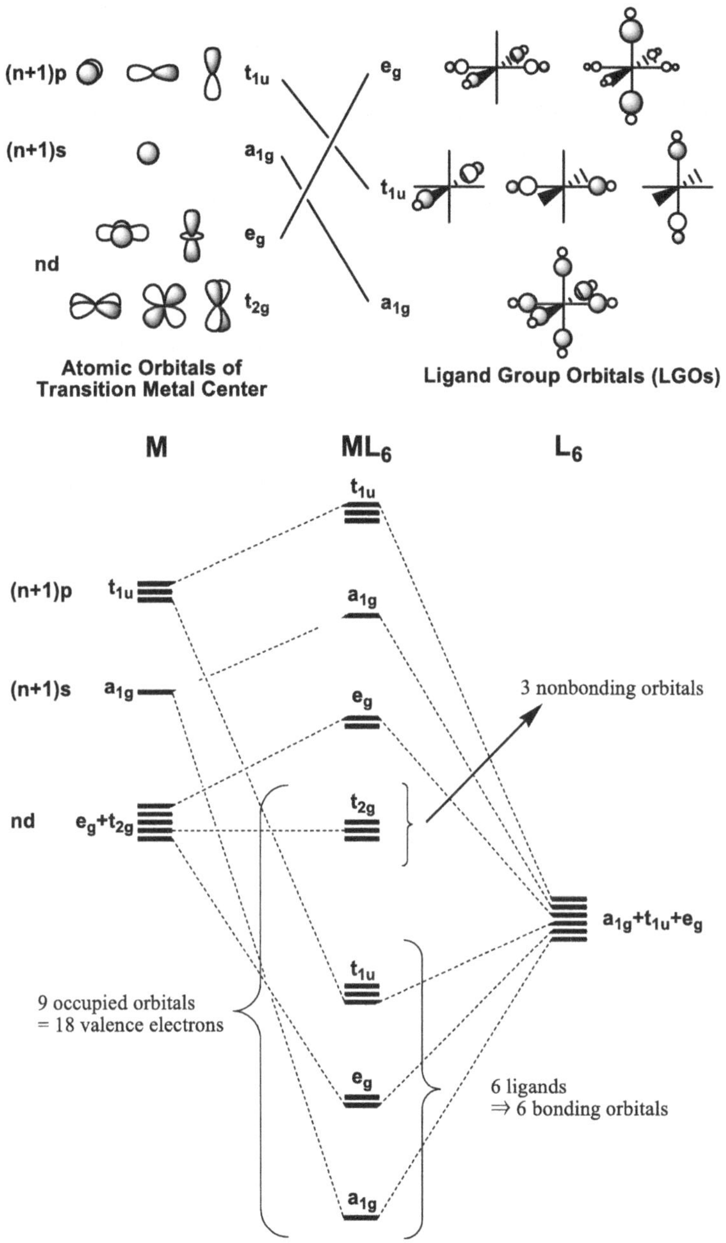

Fig. 3 Qualitative description of metal–ligand orbital interactions in octahedral transition metal complexes. On the *top part* of the figure, the valence orbitals for both the transition metal center and the ligands are shown. The valence orbitals provided by the transition metal center are the 5 *n*d orbitals, 1 (*n*+1)s orbital, and 3 (*n*+1)p orbitals. The ligand group provides a set of ligand group orbitals (LGOs) governed by the molecular symmetry. Note that there are certain LGOs with the same symmetry as the atomic orbitals in the transition metal center (denoted by *lines*), and these pairs will interact with each other and form bonding/antibonding pair in the overall molecular orbitals (*bottom part* of the figure). Note that there are no corresponding LGOs for the t_{2g} set of

overlap between the a_{1g} sets, then the next set is formed by the overlap between the e_g sets, and the last set of bonding molecular orbital is formed by the overlap between the t_{1u} sets. Note that in the process, although all the metal's (*n*+1)s and (*n* +1)p orbitals are already paired up with their corresponding LGOs and form bonding/antibonding molecular orbitals, only two d orbitals (d_{x2-y2} and d_{z2}) are involved in bonding, and there are still three d orbitals not involved (i.e., they cannot find their corresponding LGO partners), namely, the d_{xy}, d_{yz}, and d_{xz} orbitals, which remain purely as nonbonding orbitals.

2.2 Lewis Description

The six metal–ligand σ-bonding molecular orbitals, however, can be reconsidered via a Lewis approach. Note that in a classical Lewis-like consideration of an octahedral complex, we can describe the complex as having six metal–ligand σ bonds by considering the ligand donor electron pairs as the bonding pairs of electrons between metal and ligands. Thus, we argue that there actually exists a localization scheme that serves as the 6-to-6 mapping from the six bonding molecular orbitals to the six classical bonds. This mapping might sound a bit ad hoc; however these six "classical bonds" can actually be interpreted as the "natural localized molecular orbitals," and this localization is equivalent to a unitary transformation of the bonding molecular orbitals [24]. When we put it this way, it can be seen that the "molecular orbital" picture in the transition metal complex is actually not much different from a classical Lewis treatment. However, due to the unitary transformation during the localization in our treatment, the orbital energies no longer follow the usual interpretations based on Koopman's theorem, though it can be seen from later examples that with this sacrifice, the localized model can effectively serve as a building block for constructing larger molecules.

Another way to interpret this localization process is via a Pauling's hybridization approach, where we consider each M–L bond as an overlap between a metal hybridized orbital and a localized ligand orbital. As seen in Fig. 4, we can "hybridize" the valence atomic orbitals on the metal center into orbitals that are localized on each M–L axis [25]. These "hybridized orbitals" can then overlap with the corresponding localized ligand orbitals (Fig. 2, right-hand side) to form the six M–L bonds.

Interestingly, from Pauling's hybridization scheme, there are still orbitals that are not utilized and remain unhybridized, and these orbitals also do not contribute to the M–L bondings [28]. We will therefore view these orbitals as nonbonding orbitals, and the question now becomes how to localize these nonbonding orbitals.

Fig. 3 (continued) orbitals in the transition metal center (corresponding to the d_{xy}, d_{yz}, and d_{xz} orbitals), which remains as nonbonding

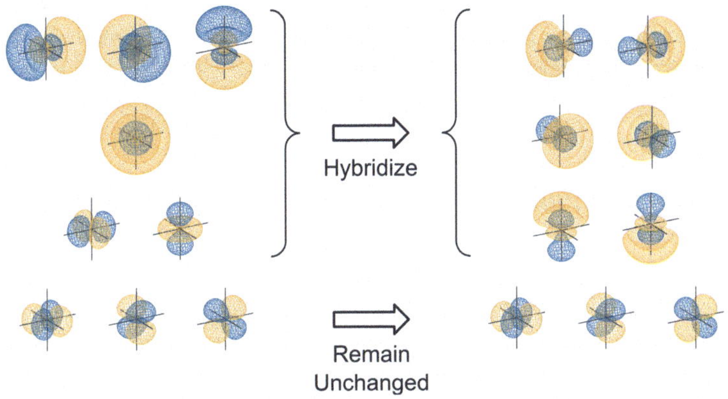

Fig. 4 Hybridization of transition metal atomic orbitals into hybrid orbitals for octahedral complex. The *left column* denotes the set of valence atomic orbitals and the *right column* denotes the hybrid orbitals, which can to a certain extent be considered as a reverse process of the delocalization in Fig. 2. All orbital plots or density plots in this chapter are based on hybridizing hydrogenic orbitals with hybridization factors taken from [25] and plotted with matplotlib [26] and visvis [27] visualization libraries. The density levels in all these plots are manually chosen for ease of visualization of density peak features

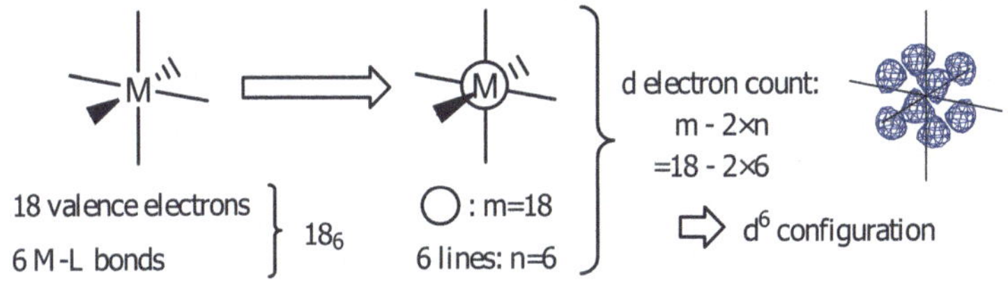

Fig. 5 Detailed description on the notation adopted in this chapter. The *circle* on the metal will indicate that the metal center follows 18-electron rule. The density plot on the *right* illustrates the combined densities of the six nonbonding electrons in an octahedral complex

Here, we will bring out one core view of this work: these nonbonding orbitals can be kept as is, but instead, we ignore the usual one-to-one relationship between a lone pair and an electron density peak/domain. For this octahedral case, the electron density contributed by the occupation of the three nonbonding orbitals occupies the eight corners of a cube (see Fig. 5 right-hand side). Note that in the classical view of "lone pairs" used in the VSEPR model, three "lone pairs" can only contribute to three "electron domains" instead of eight. We argue that by allowing this "multifurcation" of lone pair (meaning we allow a lone pair to have multiple density peaks), we can simplify a lot of our discussions and can get an interesting view for Lewis description for transition metal complexes.

With these analyses in hand, we have basically simplified an octahedral coordination complex into six "bonded pairs" (for the six classical M–L bonds) and three "lone pairs" (as in a d^6 electron configuration from the transition metal center), yielding a total of 18 valence electrons around the metal center. The feature that makes our model differ from the classical model is that the three lone pairs are localized toward the eight vertices of a cube.

2.3 Valency Counting Rule

One interesting point to note from the analysis given above is that similar to the octet rule for main group compounds, we can also derive an analogous "valency counting rule" for transition metal complexes. In particular, similar to the hybridization of 1 *n*s orbital and 3 *n*p orbitals in a main group compound to give sp^x hybridization and octet rule, we can similarly apply this to 9 orbitals (5 *n*d, 1 (*n*+1)s, 3 (*n*+1)p) so that they follow the "18-electron rule" (although the electron pairs can be contributed by either the metal or the ligands). Note that the choice of these nine orbitals is invariant to the choice of ligands (similar to the case in main group compounds, where the choice of s and p orbitals does not depend on the bonded element), but instead they can depend on the "dimension" of the compound (like BF_3 is planar and the boron center does not follow the octet rule). Moreover, as will be discussed in later examples, these nine orbitals might not always be filled in some geometries (even though the complexes are, say, three-dimensional).

In this work, we will adopt the notation 18_6 to indicate a coordination center with 18 valence electrons and 6 coordination bonds. The notation is similar to that of Sidgwick's in the presentation of effective atomic number [2]; however, instead of considering both core and valence electrons (for the determination of an "effective atomic number"), the 18 here only includes the valence electrons rather than all electrons. Because of the abundance of examples that follow 18_n, we will pictorially indicate this by drawing a circle on the metal center, and the number of M–L bonds present automatically denotes the number n in the notation, as seen in the example shown in Fig. 5. Note that because we are considering the contribution of the electrons, it will be convenient for us to consider only the electron densities based on the corresponding orbital contributions. To make all density contributions from different orbitals comparable, in this work we only consider cases where orbitals are either empty or doubly filled.

With the notation we defined, we can easily count the number of "nonbonding pairs of electrons" that contribute to the nonbonding densities, for example, in the octahedral case (18_6), each of the six M–L bonds is viewed as a classical 2-center-2-electron bond that contributes two electrons ($2 \times 6 = 12$ out of 18), and there will be three "lone pairs" (or six electrons, $18 - 2 \times 6$) left. These three lone pairs can then reside in the three d orbitals that we indicated in Fig. 3 as "nonbonding." In general, for an m_n complex (where $m = 18$ in many cases), there will be 2n M–L bonding electrons (or n bonding pairs) and the d electron count can be determined by $(m - 2n)$. This can then bridge our discussion between simple electron counting and the geometry of the coordination complex and will be very important in the determination of the number of nonbonding electrons in metal complexes with other coordination numbers.

3 Transition Metal Complexes with Seven, Eight, and Nine Coordination

To apply the valency counting rule we deduced above, we note that for a three-dimensional (i.e., neither planar nor linear), 18-electron ML_n coordination complex, we expect there would be $(9 - n)$ lone pairs (or $18 - 2n$ nonbonding d electrons). In particular, if we take the valency 18_n as a general rule of thumb, we would expect that the seven-coordinate complexes are most likely to have d^4 configuration (regardless of the actual coordination geometry), because we expect the nine valence pairs are filled by seven bond pairs and two remaining "nonbonding lone pairs" (or four electrons). Similarly, we would expect eight-coordinate complex to have a d^2 configuration and nine-coordinate complex to have a d^0 configuration. Indeed, the 18_n model applies well in all these coordination modes (as illustrated in Fig. 1). The underlying electronic reasons behind the 18-electron rule have been discussed in various works [22, 29]. In this section, however, we will first treat this as a rule of thumb and focus our effort into locating the "nonbonding orbitals," which we can seek help from the CSED model developed by Mingos [21, 22].

3.1 Pentagonal Bipyramidal Complexes

From the localized treatment mentioned above, for a pentagonal bipyramidal 18-electron ML_7 complex, there should be two "lone pairs" and seven M–L bonding pairs. In other words, we expect that there are exactly two "nonbonding orbitals" that will host the two "lone pairs," and our job here will be to locate these two "nonbonding orbitals."

If we directly apply the CSED model to determine the nonbonding orbitals, we will see that the d_{xz} and d_{yz} orbitals (the e_1'' set) are available but did not contribute to the metal–ligand bonding in the pentagonal bipyramid geometry. We can therefore declare that these two orbitals should each be filled with two electrons, consistent with the valency counting picture that the metal center should have a d^4 configuration.

We might also view these 4 d electrons from an alternative angle: the major feature of a pentagonal bipyramidal coordination geometry that distinguishes it from the octahedral coordination geometry is the extra ligand on the equatorial plane, and this incorporation of an extra ligand allows the d_{xy} orbital from the metal center to also participate in the metal–ligand bonding. At the same time, due to the

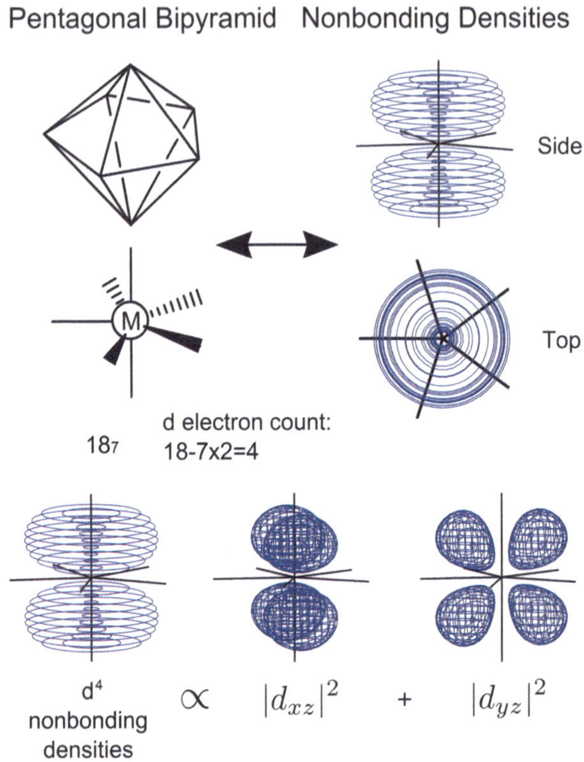

Fig. 6 Valency counting picture and the nonbonding densities for pentagonal bipyramidal geometry. Note that the nonbonding densities are contributed by occupation of two different orbitals: d_{xz} and d_{yz} (each with an occupation number of 2)

change in symmetry, the d_{x2-y2} orbital on the equatorial plane no longer degenerates with the d_{z2} orbital. Instead, d_{x2-y2} degenerates with d_{xy} to give an e_2' set (and now they both participate in the M–L bonding with the equatorial ligands). Note that, with this introduction of an extra ligand to the xy plane, all the s, p, d orbitals having maximum amplitude on the xy plane are all utilized for the M–L bonding, whereas for an octahedral complex, the d_{xy} orbital is not utilized.

Moreover, due to the participation of the d_{xy} orbital in equatorial M–L bonding, it is no longer a nonbonding orbital. If we continue our comparison with the octahedral complex, it can be easily seen that the only two orbitals that remain nonbonding are the d_{xz} and d_{yz} orbitals (the e_1'' set). Note that in total there are seven different atomic orbitals involved in M–L bonding with the seven different LGOs, which can again be transformed into seven classical M–L bonds. Here, we can see that the number of M–L classical bonds and number of "lone pairs" on the metal center are all consistent with the simple valency counting picture (Fig. 6).

3.2 Square Antiprismatic and Dodecahedral Complexes

If we step up another notch to an eight-coordinate complex, we will see that one widely found coordination geometry is the square antiprism. Before we even consider the orbital contributions based on the geometry, we should expect that if all nine valence orbitals were utilized, we will have the nonbonding pair contributed solely by electrons on one orbital (which from the subsequent analysis we know it is a d orbital).

To determine the specific orbital that accommodates the lone pair in an 18_n complex (all *n*d, (*n*+1)s, and (*n*+1)p orbitals being available), we can apply the CSED model [21]. For a square-antiprismatic complex, the d_{z2} orbital should hold the 2 d electrons. On the other hand, a dodecahedral complex (mathematically snub disphenoid, although in the context of coordination geometry, it is usually called "dodecahedron") should have the d_{x2-y2} orbital holding the 2 d electrons. It should be noted that there is actually certain level of s-d mixing in the case of square antiprism; however, the effect of this mixing is small and we will delay our discussion of these mixing effects to a later part of this text (Figs. 7 and 8).

3.3 Tricapped Trigonal Prismatic Complexes

The final example is a nine-coordinate tricapped trigonal prismatic complex. From either CSED or molecular orbital analysis, all the nine valence orbitals on the metal center (1 (*n*+1)s, 3 (*n*+1)p, 5 *n*d) are used to form bonds with the ligands, and no nonbonding orbitals are present, again consistent with the valency counting result.

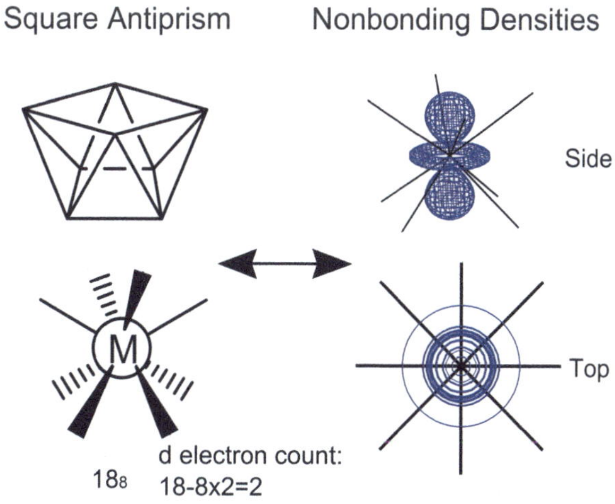

Fig. 7 Valency counting picture and the nonbonding densities for square-antiprismatic complexes. Similar to [25], s orbital contribution is omitted from the plot for clarity

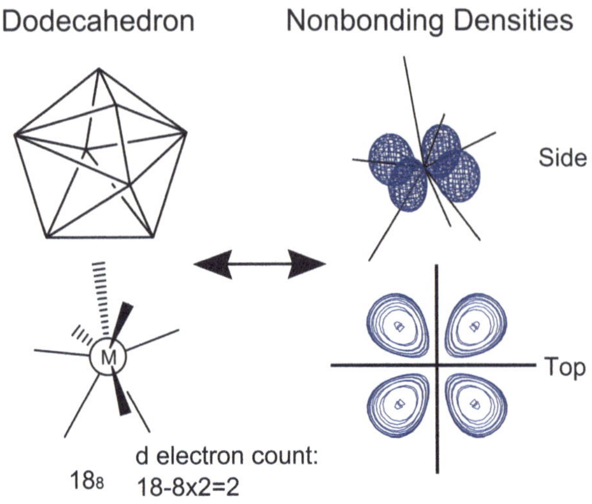

Fig. 8 Valency counting picture and the nonbonding densities for dodecahedral complexes

4 How Do the Nonbonding Electrons Affect the Coordination Geometries?

4.1 A Reference to the VSEPR Model

From the analysis and discussion above, we can clearly see that metal–ligand bonding in many different transition metal complexes can in fact be conveniently represented with Lewis structures, and the nonbonding electron counts can then be deduced based on various approaches like CSED or molecular orbital analysis. On the other hand, even though from these analyses we can determine favorable electron counts from a given geometry, the detailed connection between the electron counts and the geometries remains vague. More specifically, we have not yet been able to back-determine the coordination geometries from electron counts. Note that this is also a limitation for the original Lewis theory, in the sense that counting the valency of compounds does not directly allow us to determine their possible geometries.

Despite this limitation in determining molecular geometry, we can use main group compounds as a paradigm and thereby connect valencies to geometries via the VSEPR model [12, 13]. An intuitive generalization for 18-electron transition metal complexes will then be to generalize the VSEPR model directly to coordination number of 9 and give an interpretation for the three-dimensional coordination complexes.

If we take the basic rationale of the VSEPR model, and arrange the 9 valencies (or electron domains) according to the $N = 9$ solution [30, 31] of Tammes problem [32] (i.e., arrange points on a sphere to maximize the distance between the closest pair), we will get the tricapped trigonal prismatic arrangement. This is indeed consistent with our discussion above if we consider that all the domains are occupied by bond pairs (as we have stated, 18-electron ML_9 indeed takes up the tricapped trigonal prismatic geometry).

We can then make an analogy to tetrahedral main group compounds by first noting that in the case of CH_4, all four domains are occupied by bond pairs. From that we replace one bond pair with a lone pair (such that only three out of four domains are occupied by bond pairs while the remaining one is occupied by a lone pair), and we will see a trigonal pyramidal structure is obtained, which is indeed the representative structure for compounds with one lone pair and three bond pairs (just like NH_3). This means that the three bonds in a trigonal pyramidal structure can actually be viewed as a subset of the bonds in a tetrahedral structure.

However, in the transition metal case, if we start replacing bonding electrons in ML_9 by d electrons (and removing ligand in the process to keep the overall 18 electron count), it will be difficult for us to get geometries like square antiprism and/or pentagonal bipyramid directly if no further rearrangement is undertaken, because their M–L bonds do not resemble a subset of those found in a tricapped trigonal prismatic complex.

Because of this, VSEPR was traditionally only known to be directly applicable to coordination geometries that fall into the solution of the Tammes problem (and as we will show in later examples, this is more of a coincidence) and does not directly explain the existence of pentagonal bipyramidal, dodecahedral, and, most importantly, square planar complexes. Interestingly though, most of the main group compounds (when taking both the bond pairs and the lone pairs into account) fall into geometries that are predicted by VSEPR [13]. It is only for transition metal complexes where numerous counterexamples exist (note that the "lone pair" concept is even not very well defined for transition metal complexes).

This discrepancy arises because the VSEPR model was originally formulated as a classical exclusion model that does not take into account the presence of atomic orbitals. However, we know from later development of quantum mechanics that the behavior of these electrons does not resemble classical particles, but rather their "populations" are governed by wavefunctions that have particular forms. Here, to further analyze the success and failure of the VSEPR model via a more "quantum-mechanical" treatment, we will borrow the "hybridization" concept from the valence bond theory. In such analysis, main group compounds can be described with the hybridization model considering mixture (hybridization) of s and p orbitals, such that all "bonding orbitals" are actually sp^x hybridized (even in the case of hypervalent compounds, because the involved orbitals are ns, np, and nd as oppose to nd, (n+1)s, (n+1)p in transition metal complexes, the d orbital still has relatively high energy and have minimal impact to the lone pairs in such main group compounds), noting that such hybridization will still keep the axial symmetry of the p orbitals (though the sizes of lobes can be scaled in the hybridization process) that there are usually only a single maximal angular lobe (unless it is a pure p orbital). This means that, when we convert the orbitals into electron densities (or "domains" in VSEPR), the major electron domain (if we ignore the radial nodes) is always continuous. This assumption is actually the key to the success of this simple model, because no "lone pair" can be bifurcated/multifurcated and all the "electron domains" to a large extent resemble point particles (by collapsing the continuous domain into a point) and follow the simple exclusion rule.

In transition metal complexes, however, if we apply the Pauling scheme of hybridization, we can see that the resulting hybridized bonding orbitals from the metal center can be d-s-p hybridized [25] (example of which has been shown in Fig. 4), the results of such hybridization will give orbitals that have as much as four angular lobes (two positive lobes and two negative lobes as seen in a d_{xy} orbital). With the introduction of these extra angular lobes, the hybridized orbital is no longer axially symmetric, and there can be multiple major "electron domains" attributed to one orbital. All these will make simple "orbital exclusion" schemes less intuitive, because other orbitals may, instead of moving away from the orbitals, move in between the bifurcations/multifurcations (which was not possible in the main group case because the major electron domain has no angular node and is continuous in the spherical projection), assignments of which can cause complications and thus historically this is not a recommended way to handle transition metal complexes.

4.2 An Alternative View: Face-Dual Relationship

In view of this limitation, an exclusion-based model is seldom used for transition metal complexes. However, in this work, we will illustrate that, with only simple modifications, we can give a simple localized picture that can also account for the geometries of various coordination complexes. Yet before we proceed to an alternative view, we will first take another examination of the orbitals in our orbital analysis. Again, instead of paying attention to the sign structure of orbitals, we will only consider the electron density by taking the square of a wavefunction (note here we again ignore a proportionality constant of 2, which is the occupation number of each occupied orbital).

Considering an octahedral complex as the example, we can take the six M–L bonding molecular orbitals (both the atomic orbitals or the localized orbitals give the same densities) and plot their corresponding densities, and we can expect that these densities form six peaks localizing on the vertices of an octahedron (this is more obvious if we consider the "densities" contributed by the localized hybrid orbitals in Fig. 4). Here, by considering all bonding density peaks as a whole, we can avoid the inconvenient fact that molecular orbitals are mainly based on symmetry constraints rather than bond localization (e.g., the overlap between the s orbital and the a_{1g} set of LGOs is contributed by all six "classical bonds"), which can yield certain complications in understanding the feature of these bonding densities.

For the nonbonding electrons, if we go directly ahead and plot the electron densities arisen from these nonbonding pairs, we can clearly see that the density peaks resulted from these three nonbonding orbitals (d_{xy}, d_{xz}, d_{yz}) form the vertices of a cube (Fig. 9).

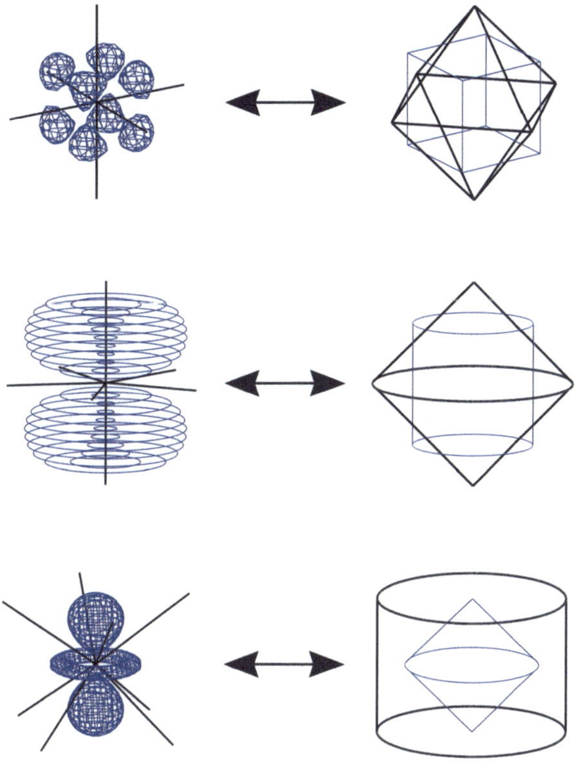

Fig. 9 Face-dual relationship of bonding and nonbonding densities in different coordination complexes. The nonbonding densities of octahedral, pentagonal bipyramidal (simplified as bicone), and square-antiprismatic (simplified as cylinder) geometries can be approximated as cube, cylinder, and bicone, respectively, illustrating a face-dual relationship. The face-dual plots are made with visvis [27] visualization library

To relate the bonding densities and the nonbonding densities, we note that this relationship between cube and octahedron is known as the face-dual relationship in mathematics, and we argue that this is an extremely helpful concept in understanding the structure and bonding of transition metal complexes in many cases.

4.3 Face-Dual Relationship for Various Coordination Geometries

Continuing the process, we might proceed in identifying the nonbonding densities of other coordination complexes. We can see that the nonbonding densities are all localized (maximized) in between the axial and the equatorial regions, fulfilling our prediction above on the "exclusion" between the bonding and the nonbonding densities (Fig. 9). One might argue that if we consider the face dual of the pentagonal bipyramid, we should get the pentagonal prism. However, if we consider only the contributions of s, p, and d orbitals, we cannot have enough angular resolution to identify densities with five density peaks on a plane (analogous to the case where we have exhausted all of the available orbitals on the xy plane for bonding with the equatorial ligands). In this case, it might be more appropriate for us to consider the face-dual relationship between bicone and cylinder. Another quick note on this is that the d_{xz} and d_{yz} orbitals together can already generate densities on a cylinder, which is analogous to the contribution of the p_x and p_y orbitals in ethyne to give a circular π-density, a fact that will be important in later examples.

For a square-antiprismatic complex, we can approximate the coordination geometry to be cylindrical, which has a face dual as bicone (to be exact, the face dual of square antiprism is tetragonal trapezohedron, but like the pentagonal bipyramid case above, cylinder and bicone will be a good enough approximation for our purpose). A bicone resembles the density peaks of a d_{z2} orbital, so we can predict that the nonbonding electron densities are predominantly contributed by the electrons on a d_{z2} orbital.

4.4 Exclusion of Bonding–Nonbonding Densities in Dodecahedral Geometry

For the dodecahedral geometry, however, one might get confused if we directly take the "face dual" of the complex to approximate the nonbonding densities. From the valency counting rule, it should only have two nonbonding electrons residing on one nonbonding orbital; however, the face dual has more vertices (which correspond to maxima in nonbonding densities) than any orbital with d-s-p hybridization. Moreover, because we only have one nonbonding orbital, unlike the case of pentagonal bipyramid, we cannot use a degenerate pair of orbitals to contribute to a degenerate set of maxima (while in the case of pentagonal bipyramid, we can represent its dual by a degenerate pair of orbitals).

We therefore need to take a closer look on the possible geometries of electron densities derived from these orbitals and have a more careful analysis. One approach we propose here is to try to view along the S_4 axis (the z axis) of the complex, in which we can notice all the ligands (when projected to the xy plane) lie on either the x axis or the y axis; thus, no orbitals contributing to M–L bonding have density peaks on any of the four quadrants. If we take the "ligand exclusion" argument here (on a two-dimensional projection, however), we might see that a nonbonding orbital of d_{xy} can be assigned here, and no other orbitals can be assigned this way such that there is no overlap between the nonbonding orbital and the ligands. Indeed, if we go back to molecular orbital theory, one might note that this d_{xy} orbital is indeed the nonbonding orbital (Fig. 8).

4.5 The "Exclusion," "Face-Dual" Relationship, and Complementary Model

Simply taking an exclusion of electron densities via a simple face-dual relationship might sound like a surprising move, and it is natural to question the validity of performing such treatment even though it seems to be very helpful in many cases. Here, one important requirement of molecular orbitals that can give us a great help here is the orthogonality. The orthogonality between orbitals mandates each pair of molecular orbitals to have zero overlap, and we would like to argue that this by itself gives a great deal of insight on the "electron domain exclusion" process.

To illustrate this, we will take sin(x) and cos(x) as examples; this pair of functions is orthogonal to each other if we view them as "wavefunctions" or "orbitals" (indeed they can be wavefunctions for periodic systems), yet when we consider the corresponding densities of "electrons," by taking square of each function, we can clearly see that the peaks of $\sin^2(x)$ and $\cos^2(x)$ are actually alternative (or "exclusive") to each other (see Fig. 10); this serves as an analogous picture that gives rises to the "exclusion picture," and we argue that such "exclusion" effect does not arise from the so-called electron repulsion or steric repulsion as one might have believed. Here, we note that this simple "orthogonality" can help us a lot in the previous or later context by mandating all "nonbonding orbitals (lone pairs)" to take positions that are "dual" to the ligands.

Putting this back to the octahedral case, we note that a similar case happens where the bonding and nonbonding orbitals are mandated to be orthogonal to each other, and analogous to the case of sine and cosine functions, their densities appeared to have an "exclusion effect" that gives the face-dual relationship, as seen in the octahedral bonding densities and the cubic nonbonding densities (Fig. 9). In fact, as seen in Table 1, the spherical harmonics can also be expressed in the form of sine and cosine functions, and so a similar argument can be used to describe the "exclusion"/"orthogonality" concept discussed above.

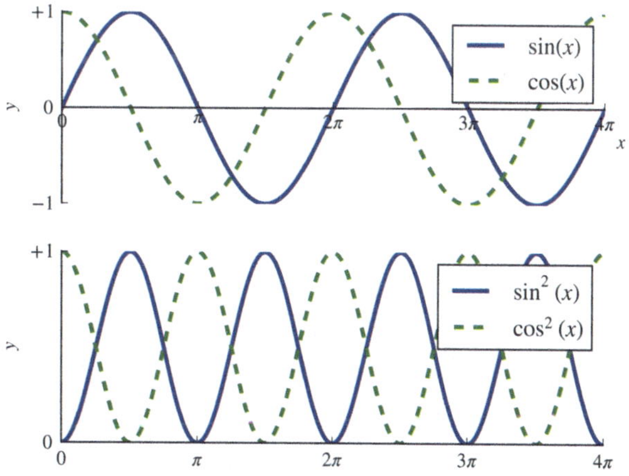

Fig. 10 Plots for sin(x), cos(x), $\sin^2(x)$, and $\cos^2(x)$. It can be seen that $\sin^2(x)$ and $\cos^2(x)$ have alternating peaks, analogous to an "exclusion" process. The plots are prepared with matplotlib [26]

Table 1 Real spherical harmonics

"Orbital"	Polar form
s	$\sqrt{\frac{1}{4\pi}}$
p_x	$\sqrt{\frac{3}{4\pi}}\sin\theta\cos\phi$
p_y	$\sqrt{\frac{3}{4\pi}}\sin\theta\sin\phi$
p_z	$\sqrt{\frac{3}{4\pi}}\cos\theta$
d_{xy}	$\sqrt{\frac{15}{16\pi}}\sin^2\theta\cos 2\phi$
d_{yz}	$\sqrt{\frac{15}{16\pi}}\sin^2\theta\sin 2\phi$
d_{xz}	$\sqrt{\frac{15}{4\pi}}\sin\theta\cos\theta\cos\phi$
d_{x2-y2}	$\sqrt{\frac{15}{4\pi}}\sin\theta\cos\theta\sin\phi$
d_{z2}	$\sqrt{\frac{5}{16\pi}}(3\cos^2\theta - 1)$

It should be well noted that this view is highly related to the spirit of CSED model proposed by Mingos [21, 22], in the sense that the nonbonding orbitals on the metal are similarly orthogonal to the metal–ligand linear combinations of orbitals which have been defined by complementary spherical harmonic functions, and these two sets when taken together provide a complete spherical set. In the model presented here, we assign specifically all these orbitals to specific atomic orbitals and take a step further to analyze more detailed contribution of bonding or nonbonding orbitals from s or p or d orbitals to take into fine account of the possible hybridization as well as all their effects on energy levels of the resulting orbitals. The importance of these energy-level analyses will be made clear in later sections, when we try to compare different geometries with the same coordination number.

4.6 Determining Coordination Geometries from d Electron Counts

So how exactly does the "face-dual" relationship help us determine the geometry of a molecule? One should first note that the "face-dual" argument applies in two ways: when we say the nonbonding densities are "face dual" to the ligand bonding directions, we are also implying that the ligand bonding directions are "face dual" to the nonbonding densities, and here we will make use of the latter argument. Note that for transition metal complexes, the d orbitals are most low-lying in energy (among the valence orbitals) and have multifurcations that allow minimal overlap with ligands, and therefore, they are actually the perfect choices for the d electrons to reside. Moreover, just like the case in the VSEPR model, where the "lone pair" plays a more important role in determining the molecular geometry (in the sense that "lone pairs" are more "exclusive"), we argue that in transition metal

Table 2 Lone pairs (filled d orbitals), coordination number, and the corresponding coordination geometries

Filled d orbital	Coordination number	Coordination geometry
(None)	9	Tricapped trigonal prism
d_{z2}	8	Square antiprism
d_{xy}	8	Dodecahedron
d_{xz}, d_{yz}	7	Pentagonal bipyramid
d_{xy}, d_{z2}	7	Capped trigonal prism
d_{xy}, d_{x2-y2}	7	Capped octahedral
d_{xy}, d_{yz}, d_{xz}	6	Octahedral

complexes, the d electrons also play a crucial role in the molecular geometry, and the fact that d orbitals can be multifurcated is therefore extremely important in this case.

Here, we will actually embrace the seemingly unreasonable "bi- or multifurcated lone pairs," because in our view, this will help our later determination of molecular geometry. In particular, we will take a greedy approach to first mandate the d electrons to reside on particular orbitals as "lone pairs" and then use the face-dual argument to rearrange all the bonding directions for the specified number of ligands. This approach is listed as follows:

1. Determine all possible combinations of d orbitals that could be occupied by the specified number of d electrons (or "lone pairs").
2. For each of these combinations, determine the overall nonbonding densities and identify the "face-dual" or "excluded" sites as possible vacancies for ligands.
3. Try to put the ligands back to these vacancies, if the number of vacancies is not enough, the "lone pair" combination is unlikely to exist and another combination from step 1 should be selected.

We may apply this protocol to understand the 18_8 complexes discussed above. Note that there are only two choices of the d orbital for the lone pair: d_{z2} or any four-lobe d orbital (e.g., d_{xy}). For the d_{z2} nonbonding orbital, with a nonbonding density that resembles a bicone, the eight ligands can be arranged in the dual positions in the square-antiprismatic manner. Similar analysis can be done for the four-lobe d_{xy} orbital, which will easily result in dodecahedral complexes.

This way, we only need to use atomic orbitals of a metal center to accommodate the "lone pairs" (or nonbonding electron densities), whereas all the M–L bonding pairs are still handled in a "Lewis bonding pair" manner. This will allow us to have a more modular handling that can be extended to the understanding of complexes with various geometries (e.g., see later text) while at the same time give a correct prediction for geometries and nonbonding electron densities. We can easily apply this approach to predict the preferred geometries of various 18-electron complexes, as predicted by the CSED model [21, 22] and seen in Table 2. Note that even with the same coordination number and electron count, there could still be multiple possible coordination geometries [33, 34].

One point to note is that up till now all our examples are focused on 18-electron complexes, and we clearly know that there are also complexes that do not fulfill the 18-electron rule. This actually does not pose a hurdle to the protocol presented above. In fact, as mentioned in the CSED model, it is entirely possible for the 18-electron rule to be violated in certain structures when the relevant orbitals that will hold the "lone pair(s)" get too high in energy and become "unavailable" [22]. The protocol we presented here is flexible enough to accommodate even such cases, as will be seen in the trigonal prismatic coordination geometry discussed in a later section.

Another important note is that even though this protocol works especially well for structures with small d electron counts (or those with large coordination number), when the d electron count increases, the number of "vacant sites" for ligands will start to decrease and we might soon find out that there are no possible structures we can determine (e.g., for 18_4 complexes, from the analysis there are 10 d electrons which will fill up all d orbitals and so no "dual" position will be left). To resolve this issue, we need to either include the hybridization concept to allow a more comprehensive understanding or allow violation of the 18-electron rule by filling in less d electrons. These two cases will be detailed in the following section.

Another aspect we would like to briefly elaborate on is that the nonbonding d orbitals listed in Table 2 might interact with the π-orbitals on the ligands and can lead to certain preferred geometric isomers for a given geometry. For example, for a pentagonal bipyramidal complex with 1 or 2 π-accepting ligand(s), these ligands will prefer to occupy the axial positions in order to maximize their π-bonding interactions with the metal center via overlapping with the d_{xz} and d_{yz} orbitals (Table 2).

5 Same Coordination Number, Different Electron Counts

In the previous sections, we have provided many examples of coordination compounds that fall into the 18_n category. However, there are also geometries that do not in general follow the 18-electron rule. In this section, we will take a closer look on some of these geometries and analyze the relationship between their geometries and electron counts.

5.1 *Effect of Hybridization on Orbital Energies*

Before we begin our in-depth discussion on various examples, we will first step back a bit to pick up the discussion we have skipped through in the discussion of square antiprism and see how hybridization will affect the geometry and energies of nonbonding densities. Even though its role is not apparent in our previous discussion on 18-electron complexes, hybridization of nonbonding orbitals is actually critical to complexes having a coordination number smaller than 6 and other complexes that do not follow the 18-electron rule. Such effect of hybridization will be discussed individually on various types of complexes in the following subsections. Here, we will only give a general discussion on how hybridization affects orbital energies.

The effect of hybridization to overall orbital energies could be roughly understood as follows. From the usual orbital analysis, atomic orbital energies follow the order of nd < (n+1)s << (n+1)p for a transition metal center. It should be noted that because p orbitals extend out much more than d orbitals, if d-p mixing occurs in a coordination complex, a "hybridized orbital" that is used for interaction with a ligand is mainly contributed by metal's p orbital(s), whereas a "nonbonding orbital" that contributes to the nonbonding densities is mainly contributed by metal's d orbital(s). Because of their order in orbital energies, if p-character is mixed into a "d-like" nonbonding orbital, the energy of that nonbonding orbital will be significantly increased.

This is somehow related to the hypothetical case of a trigonal planar NH_3 versus a trigonal pyramidal NH_3. If we only consider the bonding electrons, we will see that a trigonal planar NH_3 is more favored (c.f. "BH_3" is planar). Yet when we put back the lone pair into consideration, we will see that the lone pair occupies a pure p_z orbital (can be explained by hybridization scheme or again face-dual relationship). When we then perform a slight perturbation to this trigonal planar structure, we can notice that the p_z orbital in this case can be hybridized with the s orbital (that is usually used on the N–(H3 LGO) bonding interaction) and this hybridization will cause a stabilization of the nonbonding orbital (by mixing in the s-character) but cause destabilization of the N–(H3 LGO) bonding (by mixing in more p-character). The overall net stabilization, due to a more pronounced stabilization to the nonbonding orbital(s) [28, 35, 36], explains the difference in geometry between NH_3 (trigonal pyramidal) and "BH_3" (trigonal planar) and illustrates the important role of lone pairs in governing the overall geometry of a compound. The above discussion is consistent with the conclusion made in the context of the CSED model that empty orbitals maximize the amount of p orbital character because p is higher in energy than s.

A side note on the necessity to consider hybridization when considering lone pairs in main group compounds is that if we follow our approach in Sect. 4.6 and fill the "lone pair" on the lowest valence orbital (in main group case the s orbital), the coordination sphere will be completed immediately and no other bonds can form. This means that lone pairs in main group compound cannot naturally form without hybridization or other stabilizations, and marks an important difference between main group compounds and transition metal complexes.

5.2 Five-Coordinate Geometries

According to the valency counting rule, an 18_5 complex has a d^8 electron configuration, which we view as four "lone pairs." If we follow our argument and mandate these eight electrons to occupy only d orbitals (forming four multifurcated lone pairs), we will see that the "face dual" of such will either only have two (if we leave out the d_{z2} orbital) or four (if we leave out a d_{xy}-type orbital) vacancies for M–L bonding, none of which can fit five coordinating ligands. This means either we need to allow a complex to violate 18-electron rule or need to rearrange these nonbonding orbitals via hybridization to make up new vacancies for ligands. A detailed examination on the involved orbitals is clearly necessary. Here, we consider two representative geometries of five-coordinate complexes: the square pyramidal and the trigonal bipyramidal geometries.

5.2.1 Square Pyramidal Complexes

Square pyramidal structure is an important structure that is often proposed to be involved in organometallic reactions. Its formation often involves the dissociation of a ligand from an octahedral complex, and we will start by imitating this process in our analysis.

Note that from our arguments, removal of ligands should not adversely affect the contributing orbitals of the nonbonding densities, because orbitals that did not participate in the original complex will also not participate in the new one. Of course, this is just a very crude approximation, and more sophisticated mathematical analysis can be performed to deal with the change in hybridization with respect to removal of a ligand [25]. Still, if we assume that there is no major change in geometry and hybridization, then from our previous argument, the d_{xy}, d_{yz}, and d_{xz} orbitals should all still behave like "lone pairs." The remaining question is how we can fill up the "hole" with nonbonding densities created by the removal of a ligand.

From the original octahedral complex, the removed ligand was originally coordinated to a metal d-s-p hybrid orbital (localized picture), and removing the ligand yields a 16-electron complex (16_5). To "fill up" this orbital to get back an 18-electron complex (18_5), we just need to put two electrons into the hybrid orbital that was originally used for bonding with the removed ligand (Fig. 11). One important note of this "nonbonding" orbital is that this orbital is closely related to the bonding orbital toward the apical ligand, in the sense that these two orbitals "share" the p_z and d_{z2} orbitals. From our discussion in the previous section, if a bonding and a nonbonding orbital "share" the p and d contributions, the bonding one will be more contributed by the p orbital, whereas the nonbonding one will be more contributed by the d orbital. This means that in this case, apart from the "t_{2g} set" (quotation because we actually broke the symmetry), we will have a nonbonding orbital mainly contributed by the d_{z2} orbital.

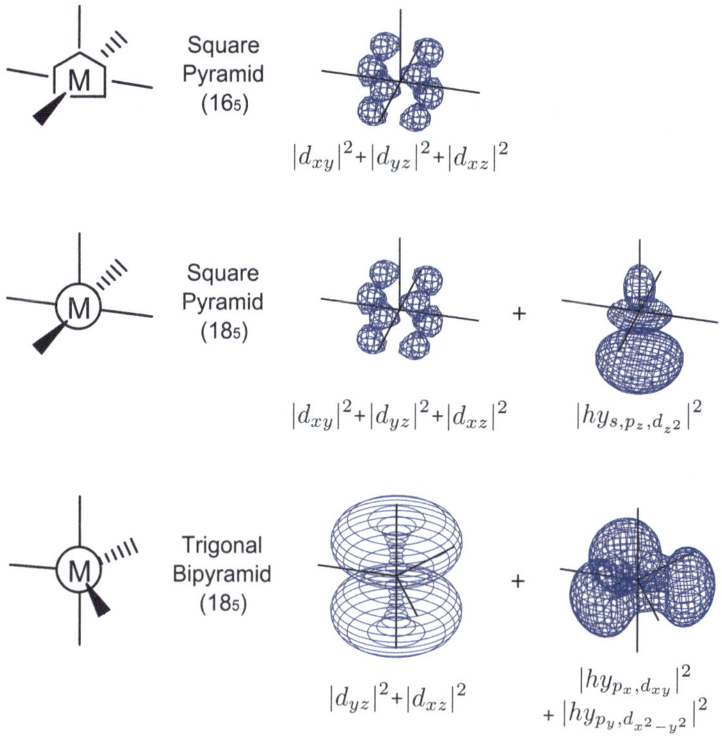

Fig. 11 Valency counting picture and nonbonding densities of square pyramidal (16_5 and 18_5) and trigonal bipyramidal geometries. The pentagon in the first Lewis structure indicates a "lone pair" was removed from the flat side, so that the complex does not follow the 18-electron rule but instead has a 16_5 configuration

However, we should also note that the orbital corresponding to the vacant site (with respect to an octahedral complex) is a d-s-p hybrid. From the discussion in Sect. 5.1, we should note that mixing in p-character to a d orbital will significantly raise the corresponding orbital energy; thus, we would expect this orbital corresponding to the vacant site will have higher energy than the nonbonding d_{xy}, d_{yz}, or d_{xz} orbital. Thus, it is possible that this orbital remains empty (in terms of the language of CSED, this orbital is termed "unavailable"), giving rise to a 16_5 complex with d^6 ($=16 - 5 \times 2$) electron configuration. This possibility allows the octahedral complex to simply dissociate one ligand and leave the vacant site unfilled, making square pyramidal complexes very important intermediates in organometallic reactions.

In fact, the two possible electron counts (18_5 and 16_5) for square pyramidal geometry illustrate well two possible solutions of the inadequate vacancies for ligands we mentioned in the end of the last section: giving up the 18-electron rule or allowing hybrid orbitals for lone pair(s). In particular, if there are only 6 d electrons (three "lone pairs," meaning 18-electron rule is violated), only three orbitals need to be occupied, where we can obviously see by occupying the d_{xy}, d_{yz}, d_{xz} orbitals, we can have six vacant sites arranged in an octahedral geometry (from face-dual relationship), and we can simply choose five of them for ligand coordination to give a 16_5 complex. On the other hand, if we insist eight electrons to be occupied in a five-coordinate complex (to conform to 18-electron rule), we can allow hybridization of s, p_z, and d_{z2} orbitals to make the electron density peak at only the basal site, thus leaving the apical site for ligand coordination to give an 18_5 complex. More detailed analysis on how the electron counts (18_5 and 16_5) affect the apical-metal-basal bond angles has been discussed in the literature [37] and will not be discussed in detail here.

5.2.2 Trigonal Bipyramidal Complexes

Apart from square pyramid, the other well-known five-coordinate geometry is the trigonal bipyramid. This geometry is analogous to pentagonal bipyramid and octahedral geometry in the sense that it again has an axial–equatorial arrangement of ligands. In this case, however, instead of having one extra ligand compared to octahedral coordination as in the pentagonal bipyramid case, we have one less ligand and all three equatorial ligands are again evenly spaced. We can notice that in this case the d_{xy} orbital is no longer avoiding all ligands (as opposed to the case of square pyramid), and from our argument there should be two d-like orbitals contributing to the nonbonding densities on the equatorial plane (because we have one less ligand than in octahedral case). However, it can be easily seen that neither d_{xy} nor d_{x2-y2} should directly contribute to the nonbonding densities or otherwise severe overlap between bonding densities and nonbonding densities will occur.

To resolve this issue, we need to allow d-p mixing on the equatorial plane. Again we can make use of the results of the CSED model, which states that the metal d electrons will be filling the d_{xz} and d_{yz} orbitals, as well as a pair of d-p hybrid orbitals [22]. In particular, if we allow p_x to hybridize with d_{x2-y2} and p_y with d_{xy}, we will get two orbitals that point directly toward the ligands (more contribution from p orbitals, bonding) and two orbitals with major lobes in between the ligands (more contribution from d orbitals, nonbonding). From this, we get two "lone pairs" on the equatorial plane with the total densities "complementary" to the equatorial ligands (Fig. 11).

On the other hand, the nonbonding orbitals (based on the octahedral geometries) d_{xz} and d_{yz} are unaffected by the removal of a ligand on the equatorial plane, and so these two orbitals will hold the other two pairs of nonbonding electrons. When we combine these with the two "lone pairs" on the equatorial plane, we will have four nonbonding "lone pairs" as predicted by the valency counting rule.

5.3 *Four-Coordinate Geometries*

Four-coordinate geometries are arguably the most important geometry second to the octahedral geometry, and there are two widely seen examples of such: the square planar geometry and the tetrahedral geometry. As mentioned in the Introduction (Sect. 1), square planar complex is one of the important examples that pose challenges to apply the classical VSEPR model to understand the molecular geometry of transition metal complexes. Even though square planar coordination complexes are widely known (e.g., *cis*-platin being a famous example), a VSEPR picture (on ligand pairs) would have predicted that a four-coordinate complex should be tetrahedral in shape. We would therefore take this discrepancy as a valuable example for illustrating the power of taking the classical Lewis picture and add in molecular orbital components to analyze the origin of square planar and tetrahedral complexes.

5.3.1 Square Planar Complex

For a square planar complex, its planar arrangement might cause issues in the definition of its face dual, which will make it less convenient for us to define its dual densities. However, we can consider that the ligands in a square planar geometry form a tetragonal dihedron, with the dual being the tetragonal hosohedron (Fig. 12).

From this definition, we argue that the densities of the nonbonding orbitals should reside in the vertices and edges of a tetragonal hosohedron. If we try to map them to the possible d orbitals (or spherical harmonics to be exact), we can see that the d_{xy}, d_{xz}, d_{yz}, and d_{z2} orbitals can all contribute to the density. It should then be noted that all these orbitals are not used in M–LGO bonding in the molecular orbital picture (to be exact, d_{xy}, d_{xz}, d_{yz} are indeed not participated, where d_{z2} has

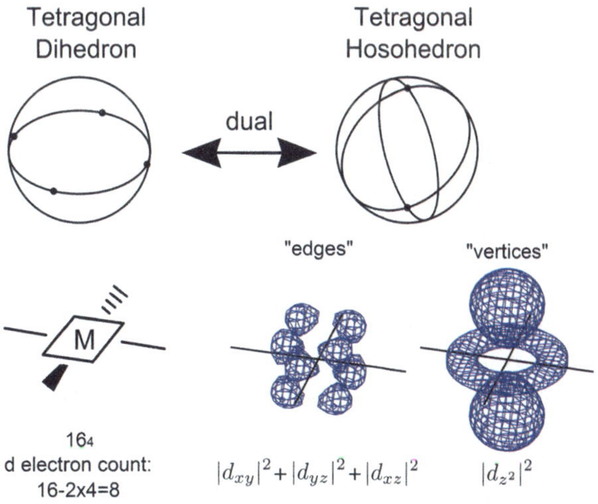

Fig. 12 The face-dual relationship of tetragonal dihedron and tetragonal hosohedron, together with the bonding picture and nonbonding densities they represent. A square instead of a circle was put around the metal center to indicate the square planar geometry (which is essentially two-dimensional) does not follow the 18-electron rule but instead has a valence electron count of 16

only minimal contribution to M–LGO bonding) and thus give rise to four nonbonding orbitals. When these nonbonding orbitals are combined with all the bonding orbitals in the molecular orbital picture (four in total), we can obtain the conventional understanding of the 16-electron rule for square planar complexes (16_4). Another view of this is that due to the fact that the square planar geometry has a two-dimensional arrangement and there is no stabilization for the p_z orbital via M–LGO bonding, and because of the high energy of p_z (unlike d_{z2}), it will not be filled in the final electron configuration (i.e., it is "unavailable" [22]) giving only a total of 16 electrons. Such orbital assignments are also in general consistent with the predictions from the Mingos' CSED model, in which these orbitals are "complementary" to the d_{x2-y2}, s, p_x, and p_y orbitals used for actual bonding [22].

The square planar geometry also serves as another example other than the 16_5 square planar complex that 18-electron rule is relaxed. When we put 8 d electrons into a four-coordinate complex, we will soon notice that we can only place the four "lone pairs" into d_{z2} and three four-lobe d orbitals (say d_{xy}, d_{yz}, and d_{xz}), because if we fill all lone pairs into four-lobe d orbitals, the "dual" d_{z2} will only have two vacancies which is inadequate for four coordinations. With d_{z2}, d_{xy}, d_{yz}, and d_{xz} being filled, exactly four vacancies will remain (which resembles the density peaks of d_{x2-y2}), when we place four ligands on these four sites, we will naturally get a square planar geometry.

5.3.2 Tetrahedral Complexes

A regular tetrahedron has a self-dual property, meaning its face dual is also a regular tetrahedron. Thus, when applying the scheme we have presented, we might on the first instinct expect the nonbonding electron densities for a tetrahedral coordination complex would be tetrahedral in nature. Indeed, there is a large contribution to the nonbonding density from the dual tetrahedron. However, this is not the complete picture.

To begin with, we should note that we can begin our discussion on a tetrahedral geometry by considering first the contribution of s and p orbitals; one might notice this is then analogous to the case of methane. We will then consider the contribution of d orbitals by considering d-p mixing; this mixing will on the one hand enhance the M–L bonding but on the other hand will distort the shape of the nonbonding orbitals (by mixing in p-character). From the density plot shown in the middle column of Fig. 13, we can see that the resulting nonbonding density does resemble the dual tetrahedron.

However, this is not the end of the story. It is widely accepted that a lot of tetrahedral coordination complex follows the 18-electron rule (and because tetrahedron is three-dimensional, that will be our first guess too); an 18_4 arrangement, however, will give us a prediction of d^{10} metal center, though we have only singled out the contribution of d_{xy}, d_{yz}, and d_{xz}, contributing only to six of the electrons. The remaining two orbitals are a bit more elusive: when we consider the density peaks of both the bonding and nonbonding densities, we can imagine that they lie

Fig. 13 Valency counting picture and the nonbonding densities for tetrahedral complexes

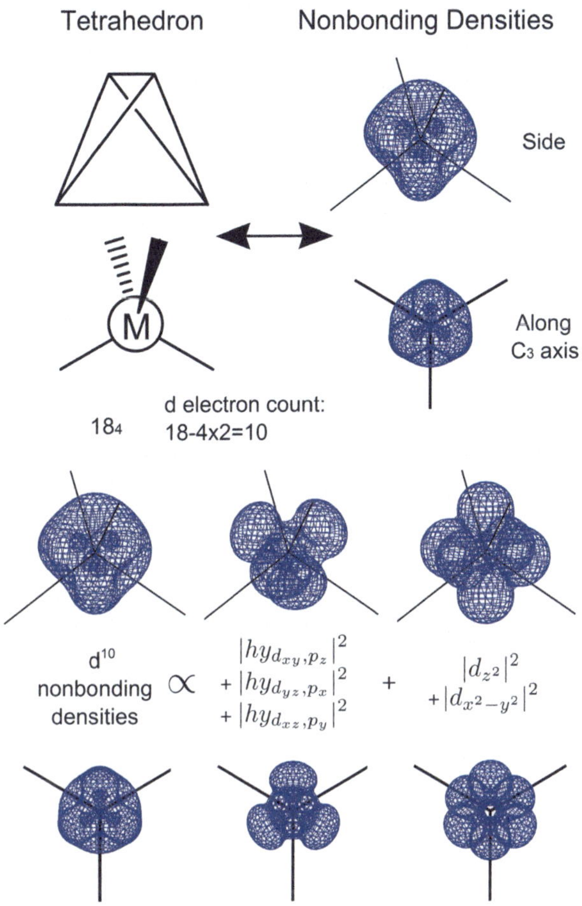

on the eight corners of the cube, meaning that the d_{x2-y2} and d_{z2} orbitals have not yet contributed, and they will therefore hold the remaining four electrons to complete the coordination sphere (note that in the right column of Fig. 13 the density peak of d_{x2-y2} and d_{z2} as a whole gives the shape of an octahedron: a dual to the densities contributed by the remaining orbitals).

5.3.3 Relationship Between d-p Mixing and Expected Valence Electron Count

The comparison on four-coordinate complexes provides us a very interesting illustration on the effect of d-p mixing on expected valence electron count. In a square planar complex, the nonbonding orbitals have minimal d-p mixings; on the other hand, for a tetrahedral complex, we would expect stabilization of the bonding orbitals and destabilization of three nonbonding orbitals (mainly contributed by d_{xy},

d_{yz}, and d_{xz}) due to d-p mixing. Thus, if the nonbonding orbitals are not filled, we would expect a tendency for the coordination complex to adopt the tetrahedral complex (due to the extra stabilization from d-p mixing which is not present in the square planar complex), which is consistent with the observation of d^0 tetrahedral complexes [38] (we can also view this as the $N = 4$ Tammes solution when there is no "multifurcated lone pairs" affecting the geometry). On the other hand, for a square planar complex, the nonbonding orbitals consist of four d orbitals and one pure p orbital. The pure p orbital is high in energy and is very unlikely to be filled, meaning eight nonbonding electrons will suffice for a square planar complex. Same thing does not hold in the tetrahedral complex, where we have two nonbonding pure d orbitals and three destabilized d-p mixed nonbonding orbitals; filling in eight electrons will cause the destabilized orbitals to be filled. Taking all the effects as a whole, for a 16-electron complex, it is much more likely to take the square planar than the tetrahedral configuration.

5.4 Six-Coordinate Geometries

The most dominant six-coordinate complexes are indeed of the octahedral geometry; however, there are also some other examples with trigonal prismatic structure. Before we conclude the section, we will take this as the last example and compare with our first example of octahedral geometry.

5.4.1 Trigonal Prismatic Complexes

Before we begin, we will have a quick recap on the octahedral geometry: all the dual sites are contributed by the d_{xy}, d_{yz}, and d_{xz} orbitals (the t_{2g} set), and these three orbitals are pure in nature and does not involve any mixing of p orbitals. The densities derived from occupation of these three d orbitals therefore form a perfect cube, which serves as the face dual of the original octahedron.

On the other hand, for a trigonal prismatic complex, we can easily tell that the face dual is the trigonal bipyramid. From the aforementioned discussion on trigonal bipyramidal complex, we know that the density lobes forming the shape of triangle are actually contributed by the mixing of the d_{xy} and d_{x2-y2} orbitals with the p_x and p_y orbitals. On top of this, the density lobes on the axial positions are contributed mainly by the d_{z2} orbital mixed with s orbital, similar to the case we see in square antiprism.

As mentioned above, during a d-p mixing, p-like orbital decreases in energy, and d-like orbital increases in energy. However, because p-like orbitals can have better overlap with ligands while d-like orbitals cannot, at the end the M–LGO bonding, orbitals are mainly contributed by p-like orbitals overlapping with ligands' orbitals. This means there will be stronger interaction between the p-like orbital and the ligands' orbitals, but at the same time the equatorial d-like orbitals have more antibonding characters (because of d-p mixing) and thus they have lower tendency to be filled.

Because of all these, in the case of trigonal prismatic complexes, only the d_{z2} orbital preserves its nonbonding character and will be favorably filled, while the d_{xy} and d_{x2-y2} orbitals become unavailable, lowering the valence electron count to 14. From our notation, a 14_6 valency predicts a d^2 electron configuration, which is indeed the case as illustrated by examples like $Os(ONO)_2$ (ONO = a tridentate ligand) [39].

On the other hand, this kind of d-p mixing is not seen in octahedral complexes, because in these octahedral complexes, all ligands are lying on the angular nodes of the three nonbonding orbitals, meaning that no d-p mixing in the nonbonding orbitals will occur (because in this case p and d orbitals are with different symmetries).

Viewing this in a reverse way, if we know that there is a six-coordinate d^2 metal complex, we can apply our analysis detailed in Sect. 4.6 and attempt to fill first a d orbital (d_{z2} in this case); all the remaining ligand pairs can be filled into the dual position (if we view that d_{z2} as a bicone, the dual will resemble a cylinder), which we can also obtain a trigonal prism by evenly distributing the ligands to resemble a cylinder. This also serves as another illustration on how our protocol can be extended to complexes that does not conform to the 18-electron rule.

This comparison between the octahedral and the trigonal prismatic structure has a more important implication, in the sense that octahedral structure is believed to be the only stable structure predicted by the VSEPR scheme; the existence of trigonal prismatic structure with the corresponding d^2 electron count shows that the nonbonding d electrons also play an important role in the geometry, yet they are also different from the usual understanding in VSEPR that the lone pair plays roughly the same role as a bonding pair. The possibility for d-p mixing to give multifurcated "nonbonding" densities, we believe, undermines the possibility for us to have trigonal prismatic and other geometries that cannot be predicted in VSEPR model.

Trigonal prismatic structures have been observed also in d^0 complexes, such as $Mo(CH_3)_6$ [40] and $Mo(S_2C_6H_4)_3$ [41]. Adoption of trigonal prism instead of octahedron can minimize the p-character and maximize d-character in the metal's hybridized orbitals used in M–L bonding [16] (note p orbitals are higher in energy than d orbitals in transition metal centers). In other words, for d^0 complexes, trigonal prismatic geometry can gain extra stability because of optimal M–L bonding interactions and thus be more commonly observed than octahedral geometry.

6 Real-World Examples

After presenting results on different geometries, we will end this chapter by providing examples in the literature and show how we can apply the valency counting approach to elucidate the structure and bonding of such complexes.

6.1 Simple Transition Metal Complexes

A large part of the discussions in this chapter has been spent on discussion of the face-dual relationship between bonding and nonbonding densities, yet we have only presented a theoretical argument without further illustrations. Though when we take a close look back to the list of the geometries we have covered, we can actually find crystals of compounds that resemble each of these geometries. In fact, on top of the geometries we listed in Fig. 1 that conforms to the 18-electron rule, we have also covered some other geometries in the process that can also be explained with the same set of rules, which illustrate that the valency counting process still works when we have correctly determined either the d electron count or the "unavailable orbitals." All these examples are detailed in Fig. 14.

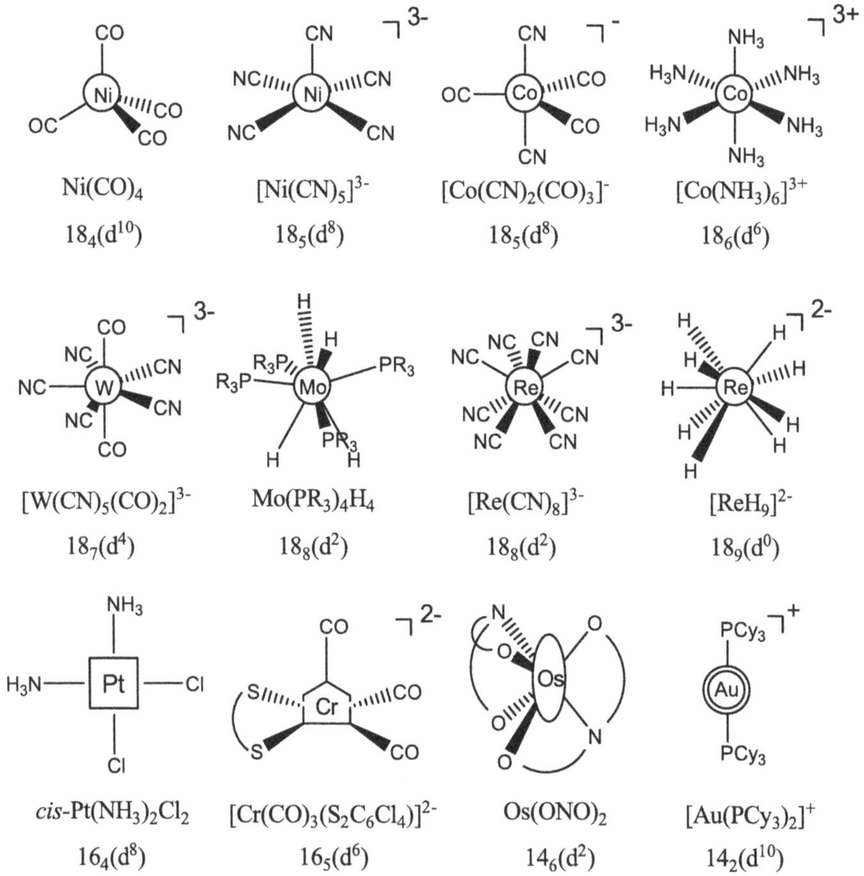

Fig. 14 Lewis description of various transition metal complexes [4–11, 39, 42–44]. *In the last row*, different symbols were used to denote the "lone pairs" specific to particular geometries that does not conform to the 18-electron rule

6.2 Z-Type Ligands

In many discussions in the field of coordination geometry, attention is usually concentrated on (in Langmuir's classification [3]) the negative valence (caused by L-type ligand) or covalence (caused by X-type ligand). Indeed, when determining the d electron count in all the previous cases, we only consider the m_n valency and count these two types of valance toward n.

A much later discovered type of ligands, known as the Z-type ligands, fills up the remaining "positive valence" category. This type of ligands, instead of serving as electron donors like the earlier discovered counterparts, accepts electrons from the metal center. This type of ligands gives even less intuitive molecular orbital diagram and can complicate the analysis [45].

Even so, with our revised view on handling bonding and nonbonding densities, the nature of Z-type ligand can also be well explained. Here, we consider Z-type ligands as all those ligands with empty orbital pointing to the "nonbonding densities" of the coordination center. For example, in a square pyramidal complex 18_5 (without considering the Z-type ligand), we would expect the Z-type ligand should interact with either the cubic nonbonding densities or the densities on the basal site. The choice between these two depends on their hybridizations. Recall that based on our discussion in Sect. 5.1, a p-mixing into a d orbital will increase its orbital energy. Also, from our chemical intuition, we would expect electrons on orbitals with higher energy will serve as a better donor. Combining these two arguments, we would expect that the Z-type ligand should occupy the basal site (so that the overall geometry looks like octahedral). Similar argument can help us explain another example with L-type ligand arranged in seesaw geometry and the two Z-type ligands filling the two remaining sites to give an overall octahedral geometry, or the L-type ligand arranged in a linear geometry and a Z-type ligand perpendicular to them (Fig. 15).

Another example of which is based on the tetrahedral structure, a Z-type ligand will have a choice to interact with the densities based on the t_2 set (d_{xy}, d_{yz}, d_{xz}) and the e set (d_{z2}, d_{x2-y2}). From our previous discussion, the orbitals in the t_2 set are hybridized with p orbitals, while the orbitals in the e set are not; thus, we expect that the Z-type ligand should interact with the density peaks on the t_2 set but not the e set. Indeed, examples of such show that the Z-type ligand always lies on the face instead of the edge of the coordination tetrahedron.

There is another more special class of ligands called "ambiphilic ligand," which can serve as either an L-type or a Z-type ligand depending on coordination mode [50–52], where our analysis on either type of ligands can be applied correspondingly. The computational results from literature on $\{(PH_3)_2Pd(SO_2)\}$ give both planar and nonplanar geometry around the sulfur center [52]. In our notation, the planar one refers to a standard 16_3 complex with SO_2 serving as an L-type ligand; and the nonplanar one is a 14_2 complex with SO_2 serving as a Z-type ligand. In fact, the planar isomer resembles more of a trigonal planar complex, and the nonplanar isomer resembles more of a linear complex with SO_2 coordinating along the

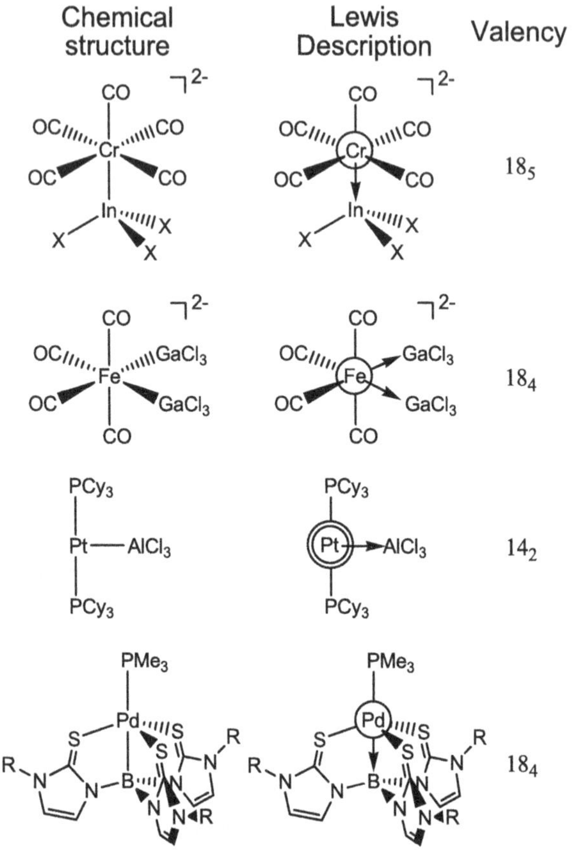

Fig. 15 Various examples of transition metal complexes with Z-type ligand(s) [46–49]. In the third example, a *double circle* was put around the Pt center to indicate a 14-electron count well known for a linear structure

perpendicular direction (similar to the third example in Fig. 15), consistent with our distinction of L-type and Z-type ligands.

6.3 *Linear Pd Clusters*

One main objective of this work is to present a localized picture for different transition metal complexes, and one might feel that none of the examples discussed illustrates a clear advantage of the localized Lewis-like approach against the delocalized treatment. Indeed, for mononuclear transition metal complexes, the advantage might not be obvious due to the existence of the n-to-n mapping between localized bonds and M–LGO bonding interactions. We will therefore present cases where a localized treatment can be more intuitive than a delocalized treatment.

We will start by presenting a case of sandwiched linear palladium clusters [53]. If a delocalized view is taken, it might not be easy to explain the reason behind the fact that such clusters can be found with variable length. However, with the use of a more localized model, this observation can actually be easily explained.

Before getting into the details of the complexes, we will first argue that in a Lewis-type treatment, "valencies" can be of different kinds, either provided by the core atom, a donor, or equally shared. In fact, in the valency counting process, we often ignore the origin of electrons, and this gives us a high flexibility in defining the so-called ligand. In particular, even a dangling electron pair can be defined as an "L-type ligand" without affecting the valency counting.

If we put this concept back to the Pd clusters (Fig. 16), it is obvious that all the Pd centers are lying on a square planar geometry (with a dangling electron pair serving as an μ^2 "L-type" ligand between two Pd centers), and we can predict from previous arguments that each center has a 16_4 configuration or an electron count of d^8. Of course, because the dangling pairs are not yet assigned to any atoms, we should then

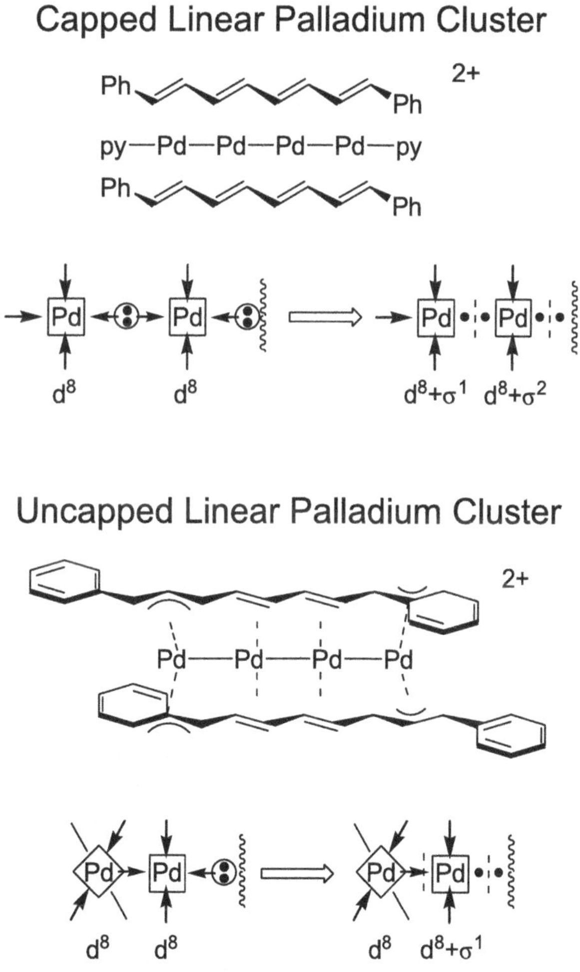

Fig. 16 Examples of sandwiched linear palladium clusters [53] and their corresponding valency counting pictures [54]. Each square around Pd denotes that the center has a 16_4 configuration as defined in Sect. 5.3.1

equally partition them between two neighboring Pd centers, all the Pd centers can then be assigned $d^8+\sigma^1$ or $d^8+\sigma^2$ accordingly (Fig. 16). We should then note that each repeating unit of the linear cluster (the $d^8+\sigma^2$ metal center and the π-donor) is formally neutral, and so these clusters can be easily found with variable length [55].

There is also another list of examples of uncapped linear palladium clusters (Fig. 16), where the terminal palladium centers should be treated in a different manner. One should note that, due to the change in the hapticity, each conjugated chain provides an L-type as well as an X-type ligand (unlike the termini in a capped chain, where each π-system serves only as an L-type ligand), which means that a terminal palladium center already by itself serves as a square planar coordination complex. Interestingly, there is a dual orbital (or "lone pair") with one of its maxima pointing toward its neighboring palladium, meaning that we can actually treat the neighboring palladium as a Z-type ligand; on the other hand, if we focus instead on the neighboring palladium center, the terminal Pd center will serve as a classical L-type ligand, allowing all internal centers to be properly considered as having square planar geometries. We can thus determine the nature of the Pd–Pd bond qualitatively without resorting to quantum chemical calculations.

Detailed analysis of this type of clusters with more examples included can be seen in [54].

6.4 $[Zn_3Cp^*_3]^+$

Another example that can also be much easily handled by a localized picture is a recent work by Fischer et al., where several "σ-aromatic" compounds are presented together with quantum-mechanical analysis of their structures and bondings [56]. Here, however, we will take an alternative view to such complexes. Similar to the case of linear Pd clusters, we will also make use of the concept of considering an electron pair as an "L-type" ligand.

In $[Zn_3Cp^*_3]^+$, the three [Cp*Zn] units form a triangle (Fig. 17). Following a usual treatment of $[Cp^*]^-$, we can consider it as 3 L-type donors to a Zn metal center. We then consider there is a dangling lone pair on the center of the Zn_3 triangle that serves as a μ^3 "L-type" ligand. We can then categorize these Zn centers to be of the type 18_4 and expect a d^{10} configuration on zinc. Indeed, if we put all Zns

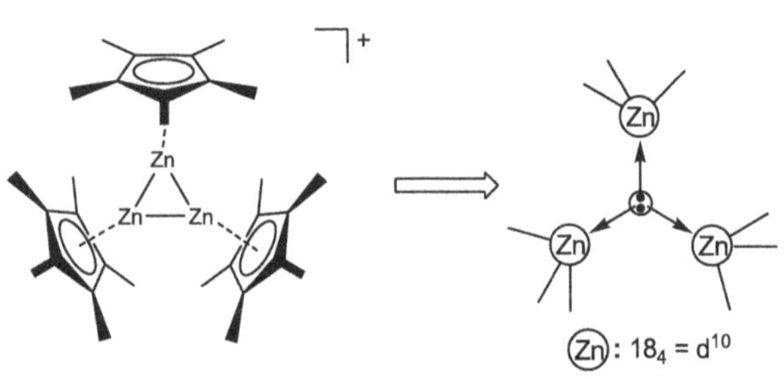

Fig. 17 Structure and Lewis description of $[Zn_3Cp^*_3]^+$ [56]

to have a charge of +2 (to give a primary electron count of d^{10}) and each Cp with a charge of −1, together with the dangling electron pair in the middle, we will have an overall charge of +1 for the complex, consistent with the experimental observation, as seen in Fig. 17. Note that if we then assign back the charge of the dangling electron pair to the metal centers, we can have a fractional charge of 4/3 to each Zn center.

6.5 Mixing the Valency Counting Picture with Delocalized Treatment

A final example concerns how the localized model can be used together with a delocalized model to give rise to the understanding of structures that are much more difficult to analyze.

In a work by Kirchmann et al., a nickel complex with six stanna-*closo*-dodecaborate clusters as ligands was reported [57] (Fig. 18). A usual understanding on borane-like compounds is that they cannot be properly described in localized treatment. Indeed, there are not yet widely accepted localized model to treat the skeletal (tangential) bonding of such clusters. However, with respect to the radial valence pair, it can still be considered as localized on each unit (the radial valence pair serves as a B–H bonding pair in a BH unit and a lone pair in a Sn center).

With this localization of "lone pair" on each Sn, they can simply be considered as L-type ligands and so the octahedral nickel complex follows the 18_6 arrangement in a straightforward manner.

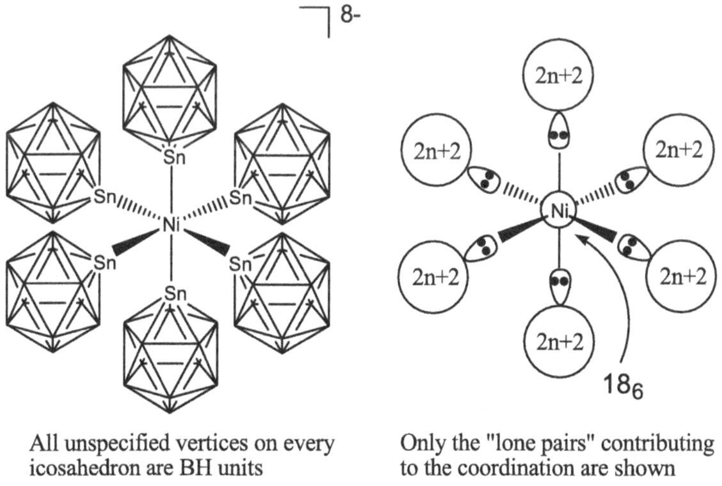

Fig. 18 Structure of $[Ni(SnB_{11}H_{11})_6]^{8-}$ and its corresponding hybrid valency counting picture [57]

7 Summary

In this work, we have reviewed the fundamental differences between the transition metal complexes and main group compounds. We have then exploited such differences to extend the classical Lewis description and formulate a revised picture to correlate electronic structure to molecular geometry for transition metal complexes.

In particular, we argue that the most prominent difference that distinguishes the transition metal complexes from main group compounds is the presence of low-lying d orbitals. Noting that nonbonding electrons preferentially populate low-lying orbitals, we can see that d electrons play a crucial role in affecting the overall molecular geometry. On top of this, unlike the lone pairs contributed by s-p hybrids in main group compounds, the possible bifurcation or multifurcation of electron domains for "lone pairs" contributed by the d orbitals allows the ligands to go in between multifurcated lobes instead of being excluded away. All of these undermine the apparent failure of classical VSEPR model in transition metal complexes. In view of this, we have proposed the adoption of the revised concept of "multifurcated lone pairs," which can allow the relationship between structure and bonding to be described in a more localized manner.

With the emergence of newer and more complicated transition metal complexes being discovered, we expect this local approach will be more useful in a general understanding of compounds without always resorting to quantum chemical calculations. The applicability of this approach has been demonstrated via a wide range of examples, and we hope that we have provided the readers an alternative view when encountering new transition metal complexes.

Acknowledgments We would like to acknowledge the financial support from the Hong Kong Research Grants Council (Grant No. GRF16303014 and CUHK7/CRF/12G). F.K.S. acknowledges support from the Hong Kong Ph.D. Fellowship Scheme 2012/13 (PF11-08816).

References

1. Lewis GN (1916) The atom and the molecule. J Am Chem Soc 38:762–785. doi:10.1021/ja02261a002
2. Sidgwick NV (1927) The electronic theory of valency. Oxford University Press, London
3. Langmuir I (1921) Types of valence. Science 54:59–67. doi:10.1126/science.54.1386.59
4. Hedberg L, Iijima T, Hedberg K (1979) Nickel tetracarbonyl, $Ni(CO)_4$. I. Molecular structure by gaseous electron diffraction. II. Refinement of quadratic force field. J Chem Phys 70:3224–3229. doi:10.1063/1.437911
5. Spiro TG, Terzis A, Raymond KN (1970) Structure of $Ni(CN)_5^{3-}$. Raman, infrared, and x-ray crystallographic evidence. Inorg Chem 9:2415–2420. doi:10.1021/ic50093a006
6. Lo W, Hu C, Lumeij M, Dronskowski R, Lovihayeem M, Ishal O, Jiang JF (2013) $[Co^{I}(CN)_2(CO)_3]^-$, a new discovery from an 80-year-old reaction. Chem Commun 49:7382–7384. doi:10.1039/C3CC43269F

7. Wang X-Y, Justice R, Sevov SC (2007) Hydrogen-bonded metal-complex sulfonate (MCS) inclusion compounds: effect of the guest molecule on the host framework. Inorg Chem 46:4626–4631. doi:10.1021/ic070324p
8. Karunadasa HI, Long JR (2009) Synthesis and redox-induced structural isomerization of the pentagonal bipyramidal complexes $[W(CN)_5(CO)_2]^{3-}$ and $[W(CN)_5(CO)_2]^{2-}$. Angew Chem Int Ed 48:738–741. doi:10.1002/anie.200804199
9. Guggenberger LJ (1973) Structure of tetrahydridotetrakis(methyldiphenylphosphine)molybdenum(IV). Inorg Chem 12:2295–2304. doi:10.1021/ic50128a017
10. Bennett MV, Long JR (2003) New cyanometalate building units: synthesis and characterization of $[Re(CN)_7]^{3-}$ and $[Re(CN)_8]^{3-}$. J Am Chem Soc 125:2394–2395. doi:10.1021/ja029795v
11. Abrahams SC, Ginsberg AP, Knox K (1964) Transition metal-hydrogen compounds. II. The crystal and molecular structure of potassium rhenium hydride, K_2ReH_9. Inorg Chem 3:558–567. doi:10.1021/ic50014a026
12. Gillespie RJ, Nyholm RS (1957) Inorganic stereochemistry. Q Rev Chem Soc 11:339–380. doi:10.1039/QR9571100339
13. Gillespie RJ, Hargittai I (1991) The VSEPR model of molecular geometry. Allyn & Bacon, Boston
14. Pauling L (1931) The nature of the chemical bond. Application of results obtained from the quantum mechanics and from a theory of paramagnetic susceptibility to the structure of molecules. J Am Chem Soc 53:1367–1400. doi:10.1021/ja01355a027
15. Pauling L (1960) The nature of the chemical bond and the structure of molecules and crystals: an introduction to modern structural chemistry, 3rd edn. Cornell University Press, Ithaca
16. Kimball GE (1940) Directed valence. J Chem Phys 8:188–198. doi:10.1063/1.1750628
17. Landis CR, Firman TK, Root DM, Cleveland T (1998) A valence bond perspective on the molecular shapes of simple metal alkyls and hydrides. J Am Chem Soc 120:1842–1854. doi:10.1021/ja9710114
18. Mulliken RS (1967) Spectroscopy, molecular orbitals, and chemical bonding. Science 157:13–24. doi:10.1126/science.157.3784.13
19. Göthe MC, Wannberg B, Karlsson L, Svensson S, Baltzer P, Chau FT, Adam MY (1991) X-ray, ultraviolet, and synchrotron radiation excited inner-valence photoelectron spectra of CH_4. J Chem Phys 94:2536–2542. doi:10.1063/1.459880
20. Hoffmann R (1982) Building bridges between inorganic and organic chemistry (Nobel lecture). Angew Chem Int Ed Engl 21:711–724. doi:10.1002/anie.198207113
21. Mingos DMP, Hawes J (1985) Complementary spherical electron-density model. Struct Bond 63:1–63
22. Mingos DMP (2004) Complementary spherical electron density model and its implications for the 18 electron rule. J Organomet Chem 689:4420–4436. doi:10.1016/j.jorganchem.2004.07.020
23. Verkade JG (1986) A pictorial approach to molecular bonding. Springer, New York
24. Glendening ED, Landis CR, Weinhold F (2012) Natural bond orbital methods. Wiley Interdiscip Rev Comput Mol Sci 2:1–42. doi:10.1002/wcms.51
25. Mingos DMP, Lin Z (1990) Hybridization schemes for co-ordination and organometallic compounds. In: Bioinorganic chemistry. Springer, Berlin/Heidelberg, pp 73–111
26. Hunter JD (2007) Matplotlib: a 2D graphics environment. Comput Sci Eng 9:90–95. doi:10.1109/Mcse.2007.55
27. visvis (2013) The object oriented approach to visualization. Google Project Hosting. https://code.google.com/p/visvis/. Accessed 31 May 2015
28. Mingos DMP, Lin Z (1989) Non-bonding orbitals in co-ordination, hydrocarbon and cluster compounds. In: Stereochemistry and bonding. Springer, Berlin/Heidelberg, pp 1–56
29. Pyykkö P (2006) Understanding the eighteen-electron rule. J Organomet Chem 691:4336–4340. doi:10.1016/j.jorganchem.2006.01.064

30. Schütte K, van der Waerden BL (1951) Auf welcher Kugel haben 5, 6, 7, 8 oder 9 Punkte mit Mindestabstand Eins Platz? Math Ann 123:96–124. doi:10.1007/BF02054944
31. Weaire D, Aste T (2008) The pursuit of perfect packing, 2nd edn. CRC, New York
32. Tammes PML (1930) On the origin of number and arrangement of the places of exit on pollen grains. Recl Trav Bot Néerl 27:1
33. Lin Z, Bytheway I (1996) Stereochemistry of seven-coordinate main group and d^0 transition metal molecules. Inorg Chem 35:594–603. doi:10.1021/ic950271o
34. Thompson HB, Bartell LS (1968) Seven-coordination and ligand-repulsion models for bond geometry. Inorg Chem 7:488–491. doi:10.1021/ic50061a020
35. Hall MB (1978) Stereochemical activity of s orbitals. Inorg Chem 17:2261–2269. doi:10.1021/ic50186a050
36. Hall MB (1978) Valence shell electron pair repulsions and the Pauli exclusion principle. J Am Chem Soc 100:6333–6338. doi:10.1021/ja00488a007
37. Rossi AR, Hoffmann R (1975) Transition metal pentacoordination. Inorg Chem 14:365–374. doi:10.1021/ic50144a032
38. Lin Z, Hall MB (1993) A group theoretical analysis on transition-metal complexes with metal-ligand multiple bonds. Coord Chem Rev 123:149–167. doi:10.1016/0010-8545(93)85054-8
39. Cipressi J, Brown SN (2014) Octahedral to trigonal prismatic distortion driven by subjacent orbital π antibonding interactions and modulated by ligand redox noninnocence. Chem Commun 50:7956–7959. doi:10.1039/C4CC03404J
40. Roessler B, Seppelt K (2000) $[Mo(CH_3)_6]$ and $[Mo(CH_3)_7]^-$. Angew Chem Int Ed 39:1259–1261. doi:10.1002/(SICI)1521-3773(20000403)39:7<1259::AID-ANIE1259>3.0.CO;2-Y
41. Cowie M, Bennett MJ (1976) Trigonal-prismatic vs. octahedral coordination in a series of tris (benzene-1,2-dithiolato) complexes. 1. Crystal and molecular structure of tris(benzene-1,2-dithiolato)molybdenum(VI), $Mo(S_2C_6H_4)_3$. Inorg Chem 15:1584–1589. doi:10.1021/ic50161a023
42. Raudaschl G, Lippert B, Hoeschele JD, Howard-Lock HE, Lock CJL, Pilon P (1985) Adduct formation of *cis*-$(NH_3)_2PtX_2$ ($X = Cl^-, I^-$) with formamides and the crystal structures of *cis*-$(NH_3)_2PtCl_2{\cdot}(CH_3)_2NCHO$. Application for the purification of the antitumor agent cisplatin. Inorg Chim Acta 106:141–149. doi:10.1016/S0020-1693(00)87550-7
43. Sellmann D, Wille M, Knoch F (1993) Transition metal complexes with sulfur ligands. 97. Coordinatively and electronically unsaturated and saturated chromium, molybdenum, and tungsten dithiolate complexes of the type $[M(CO)_3(\text{'}S_2\text{'})]^{2-}$ and $[M_2(CO)_7(\text{'}S_2\text{'})]^{2-}$ {$\text{'}S_2\text{'}^{2-} = S_2C_6R_4^{2-}$ (R = H, Cl, Me); $S_2C_2H_4^{2-}$}. Inorg Chem 32:2534–2543. doi:10.1021/ic00063a054
44. Bowmaker GA, Brown CL, Hart RD, Healy PC, Rickard CEF, White AH (1999) Co-ordination and conformational isomerism in bis(tricyclohexylphosphine) gold(I) halides. J Chem Soc Dalton Trans 881–890. doi:10.1039/A808928K
45. Amgoune A, Bourissou D (2010) σ-Acceptor, Z-type ligands for transition metals. Chem Commun 47:859–871. doi:10.1039/C0CC04109B
46. Rutsch P, Renner G, Huttner G, Sandhofner S (2002) Ügangsmetallorganisch stabilisierte Indate und Germanate: Synthese, Struktur und Eigenschaften der Komplexe $[\{(CO)_5M\}EX_3]^{2-}$ (M = Cr, Mo, W; E = In; X = Cl, Br), $[(CO)_5Cr)InBr(\mu_2\text{-}Br)]_2^{2-}$ sowie $[\{(CO)_5Cr\}E(oxinat)_2]^{n-}$ (E = In, n = 2; E = Ge, n = 1). Z Naturforsch 57b:757–772
47. Leiner E, Hampe O, Scheer M (2002) Synthesis and structure of novel complexes of gallium. Eur J Inorg Chem 2002:584–590. doi:10.1002/1099-0682(200203)2002:3<584::AID-EJIC584>3.0.CO;2-J
48. Braunschweig H, Gruss K, Radacki K (2007) Interaction between d- and p-block metals: synthesis and structure of platinum–alane adducts. Angew Chem Int Ed 46:7782–7784. doi:10.1002/anie.200702726

49. Pang K, Quan SM, Parkin G (2006) Palladium complexes with Pd→B dative bonds: analysis of the bonding in the palladaboratrane compound [κ^4-B(mim$^{Bu^t}$)$_3$]Pd(PMe$_3$). Chem Commun 5015–5017. doi:10.1039/B611654J
50. Mingos DMP (2014) Ambivalent Lewis acid/bases with symmetry signatures and isolobal analogies. In: Mingos DMP (ed) Nitrosyl complexes inorganic chemistry biochemistry medicine II. Springer, Berlin/Heidelberg, pp 1–51
51. Mingos DMP (2014) A review of complexes of ambivalent and ambiphilic Lewis acid/bases with symmetry signatures and an alternative notation for these non-innocent ligands. J Organomet Chem 751:153–173. doi:10.1016/j.jorganchem.2013.08.033
52. Mingos DMP (2015) A theoretical analysis of ambivalent and ambiphilic Lewis acid/bases with symmetry signatures. Coord Chem Rev 293–294:2–18. doi:10.1016/j.ccr.2014.11.009
53. Murahashi T, Mochizuki E, Kai Y, Kurosawa H (1999) Organometallic sandwich chains made of conjugated polyenes and metal–metal chains. J Am Chem Soc 121:10660–10661. doi:10.1021/ja992387f
54. Sheong FK, Chen W-J, Lin Z (2015) Electron counting approach to the structure and bonding of sandwiched low dimensional palladium clusters. J Organomet Chem. 792:93–101. doi:10.1016/j.jorganchem.2015.02.026
55. Horiuchi S, Tachibana Y, Yamashita M, Yamamoto K, Masai K, Takase K, Matsutani T, Kawamata S, Kurashige Y, Yanai T, Murahashi T (2015) Multinuclear metal-binding ability of a carotene. Nat Commun 6. doi:10.1038/ncomms7742
56. Freitag K, Gemel C, Jerabek P, Oppel IM, Seidel RM, Frenking G, Banh H, Dilchert K, Fischer RA (2015) The σ-aromatic clusters [Zn_3]$^+$ and [Zn_2Cu]: embryonic brass. Angew Chem Int Ed 54:4370–4374. doi:10.1002/anie.201410737
57. Kirchmann M, Eichele K, Schappacher FM, Schappacher FM, Pottgen R, Wesemann L (2008) Octahedral coordination compounds of the Ni, Pd, Pt triad. Angew Chem Int Ed 47:963–966. doi:10.1002/anie.200704814

Struct Bond (2016) 169: 131–156
DOI: 10.1007/430_2015_202

Published online: 14 February 2016

Gilbert Lewis and the Model of Dative Bonding

Gernot Frenking and Markus Hermann

Abstract The electron-pair bonding model that was introduced by Gilbert Lewis 100 years ago is discussed in the light of modern quantum chemical methods for analysing the electronic structures of some simple molecules. It is argued that Lewis structures in conjunction with accurate quantum chemical calculations are still very useful for the description of chemical bonding. The emphasis lies on the difference between electron-sharing bonds A–B and dative bonds $A \rightarrow B$ which were suggested by Lewis as a general definition for acids and bases. The electron-pair model, if combined with quantum chemical calculations, remains a powerful guide for the search of new molecules and for understanding molecular structures.

Keywords Carbone • Chemical bonding • Dative bond • Electron-sharing bond • Multiple bond

Contents

G. Frenking (✉) and M. Hermann
Fachbereich Chemie, Philipps-Universität Marburg, Hans-Meerwein-Strasse, 35043 Marburg, Germany
e-mail: frenking@chemie.uni-marburg.de

1 Introduction

This issue of *Structure and Bonding* is dedicated to the 100th anniversary of the landmark publication "The Atom and the Molecule" by Gilbert Lewis where he introduced the model of electron-pair bonding into chemistry [1]. It was a bold suggestion that was born from the attempt to explain the wealth of chemical information which was available at that time with a model for molecular structures that was shaped by classical physics. The boldness of the suggestion lies in the fact that Lewis knew about the inability of classical physics to correctly describe chemical bonding in terms of electrostatic attraction as underpinning forces for the model of electron-pair bonding. He speculated about a possible deviation from Coulomb's law when he wrote that "Electric forces between particles which are very close together do not obey the simple law of inverse squares which holds at greater distances" [2]. Eleven years later, this foresighted postulate was proven by Heitler and London to be correct [3], although Lewis could not foresee the paradigm change which was to come by the quantum theory that was suggested by Schrödinger and Heisenberg.

The work by Heitler and London published in 1927 was the first study which correctly described chemical bonding in terms of modern quantum theory that was introduced by Heisenberg and Schrödinger two years earlier. It was the birth of quantum chemistry which received its first textbook *Einführung in die Quantenchemie* [Introduction to Quantum Chemistry (in German)] in 1937 by Hans Hellmann [4]. (The book has recently been republished with biographical notes of the son Hans Hellmann Jr. by Andrae [5].) The quantum chemical explanation of chemical bonding was a revolutionary view of the nature of the interatomic interactions. In order to understand the strong attraction between two neutral atoms which form a chemical bond, electrons have to be considered as waves rather than particles, and the covalent bond must be understood as a (translated from German by the authors) "quantum mechanical vibrational phenomena" [3] which comes from the mixing of the wave functions. Thus, chemical bonding is not due to the formation of an electron pair as suggested by Lewis. The tendency of molecules to have electron pairs is rather due to the Pauli principle which allows a maximum of two electrons in the same region of space. A chemical bond can be formed with only one electron such as in H_2^+.

It is illuminating to consider the position of Gilbert Lewis, who had an excellent knowledge of modern physics, to quantum theory. In his pioneering publication in 1916, he shortly discusses new models which were suggested to explain the apparent violation of the common laws of electricity in the atomic region. He writes that "The most interesting and suggestive of these theories is the one proposed by Bohr and based upon Planck's quantum theory" [6]. But then he dismisses Bohr's atomic model because it "…is not only inconsistent with the accepted laws of electromagnetics but, I may add, is logically objectionable, for that state of motion which produces no physical effect whatsoever may better be called a state of rest". In spite of his reservations against quantum theory which was only known in the suggestions of Planck and Bohr and prior to the work of Schrödinger and Heisenberg, Lewis devoted more space and thoughts to it in his seminal book "Valence and The Structure of Atoms and Molecules" which was published in 1923 [7]. It was the first publication by Lewis fully devoted to chemical bonding after his now celebrated 1916 study. In the meantime Irving Langmuir had published a series of papers where he further developed the theory of electron-pair bonding [8–10]. This led to the situation that the electron-pair model was widely attributed to Langmuir rather than to Lewis, and still today the name Langmuir–Lewis model is sometimes used. In fact, the term "covalency" and the octet rule are due to Langmuir and not to Lewis [11, 12].

Lewis devoted a section in his book to the topic entitled "The Quantum Theory" where he acknowledges that the discrete nature of matter introduces a fundamentally new understanding of atomic structure and light. He frequently refers to Einstein's thoughts about quantum theory, and he mentions a remark which Einstein made to Lewis that "...the quantum theory was not really a new theory, but merely a recognition of the falsity of previous theories" [13]. Lewis goes on and points out that quantum theory (which was prior to the Schrödinger/Heisenberg development of the theory) was not really capable to furnish an understanding of interatomic interactions. But then he concludes "Quantum theory has been criticized for furnishing no adequate mechanism, but presumably the root of our present problem lies deeper than this, and it is hardly likely that any mechanism based on our existing modes of thought will suffice for the explanation of the many new phenomena which the study of the atom is disclosing" [13]. Lewis shared the deep-rooted dislike of quantum theory which he refers to as "the entering wedge of scientific bolshevism" [14] with Einstein but sensed at the same time that something new was coming up to explain chemical bonding and other molecular properties. He speculated that some of the abstractions which he used in his book may in the future have to be abandoned while others "...may have to be modified, and my chief purpose in writing the present section is not so much to predict just how these modifications are to occur as it is to emphasize the necessity of maintaining an opening of mind; so that, when the solution of these problems, which now seem so baffling, is ultimately offered, its acceptance will not be retarded by the conventions and the inadequate mental abstractions of the past" [15]. These are the closing remarks in the book which are a challenge and a legacy of Gilbert Lewis to the following generations.

After reading the original works by Gilbert Lewis about chemical bonding and in particular his 1916 paper [1] and the 1923 book [7], it may be recognized that his legacy is not just the suggestion that the chemical bond shall be identified with an electron pair. It is also the appeal to future generations not to hang on to old conventions and traditional models but to continue in developing new models and to be open to new insights which become available when future methods provide more information about chemical bonding.

One of the models which Lewis introduced in his book shall be the topic of this article which is written in the spirit that is expressed in the above-cited closing statement of the author. It is the general definition for acids and bases which now

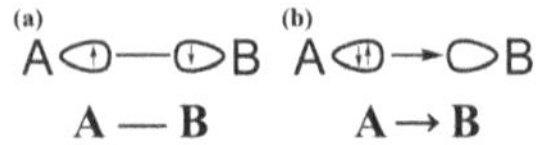

Scheme 1 Schematic representation of (**a**) an electron-sharing electron-pair bond and (**b**) a dative (donor–acceptor) electron-pair bond. Electron lone pairs are represented in this and the other figures by a *bar*

carries his name *Lewis acids* and *Lewis bases*. In the chapter "Remnants of the Electrochemical Theory", he devotes a section to "The Definition of Acids and Bases" where he introduces his model with the statement: "A basis substance is one which has a lone pair of electrons which may be used to complete the stable group of another atoms, and....an acid substance is one which can employ a lone pair from another molecule in completing the stable group of one of its own atoms" [16]. With other words, Lewis distinguishes between two types of electron-pair bonds, i.e. the shared-electron bond A–B and the dative bond A → B (Scheme 1). Fifteen years later in 1938, when quantum chemistry was already blossoming and Linus Pauling was on his way to provide a quantum theoretical underpinning of the electron-pair model in his book "The Nature of the Chemical Bond" that was eventually published in 1939 [17, 18], Lewis elaborated on the topic in his later work entitled "Acids and Bases" [19]. He mentions the model of resonance that was suggested by Pauling for describing the electronic structure which is particularly important for unsaturated species such as Lewis acids. But he points out that the development of quantum chemistry does not really alter the essential definition of acids and bases in terms of lone electron-pair donation.

The relevance of distinguishing between electron-sharing bonds A–B and dative bonds A → B for understanding molecular structures of main-group compounds has been stressed by Haaland in a review article in 1989 [20]. The model of donor–acceptor interactions is well established in transition metal chemistry since Dewar suggested in 1951 that the structure of Zeise's salt can be understood in terms of σ donation and π backdonation [21]. The donor–acceptor model was generalized to other transition metal complexes in a series of papers by Chatt together with Duncanson and other co-workers [22, 23] (the contributions of Chatt to the present understanding of chemical bonding in transition metal chemistry have been highlighted in [23]), and therefore, it is now known as Dewar–Chatt–Duncanson (DCD) model (for a discussion of the DCD bonding model in the light of quantum chemical calculations, see [24]). The DCD model which uses dative bonds is the predominant description of chemical bonding in transition metal chemistry [25]. Prior to the works by Dewar and Chatt, similar suggestions were made by Hieber [26] and later by Orgel [27] who pointed towards a synergic bonding in transition metal complexes.

The review by Haaland [20] is a good starting point for the present manuscript. It summarizes the knowledge of classical Lewis acid/base complexes mainly of group 13/15 adduct which were known at that time. It is shown that the discrimination between electron-sharing bonds and dative bonds is very useful for understanding molecular structures and stabilities. But during the last decade, it was realized that there are molecules of main-group atoms which were previously described with electron-sharing bonds that may better become discussed with dative bonds [28–39]. This led to the prediction of new adducts with unusual bonds which could become synthesized and structurally characterized by X-ray analysis [40–44]. Numerous experimental studies particularly in the area of low-valent main-group atoms reported about exotic molecules whose structures were explained with dative bonds (representative examples: [45–56]). The increasing number of molecules that were sketched with dative bonds was not undisputed [57], but it was shown that many features and properties of the newly synthesized compounds are easily understood with the model of dative bonds [58] (for a reply, see [59]). We believe that it is an appropriate contribution to the special issue of *Structure and Bonding* celebrating the 100th anniversary of Gilbert Lewis' epochal paper to show that his model of dative bonding is still a powerful tool for finding new molecules and to explain unusual structures.

2 Carbon Dioxide CO_2 and Carbon Suboxide C_3O_2

Carbon dioxide CO_2 is a well-known compound while carbon suboxide C_3O_2 is a more exotic species which has received less attention in the literature, although it has been synthesized already in 1906 [60]. The molecules are usually sketched with electron-sharing double bonds O=C=O and O=C=C=C=O which let one expect linear geometries. CO_2 has a linear equilibrium structure while gas-phase studies revealed in 1986 that carbon suboxide has a bent geometry with a bending angle of 156° at the central carbon atom [61, 62]. The bending potential was found to be very flat, and the molecule adopts a linear structure in the solid state [63]. The linear geometry of CO_2 and the bent structure of C_3O_2 can easily be understood when the two types of electron-pair bonding are considered. Figure 1 schematically displays

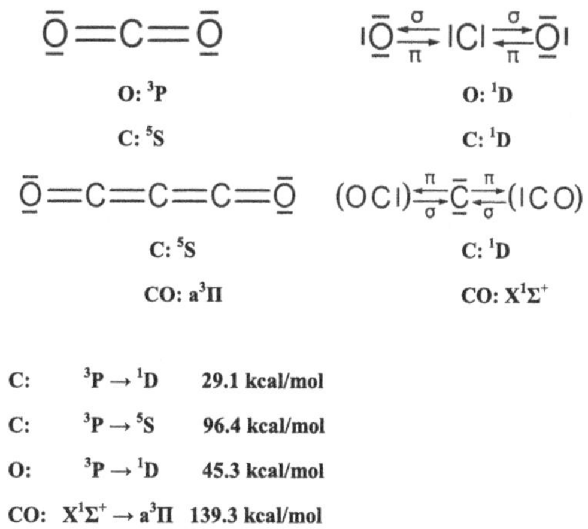

Fig. 1 Sketch of electron-pair bonds in CO_2 and C_3O_2 with electron-sharing bonds and dative bonds. Below each structure are the electronic reference states of the atoms and CO for the respective bonding interactions. At the bottom are the excitation energies from the electronic ground state to the excited state which were taken from [61]

the possible bonding situations in the molecules in terms of electron-sharing bonds and dative bonds. It also shows the electronic reference states of the relevant bonding fragments and the excitation energy which is required for promotion from the electronic ground state. (The experimental values of the atoms have been taken from [64]. The excitation energies for diatomic molecules were taken from [65].) Note that the carbon atom in the excited ^{1}D state would be a σ donor and π acceptor in CO_2 but a σ acceptor and π donor in C_3O_2.

The electron-sharing double bonds of CO_2 require oxygen atoms in the ^{3}P state which is the electronic ground state, while the carbon atom requires the excited ^{5}S state which is 96.4 kcal/mol higher than the ground state. Even more promotion energy is necessary to prepare oxygen and carbon for possible dative bonds. The $^3\text{P} \rightarrow {}^1\text{D}$ excitation energies for two oxygen atoms entail $2 \times 45.4 = 90.8$ kcal/mol and the $^3\text{P} \rightarrow {}^1\text{D}$ excitation of carbon necessitates another 29.1 kcal/mol. Thus, a total of 119.9 kcal/mol is required to promote oxygen and carbon for possible donor–acceptor interactions O⇆C⇄O, while only 96.4 kcal/mol is necessary for electron-sharing bonds O=C=O. Since electron-sharing bonds are stronger than dative bonds between the same atoms, it is clear that CO_2 should be written as O=C=O.

The situation looks different for carbon suboxide. Here, the promotion energy for dative bonds is only 29.1 kcal/mol which comes from the $^3\text{P} \rightarrow {}^1\text{D}$ excitation of carbon atom. Possible electron-sharing bonds O=C=C=C=O request excitation of the CO groups from the $X^1\Sigma^+$ ground state to the $a^3\Pi$ excited state which amounts to $2 \times 139.3 = 278.6$ kcal/mol and excitation of carbon atom to the excited ^{5}S state (96.4 kcal/mol) which leads to a total promotion energy of 375.0 kcal/mol. The difference of 345.9 kcal/mol in promotion energy is easily compensated by the strength of the donor–acceptor interactions (OC)⇆C⇄(CO). The bond dissociation energy for the reaction $C_3O_2 \rightarrow C + 2$ CO is $D_e = 136.0$ kcal/mol which gives a bond energy of 68 kcal/mol for each dative (double) bond, much less than an electron-sharing double bond. The bond dissociation energy of the electron-sharing double bond in ethylene $H_2C{=}CH_2$ is $D_e = 180$ kcal/mol (Hermann M and Frenking G. Calculated at UCCSD(T)/cc-pVTZ. Unpublished). Thus, carbon suboxide should be written as $C(CO)_2$, and the bonding situation should be sketched as (OC)⇆C⇄(CO). The σ donation OC → C ← CO is found to be stronger than the OC ← C → CO π backdonation [30] and leaves some lone-pair character at the central carbon atom. This explains why carbon suboxide has a bent geometry.

The assignment of dative bonds for C_3O_2 does not mean that the description with double bonds is wrong. Bonding models are not right or wrong; they are more or less useful. It is important to distinguish between the physical reality of the interatomic interactions and the description in terms of a model. Nature does not know electron-pair bonds nor does it discriminate between dative bonds and electron-sharing bonds. Our argument in favour of dative bonds in C_3O_2 rests on the straightforward explanation of (a) the bend equilibrium geometry of the molecule which is difficult to understanding using electron-sharing bonds,[1] (b) the rather small bond dissociation energies of the C–C bonds which do not agree with a genuine C=C double bond and (c) the trend of decreasing bond angles L-C-L′ when L becomes a weaker π acceptor (see the following section about carbones). Moreover, the high excitation energy of CO to the triplet state, which is required for building an electron-sharing double bond, makes it physically more reasonable to consider dative bonds. Thus, we find it more useful to discuss the bonding of carbon suboxide in terms of donor–acceptor interactions, because it offers an explanation and not only a mere description of the bonding situation.

The very high excitation energy of CO to the triplet state explains also the finding that the dimer ethylenedione in the singlet state is not even a minimum on the potential energy surface, although it can nicely be written with the Lewis formula O=C=C=O. The excitation energy of 278.6 kcal/mol for the required $^3\Pi$ state of CO is too high to be compensated by the bonding energy of a C=C double bond. Dative bonding in the linear form is also unfavourable, because the two fragments are both donors. Only lately has linear OCCO in the $^3\Sigma_g^-$ state been observed as quasi-bound species while singlet states of OCCO were found to be dissociative [66] (see also [67]).

Very recently, the anion $[B(CO)_2^-]$ which is isoelectronic to $C(CO)_2$ could become isolated in a low-temperature matrix experiment [68]. The analysis of the vibrational spectra and the comparison with ab initio calculations at the CCSD(T)/aug-cc-pVTZ level indicated that the boron dicarbonyl anion has a linear structure. This could be rationalized with stronger CO π backdonation $OC \leftarrow B^- \rightarrow CO$, but the structure might also be sketched with double bonds $[O{=}C{=}B{=}C{=}O]^-$. The authors investigated the nature of the bonding using an energy decomposition analysis. They compared the energy change which is associated with the formation of the molecular structure from closed-shell fragments or from proper open-shell fragments which have the frozen geometries of the molecules. The results which are shown in Table 1 provide also a detailed insight into the bonding interactions. The crucial term which indicates the relaxation of the molecular orbitals that comes from the formation of the molecular wave function is ΔE_{orb}. The calculated value when one starts with the singlet fragments which yield dative bonds is $\Delta E_{orb} = -433.4$ kcal/mol, while the open-shell fragments which lead to electron-sharing bonds give $\Delta E_{orb} = -493.9$ kcal/mol. Thus, the anion $[B(CO)_2]^-$ should be described in terms of dative bonds $(OC)\leftrightarrows B^- \rightleftarrows (CO)$ although it has a linear geometry.

The energy decomposition analysis makes it possible to visualize the charge migration which is associated with the dative interactions and to provide a quantitative estimate of the relative strength of donation and backdonation. Figure 2a

---

[1] One referee suggested that the bent geometry of carbon suboxide could be explained with the admixture of small contributions from the resonance form $O{=}C{=}C^-{-}C{\equiv}O^+$. This is a mere ad hoc description of the electronic structure rather than an explanation, because it does not answer the question why a resonance form where the more electropositive carbon atom carries a negative charge and the more electronegative oxygen atom carries a positive charge becomes so relevant.

Table 1 Energy decomposition analysis of $[B(CO)_2^{-}]$ at the BP86/TZ2P+ level

Fragments	B^- $2s^0 2p(\sigma)^0 2p(\pi)^2 2p(\pi')^2$ (OC)···(CO) singlet state	B^- $2s^1 2p(\sigma)^1 2p(\pi)^1 2p(\pi')^1$ (OC)···(CO) quintet state
ΔE_{int}	−445.3	−488.0
ΔE_{Pauli}	116.7	335.0
ΔE_{elstat} (%)[a]	−128.6 (22.9)	−329.2 (40.0)
ΔE_{orb} (%)[a]	−433.4 (77.1)	−493.9 (60.0)
ΔE_1 OC ← $E^{(-)}$ → CO (%)[b]	−127.7 (29.5)	−87.3 (17.7)
ΔE_2 OC ← $E^{(-)}$ → CO (%)[b]	−127.7 (29.5)	−87.3 (17.7)
ΔE_3 σ(+,−) OC → $E^{(-)}$ ← CO (%)[b]	−114.2 (26.3)	−151.7 (30.7)
ΔE_4 σ(+,+) OC → $E^{(-)}$ ← CO (%)[b]	−54.8 (12.6)	−151.0 (30.6)
ΔE_{orb}(rest) (%)[b]	−8.6 (2.0)	−16.9 (3.4)

Energy values are given in kcal/mol. Data are taken from (Hermann M and Frenking G. Calculated at UCCSD(T)/cc-pVTZ. Unpublished)

[a]The value in parentheses gives the percentage contribution to the total attractive interactions $\Delta E_{elstat} + \Delta E_{orb}$

[b]The value in parentheses gives the percentage contribution to the total orbital interactions ΔE_{orb}

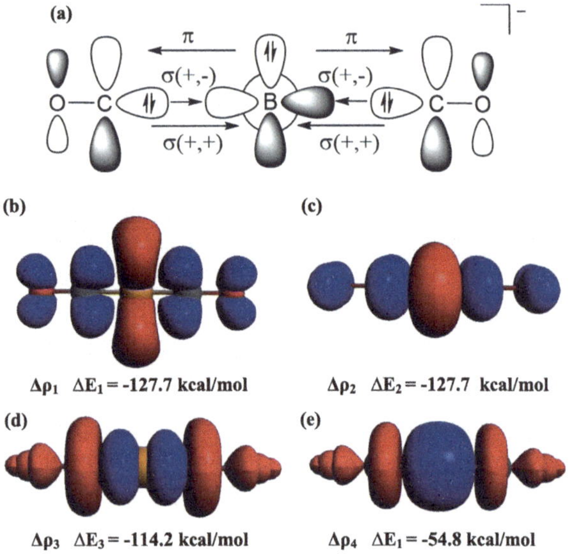

Fig. 2 (**a**) Schematic representation of the orbitals involved in the OC → $B^{(-)}$ ← CO σ-donation and the OC ← $B^{(-)}$ → CO π backdonation. Only one component of the latter is shown. (**b–e**) Plot of the charge flow which is connected with the pairwise orbital interactions in $[B(CO)_2]^-$ together with the associated interaction energies ΔE_n. The charge flow is *red → blue*

displays schematically the orbitals which are involved in the donor–acceptor interactions. Only one component of the degenerate OC ← $B^{(-)}$ → CO π backdonation is shown. Figure 2(b)-(e) shows the associated charge flows of the orbital interactions. The (+,+) component of the (OC) → $B^{(-)}$ ← (CO) σ-donation involves the 2s AO of boron while the (+,−) component implicates donation into the 2p(σ) AO. The charge flow which is associated with the dative interactions has the colour code red → blue. The strongest contributions ΔE_1 and ΔE_2 (−127.7 kcal/mol each) come from the degenerate π backdonation OC ← $B^{(-)}$ → CO. A slightly weaker stabilization $\Delta E_3 = -114.2$ kcal/mol arises from the out-of-phase (+,−) σ-donation of the CO lone-pair orbitals $OC \rightarrow B^{(-)} \leftarrow CO$ into the vacant p(σ) AO of boron. The in-phase (+,+) σ-donation $OC \rightarrow B^{(-)} \leftarrow CO$ into the vacant 2s AO of boron is much weaker ($\Delta E_4 = -54.8$ kcal/mol) than the out-of-phase (+,−) σ-donation, although the 2s AO is energetically lower lying than the 2p AO. The stronger stabilization of the donation into the latter comes from the larger overlap of the p(σ) AO than the 2s AO of boron with the donor orbitals.

The results in Table 1 and Fig. 2 demonstrate the progress which has been made since 1916 when Gilbert Lewis suggested the electron-pair model. At the same time the essential kernel of his model is retained. The development of modern quantum chemical methods did not erase the intuitive proposal of Lewis, which is complemented and can now be quantitatively expressed by advanced models of quantum chemistry.

3 Carbones CL_2 and Related Molecules

The finding that carbon suboxide may be understood as donor–acceptor complex suggests that there could be other molecules with the generic formula $L \rightarrow C \leftarrow L$ where L is a σ-donor ligand and where the carbon(0) atom retains its valence electrons as two lone pairs. A literature search shows that the replacement of CO by a phosphine PR_3 leads to well-known compounds whose structures and chemical properties agree with the model of dative bonding. Figure 3 shows a decreasing bond angle at the central carbon atom from $C(CO)_2$ (156.0°) to $C(CO)(PPh_3)$ (145.6°) and $C(PPh_3)_2$ (131.7°) which conforms with the weaker π-acceptor strength of phosphine ligands that is known from transition metal (TM) chemistry [69]. Carbodiphosphorane $C(PPh_3)_2$ is known since 1961 [70] and has been extensively studied ever since [71–74]. It easily accepts two protons at the central carbon atoms which are strongly attracted by the lone electron pairs at carbon [75, 76]. The molecule was considered in the past as diylid [70–73], and only recently it was recognized that $C(PPh_3)_2$ is better described with donor–acceptor bonds [28].

The model of dative bonds and the known behaviour of ligands in TM complexes helped to predict another type of carbon(0) compound where the ligands are *N*-heterocyclic carbenes (NHCs) [29]. There is experimental evidence that NHC ligands have similar σ-donor/π-acceptor properties as phosphines PR_3 [77]. Calculations showed that carbodicarbenes $C(NHC)_2$ have a bent geometry and a bending angle of 132° which is close to the value of carbodiphosphorane $C(PPh_3)_2$. [29, 35]. The theoretical prediction was soon verified by experiment. The first

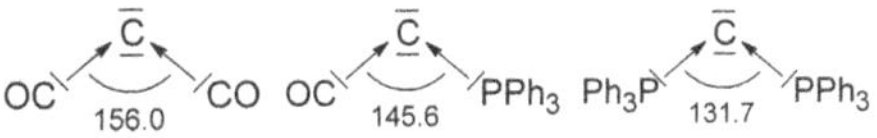

Fig. 3 Experimental bond angles at the central carbon atom of the compounds, $C(CO)_2$, $C(PPh_3)(CO)$ and $C(PPh_3)_2$

carbodicarbenes (CDCs) were synthesized and structurally characterized by X-ray analysis [78, 79] shortly after they had been calculated [29]. CDCs are now a very active and promising field of experimental research [80–83]. "The root for this very recent development lies in the suggestion of Gilbert Lewis that chemical bonds in molecules may be formed by the attraction between a fragment which has an electron lone pair and an electron deficient species that has a two-electron gap in the valence shell". The carbon atom in the ^{1}D state has two such gaps and thus, it can accommodate two donors. Compounds CL_2 which have a carbon(0) atom with two electron lone pairs are a class of compounds whose chemical reactivity exhibits characteristic differences from carbenes CR_2 that have only one lone pair at the carbon(II) atom [84, 85]. The name "carbone" has been coined for divalent C (0) compounds CL_2 [86]. Further information about the chemistry of carbones can be found in the literature [87, 88].

Carbodicarbenes are good examples to demonstrate the dichotomy of electron sharing vs. dative bonding. CDCs may also be considered as amino-substituted allenes which can be sketched with electron-sharing bonds. Figure 4 shows schematically the bonding situation in parent allene and in tetraaminoallenes (TAAs) with the two types of electron-pair bonds. Below each structure are the relevant fragments in the required electronic reference state. The crucial factor is the singlet–triplet (S/T) excitation energy of the terminal carbene fragments CR_2 (R=H, NMe_2, NEt_2) and NHC. CH_2 has a ($^3B_1$) ground state which needs no promotion for electron-sharing bonding with C (^{5}S). There is no doubt that the parent allene has electron-sharing double bonds $H_2C{=}C{=}CH_2$. The $C(NMe_2)_2$ carbene has a S/T gap of 33.5 kcal/mol which requests a promotion energy of 67 kcal/mol of the ligands to engage in electron-sharing bonds. The linear geometry suggests that the molecule may be written as $(NMe_2)_2C{=}C{=}C(NMe_2)_2$, but in the absence of EDA calculations, it cannot be ruled out that the molecule has dative bonds $(NMe_2)_2C \rightarrow C \leftarrow C(NMe_2)_2$. The latter bonding situation is clearly more

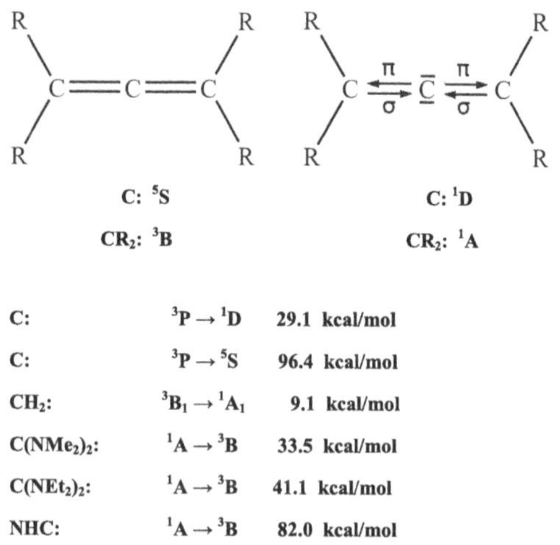

Fig. 4 Sketch of electron-pair bonds in allenes with electron-sharing bonds and dative bonds. Below each structure are the electronic reference states of the carbon atom and the carbene ligands CR_2 for the respective bonding interactions. At the bottom are the excitation energies from the electronic ground state to the excited state. The value for the carbon atom was taken from [61], and the values for the carbenes were calculated at BP86+D3(BJ)/def2-TZVPP

appropriate for the NEt_2 derivative $(NEt_2)_2C \rightarrow C \leftarrow C(NEt_2)_2$ which has in spite of the more bulky ethyl groups a calculated bonding angle at the central carbon atom of 169.5° [29]. The S/T gap of $C(NEt_2)_2$ is 41.1 kcal/mol which means that the promotion energy of the ligands to engage in electron-sharing bonds is 82.2 kcal/mol which is higher than for $C(NMe_2)_2$. An much higher S/T gap of 82.0 kcal/mol is calculated for NHC which leaves no doubt that $C(NHC)_2$ should be written with dative bonds $(NHC) \rightarrow C \leftarrow (NHC)$. There is experimental and theoretical evidence that $C[C(NMe_2)_2]$ and $C[C(NEt_2)_2]$ can both be considered as carbones CL_2, although the former species has a linear structure. The molecules readily add CO_2 and CS_2 at the central carbon atom-yielding adducts [89] that were also found for carbodiphosphorane [90]. Calculations showed that TAAs have very large first and second proton affinities which have been found as distinctive difference to carbenes [91].

The model of dative bonding proved to be a very powerful tool for related molecules EL_2 which are isoelectronic to carbones. Figure 5 shows a survey of group-15 homologues $(N^+)L_2$ and group-13 complexes $(BH)L_2$ which were calculated and in part synthesized. Some of them could become synthesized following

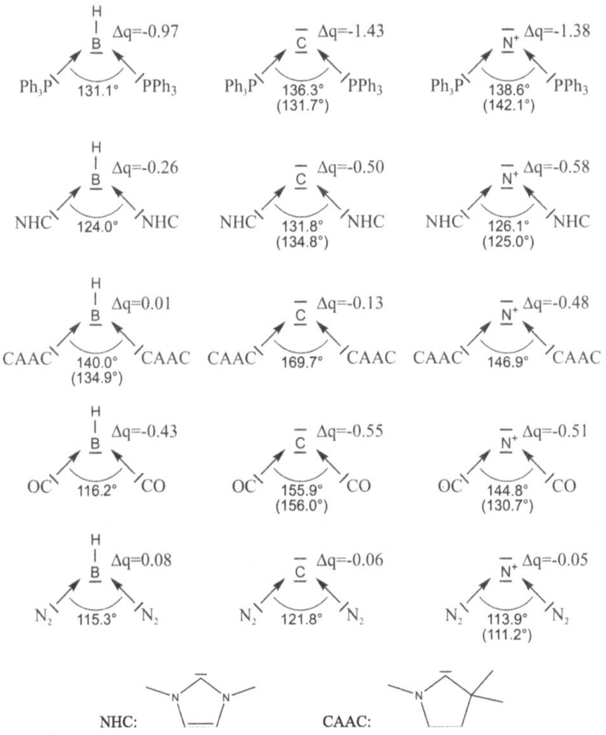

Fig. 5 List of isoelectronic molecules EL_2 showing calculated bond angles and partial charges Δq of the central fragments for E=BH, C, N^+ [89]. Experimental values of the bond angles of isolated molecules are given in parentheses

predictions that were based on the model of dative bonding. The experimental values for the bond angles of isolated molecules are given in parentheses. We want to point out that the trend of the bond angles in EL_2 nicely follows the strength of the L ← E → L π backdonation, unless steric repulsion of the bulky substituents prevents smaller angles. For example, the bond angles increases for the isoelectronic species $(N^+)(N_2)_2$ (111.2°) < $(N^+)(CO)_2$ (130.7°) < $C(CO)_2$ (156°).

The synthesis of two unusual molecules from the series shown in Fig. 5 which were previously unknown confirms the predictive value of the Lewis model of dative bonding. One molecule is the borylene complex $(BH)(CAAC)_2$ where two cyclic (alkyl)(amino)carbene ligands (see Fig. 5) stabilize a BH fragment in the highly excited $^1\Delta$ state where the lone-pair electrons occupy a p(π) AO [39]. Although the excitation energy of BH from the $X^1\Sigma^+$ ground state to the $^1\Delta$ reference state is very high (131.5 kcal/mol) [65], the strong CAAC → (BH) ← CAAC σ donation and the CAAC ← (BH) → CAAC π backdonation sufficiently stabilize the molecule that it can be isolated and structurally characterized by X-ray analysis [40]. The molecule was the first example of a tricoordinated boron compound where the boron atom is a Lewis base, and thus, it is isoelectronic with an amine. Figure 6 shows the HOMO of the molecule. The shape nicely shows the extent of the CAAC ← (BH) → CAAC π backdonation. Since the HOMO is energetically rather high lying, it can easily be protonated and can also be ionized to the radical cation which could become isolated and structurally characterized by X-ray analysis. For further details we refer to the original literature [39, 40].

The second remarkable molecule is a substituted homologue of the borylene complex where the stabilizing donor ligands are carbonyls. The borylene

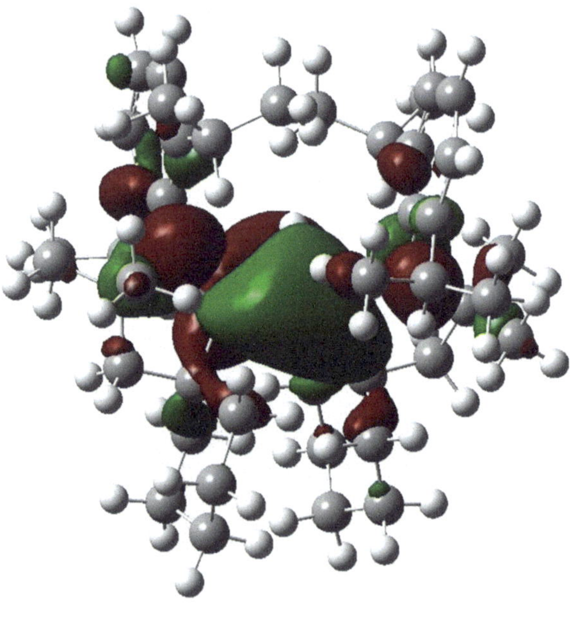

Fig. 6 Plot of the highest-lying occupied molecular orbital (HOMO) of the borylene complex (BH) $(CAAC)_2$ [90]

dicarbonyl complex $(BR)(CO)_2$ which has a bulky aryl group R has very recently been synthesized [92, 93]. The X-ray analysis confirms the trigonal planar geometry which was calculated for the parent system (Fig. 5). The bond angle OC-B(R)-CO is 104.0° which is more acute than the calculated value in the parent system OC-B(H)-CO because of steric repulsion of the carbonyl ligands with the substituent R. The borylene dicarbonyl reacts chemically like a transition metal carbonyl complex which opens the door to a new area of the p-block atoms [92, 93].

The model of dative bonding served also as useful guideline for heavy atom homologues of carbones EL_2 (E = Si – Pb). It actually seems that dative bonds are even more common in molecules of heavier main-group atoms than for the first octal-row elements. In the years 2003–2005 it was reported that the first silicon and germanium homologues of allenes had been isolated [94, 95]. However, the molecular structure did not exhibit the characteristic features of allenes which have a linear backbone $R_2C{=}C{=}CR_2$ where the terminal groups are orthogonal to each other. The isolated species have a bent geometry, and the planes of the substituents are twisted (Fig. 7a). The structural features resemble more carbodicarbenes (Fig. 7b), but the cyclic end groups do not have nitrogen atoms in α-position like CDCs.

Calculations of NHE and cycE (E = C – Pb) where NHE and cycE are group-14 homologues of NHCs and cyclopentylidene showed that the singlet/triplet gap of

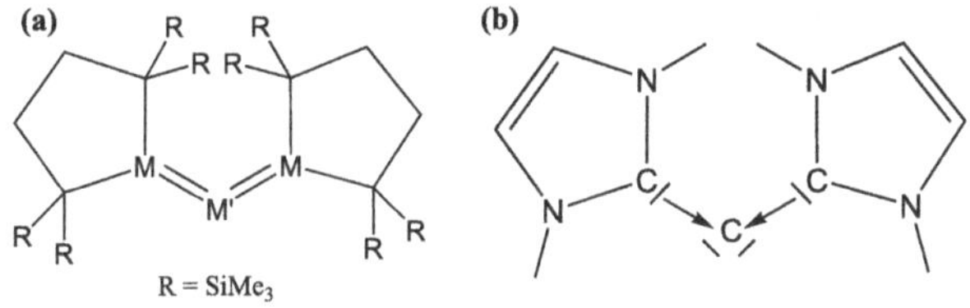

Fig. 7 Schematic representation of the bonding situation in (**a**) alleged heavy allenes and (**b**) carbodicarbenes $C(NHC)_2$

Table 2 Energy differences of the ligands NHE and cycE between different spin multiplicities at BP86/TZVPP and experimental atomic excitation energies

	E [a]		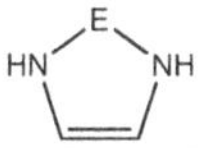		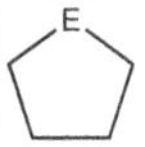	
	^{1}D	^{3}P	Singlet	Triplet	Singlet	Triplet
E=C	29.1	0.0	0.0	84.1	0.0	7.4
E=Si	18.0	0.0	0.0	60.8	0.0	27.1
E=Ge	20.4	0.0	0.0	50.5	0.0	31.0
E=Sn	24.6	0.0	0.0	37.2	0.0	31.2
E=Pb	22.4	0.0	0.0	30.5	0.0	33.8

The calculated values of NHE and cycE have been taken from [87]. All values in kcal/mol
[a]Taken from [64]

Table 3 Proposed nomenclature for divalent E (0) compounds

	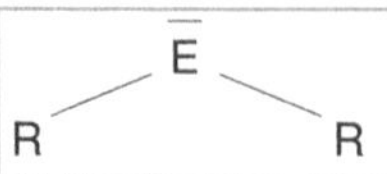	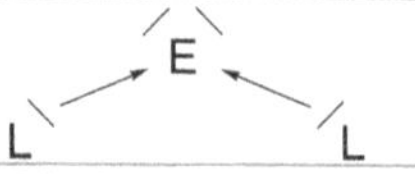
E	Divalent E(II): ylidene	Divalent E(0): ylidone
C	Carbene	Carbone
Si	Silylene	Silylone
Ge	Germylene	Germylone
Sn	Stannylene	Stannylone
Pb	Plumbylene	Plumbylone

NHE decreases but that of cycE increases when E becomes heavier (Table 2). The excitation energy $^3P \rightarrow {}^1D$ of the heavier atoms Si - Pb is smaller than for carbon which means that dative bonding for the systems (cycE) → E ← (Ecyc) becomes more favourable. For E = Si, the singlet fragments in $Si(cycSi)_2$ are favoured by (2 × 27.1 kcal/mol) – 18.0 kcal/mol = 36.2 kcal/mol, and for E=Ge the singlet fragments in $Ge(cycGe)_2$ are lower in energy by (2 × 31.0 kcal/mol) – 20.4 kcal/mol = 41.6 kcal/mol. The bonding situation of a genuine allene in $E(cycE)_2$ would only be possible if stronger electron-sharing bonds would compensate for the differences in the excitation energies. It has been shown, however, that for heavier atoms E, the E → E (E = Si – Pb) donor–acceptor interactions between singlet fragments may have a similar strength as E–E electron-sharing interactions between open-shell fragments [96]. Thus, the alleged "trisilallene" and trigermaallene" [94, 95] are rather examples for heavy group-14 homologues of carbones. Table 3 shows the names which have been suggested for tetrylones EL_2 in analogy to tetrylenes ER_2.

We close this section with an example where the different bond strength of dative bonds and electron-sharing bonds of the same fragments could be quantitatively estimated. The results provided an explanation for a puzzling experimental result. In 2009, the silylene complex NHC → $SiCl_2$ could become isolated which was the first stable silylene adduct at room temperature [97] (a stable $SiBr_2$ adduct was reported in the same issue: [98]). The complex was later reacted with the CAAC carbene which is a stronger σ donor and stronger π acceptor than NHC [99, 100]. Instead of the expected exchange reaction yielding CAAC → $SiCl_2$, the product had two CAAC ligands attached to the silylene fragment in CAAC-(SiC12)-CAAC where the Si–C bonds were significantly shorter than in CAAC → $SiCl_2$ [101]. Moreover, the latter molecule was found to have an electronic triplet state.

Figure 8 shows schematically the bonding situation in the complex NHC → $SiCl_2$ and in the triplet species $SiCl_2(CAAC)_2$ which has electron-sharing bonds between $SiCl_2$ and the CAAC ligands that accommodate the unpaired electrons.[2] It shows also calculated energies which are relevant for the system. The bond dissociation energy (BDE) of NHC → $SiCl_2$ is $D_e = 40.5$ kcal/mol. The BDE of the hypothetical adduct CAAC → $SiCl_2$ of $D_e = 42.5$ kcal/mol is as

[2] The calculation of the spin density shows that the unpaired electrons are mainly at the carbene carbon atoms and the nitrogen atoms. For details see [101].

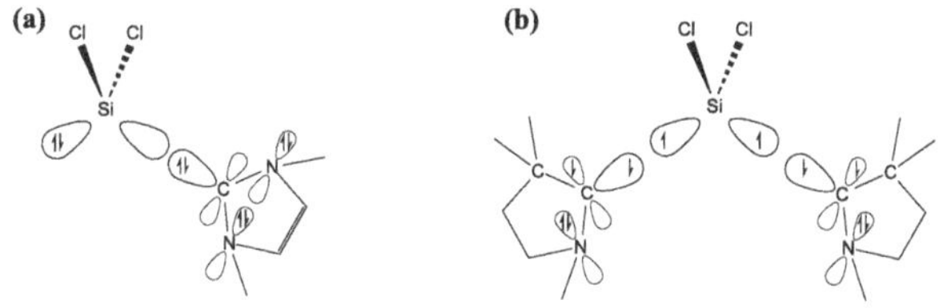

$SiCl_2(NHC) \rightarrow SiCl_2 + NHC$ $\quad D_e = 40.5$ kcal/mol

$SiCl_2(CAAC) \rightarrow SiCl_2 + CAAC$ $\quad D_e = 42.5$ kcal/mol

$SiCl_2(NHC)_2 \rightarrow SiCl_2 + 2\,NHC$ $\quad D_e = 245.5$ kcal/mol

$SiCl_2(CAAC)_2 \rightarrow SiCl_2 + 2\,CAAC$ $\quad D_e = 227.5$ kcal/mol

S/T

$SiCl_2$ = 60.1 kcal/mol

CAAC = 49.9 kcal/mol

NHC = 88.9 kcal/mol

ΔE(CAAC) = 227.5 kcal/mol - 60.1 kcal/mol - 2 x 49.9 kcal/mol = 67.3 kcal/mol

ΔE(NHC) = 245.2 kcal/mol - 60.1 kcal/mol - 2 x 88.9 kcal/mol = 7.3 kcal/mol

Fig. 8 Schematic view of the bonding situation in (**a**) the complex $NHC \rightarrow SiCl_2$ and (**b**) the molecule $SiCl_2(CAAC)_2$ in the triplet state. Below are the calculated bond dissociation energies at M05-2x/def2-TZVPP of the complexes $SiCl_2(NHC)$ and $SiCl_2(CAAC)$ and the compounds in the triplet state $SiCl_2(NHC)_2$ and $SiCl_2(CAAC)_2$ as well as the singlet–triplet gaps of the fragments. The *bottom lines* give the net stabilization energies ΔE for the formation of the triplet compounds $SiCl_2(NHC)_2$ and $SiCl_2(CAAC)_2$

expected slightly bigger. The calculated BDE of the isolated species $SiCl_2(CAAC)_2$ is $D_e = 227.5$ kcal/mol. In order to enable electron-sharing bonds, the fragments $SiCl_2$ and CAAC must be promoted to the triplet state which requires 60.1 kcal/mol + (2 × 49.9 kcal/mol) = 159.9 kcal/mol. Subtracting this value from the BDE gives a net gain of $\Delta E = 67.3$ kcal/mol that is larger than the BDE of the hypothetical complex $CAAC \rightarrow SiCl_2$ ($D_e = 42.5$ kcal/mol). Thus, the formation of the triplet species $SiCl_2(CAAC)_2$ which possesses electron-sharing bonds is energetically favoured. In contrast, the formation of the hypothetical triplet molecule $SiCl_2(NHC)_2$ is energetically unfavourable in comparison with the complex $NHC \rightarrow SiCl_2$ although the BDE of $SiCl_2(NHC)_2$ ($D_e = 245.5$ kcal/mol) is higher than that of $SiCl_2(CAAC)_2$ ($D_e = 227.5$ kcal/mol). This is because the S/T gap of NHC (88.9 kcal/mol) is much higher than for CAAC (49.9 kcal/mol). The net stabilization energy ΔE for the formation of $SiCl_2(NHC)_2$ is only 245.2 kcal/mol–60.1 kcal/mol – (2 × 88.9 kcal/mol) = 7.3 kcal/mol which is much less than the BDE of $NHC \rightarrow SiCl_2$ ($D_e = 40.5$ kcal/mol).

The data in Table 2 suggest that tetrylones with NHC ligands $E(NHC)_2$ are promising targets for the synthesis of stable divalent E(0) compounds. Very recently, the stable silylone SiL_2 and germylone GeL_2 have been isolated with

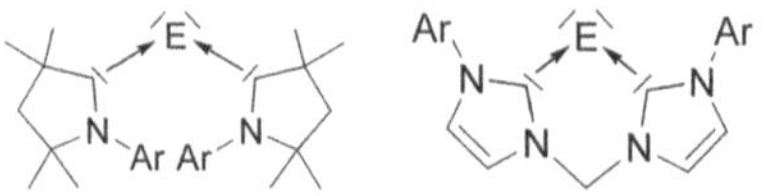

Fig. 9 Schematic representation of the silylones and germylones $E(CAAC)_2$ and E(NHC-NHC) (E = Si, Ge) which have been isolated

two NHC ligands that were bridged by a methylene group [42, 43], and the silylone $Si(CAAC)_2$ could become isolated (Fig. 9) [41].

The Lewis bonding model still remains remarkably fruitful when it becomes connected with quantum chemical calculations even 100 years after its first presentation.

4 Dative Bonding in Heavy Homologues of Acetylene HEEH (E = Si – Pb)

The fundamental difference between atoms of the first octal row and heavier homologues to form stable molecule with multiple bonds was well known when Lewis published his works on chemical bonding [1, 7]. He wrote in his book: "..... the ability to form multiple bonds is almost entirely, if not entirely, confined to elements of the first period of eight, and especially to carbon, nitrogen and oxygen" [102]. This statement was based on chemical knowledge which was available at that time. It was only in 1976 and 30 years after Lewis was deceased in 1946 that the first stable molecule whose structure could be measured by X-ray crystallography with a Sn=Sn double bond was isolated [103] followed by the syntheses of the other group-14 homologues $R_2E{=}ER_2$ ((E = Si: [104]), (E = Ge: [105], see also [106]), (E = Pb: [107])). Stable group-14 homologues of alkynes RE≡ER (E = Si – Pb) were reported between 2000 and 2004. Measurement of the X-ray analysis showed that the equilibrium geometries of ditetrylenes $R_2E{=}ER_2$ and ditetrylynes RE≡ER are very different from those of the carbon systems. $R_2E{=}ER_2$ are not planar but possess pyramidally coordinated ER_2 groups while ditetrylynes RE≡ER have a *trans*-bent arrangement of the substituents instead of a linear structure which raised the question whether they have genuine E≡E triple bonds [108–112].

Even more surprising were the results of accurate quantum chemical calculations of the parent systems E_2H_2 [113–127] (E = Si – Pb) which showed a variety of unusual equilibrium structures for several isomers (Fig. 10) that were later found in low-temperature matrix experiments [128–133]. None of them are the linear form HE≡EH which is an energetically high-lying second-order saddle point on the potential energy surface. The doubly bridged butterfly structure **A** is the global energy minimum for all heavier systems E = Si – Pb followed by the singly bridged isomer **B**. The vinylidene form **C** is the only isomer which is common for all group-14 atoms. Since the atomic connectivity differs from those of the other species, it is

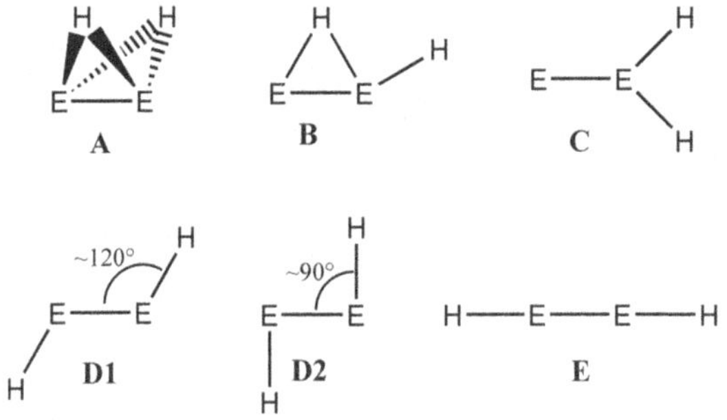

ΔE_{rel} (kcal/mol) at BP86/QZ4P

	C	Si	Ge	Sn	Pb
A	-	0.0	0.0	0.0	0.0
B	-	9.9	10.9	13.9	17.0
C	55.1	15.3	16.5	19.3	27.1
D1	-	19.9	22.3	25.3	31.0
D2	-	45.7(i=1)	37.6(i=1)	29.3(i=1)	26.5(i=1)
E	0.0	41.5(i=2)	45.7(i=2)	64.6(i=2)	92.5(i=2)

Fig. 10 Schematic view of the structures E_2H_2 which are found as energy minima on the potential energy surfaces and calculated relative energies. The values were taken from [96]

not further considered. There are two trans-bent forms **D1** and **D2** which are important for discussion. Only **D1** is an energy minimum while **D2** which has a more acute bonding angle is a transition state which is, however, for the lead system energetically lower lying than **D1**. With bulky substituents R it becomes an energy minimum [134] which could become isolated [135].

The isomers which are shown in Fig. 10 are difficult to sketch with electron-sharing bonds in an unambiguous way. It has been shown that the reason for the heavier homologues of acetylene to adopt unusual structures and the relative energies of the isomers can be explained with the dichotomy of electron-sharing and dative bonds [96]. The explanation is based on an earlier model that was suggested to rationalize the pyramidal structures of some ditetrylenes where the electron-sharing bonds for carbon $R_2C{=}CR_2$ are substituted by dative bonds $R_2E{\rightleftarrows}ER_2$ (Fig. 11) [136, 137]. This model proved in conjunction with quantum chemical calculations to be even more helpful for the exotic structures of REER.

The starting point for the discussion of the bonding situation in HEEH is the inspection of the electronic structure of the fragments EH. Figure 12 shows that the electronic ground state $^2\Pi$ has one σ electron pair and one unpaired electron in the p(π) AO of atom E. The electronic reference state which is required for electron-sharing triple bonds HE≡EH is the first excited $^4\Sigma^-$ state. The excitation energy $^2\Pi \rightarrow {}^4\Sigma^-$ is 16.7 kcal/mol for E = C but it becomes much higher for the heavier atoms E.

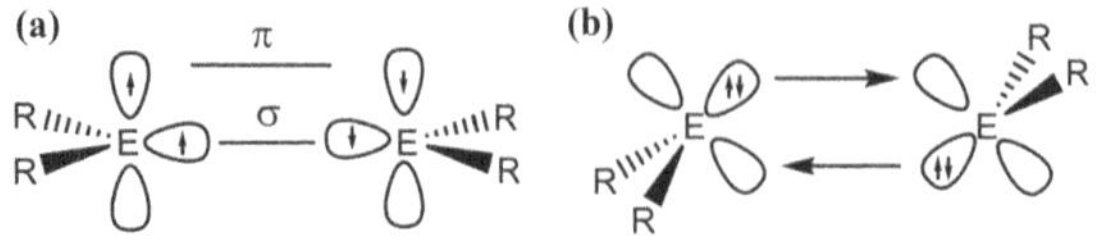

Fig. 11 Schematic representation of (**a**) electron-pair bonding in ethylene and planar analogues and (**b**) dative bonds in heavier group-14 homologues

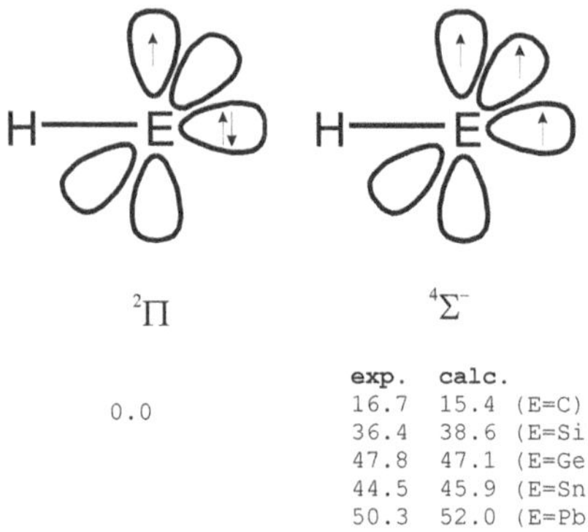

Fig. 12 Schematic representation of the electron configuration of the $^2\Pi$ electronic ground state and the $a^4\Sigma^-$ excited state of EH (E=C–Pb). Experimental [65] and calculated (BP86/QZ4P; [96]) excitation energies in kcal/mol

Table 4 Calculated bond dissociation energies D_e (kcal/mol) of linear HE≡EH → 2 EH ($a^4\Sigma^-$) and $X^2\Pi \rightarrow a^4\Sigma^-$ excitation energies ΔE_{exc} (kcal/mol) of EH at BP86/QZ4P

E	D_e	ΔE_{exc}	$D_e - 2 \times \Delta E_{exc}$
C	270.9	15.44	240.0
Si	121.6	38.56	44.5
Ge	113.3	47.09	19.0
Sn	89.4	45.87	−2.3
Pb	69.0	52.01	−35.0

The data are taken from [96]

Table 4 shows the calculated bond dissociation energies D_e for all linear species HE≡EH with E = C – Pb which exhibit a regular decrease for the heavier atoms. The decrease is particularly large from C to Si. The right column shows the net energy gain of the electron-sharing triple bonds when the promotion energies $^2\Pi \rightarrow {}^4\Sigma^-$ of the fragments EH are subtracted. Note that the $^2\Pi$ state would lead to a HE-EH electron-sharing single bond which is in competition with the HE≡EH triple bond that can be formed from the excited $^4\Sigma^-$ state. The energy gain for the carbon system to form a triple bond amounts to 240.0 kcal/mol which is much higher than the bond energy of a typical single bond. The situation is clearly different for the heavier homologues. The net energy gain for a HSi≡SiH triple bond is only 44.5 kcal/mol which is less than the value for a typical Si–Si single bond (75–80 kcal/mol) [138]. The same conclusion holds for the remaining species

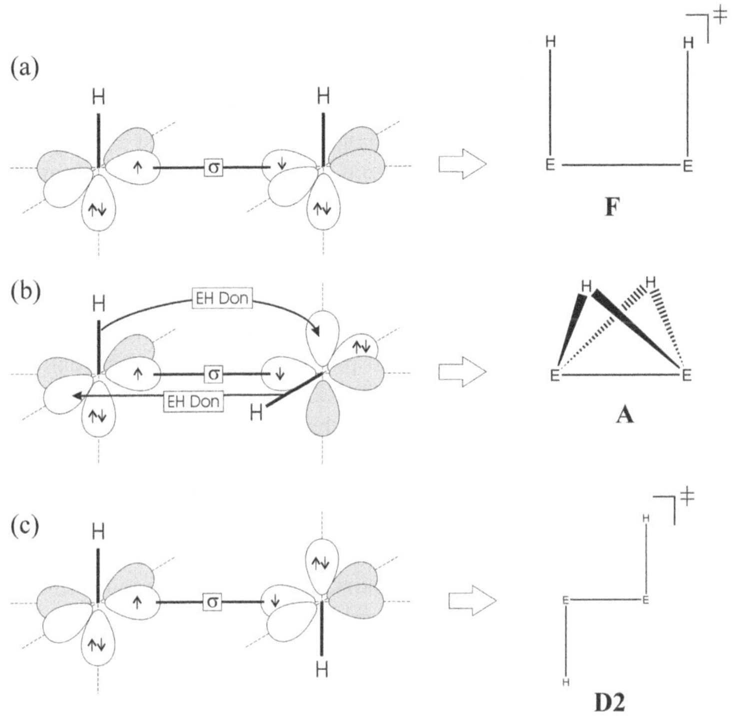

Fig. 13 Qualitative model for the orbital interactions between two EH molecules in different orientations where the unpaired electrons yield a σ bond

with E = Ge – Pb. It follows that only carbon binds through the excited $^4\Sigma^-$ state of CH while the heavier hydrides EH bind through the $^2\Pi$ ground state.

Connecting the EH fragments in the $^2\Pi$ ground state through the unpaired electrons and taking care of the octet rule straightforwardly lead to the energy minima on the potential energy surface of E_2H_2 that are shown in Fig. 10. Figure 13 shows three different arrangements where the unpaired electrons yield a σ bond HE–EH. The planar *syn*- and *anti*-forms **F** (Fig. 13a) and **D2** (Fig. 13c) which have an electron sextet in the valence shell of atom E are transition states with respect to rotation about the E–E bond. As mentioned above, **D2** becomes an energy minimum for bulky groups R and it can be isolated when E = Pb, although the lead atom has only six electrons in the valence shell [134, 135]. The octet rule is violated for very heavy main-group atoms due to relativistic effects [139]. Rotation of **F** or **D2** by 90° about the E–E axis enables mutual donation of the E–H bonds into the formally vacant p(π) AO of the other atom E, yielding structure **A** which fulfils the octet rule (Fig. 13b). The E–H bonds are tilting in order to maximize the E–H donation which straightforwardly leads to the doubly bridged butterfly structure **A**. Such kind of electron donation was already envisaged by Lewis in his book: "…. when there are not enough electrons in a molecule to provide each atom with its stable octet by the process of forming normal bonding pairs, two contiguous atoms may, to some extent, share a second or third pair of electrons, although this sharing is by no means so complete or unambiguous as in the single bond" [140]. Again, a remarkable foresight which, however, was followed by a restriction which is not correct: "....this ability to share a second or third pair is almost entirely limited to carbon, nitrogen and oxygen". The electron donation from the E–H bond into the empty p(π) orbital can be understood as a variant of the dative bond where the electron pair comes from a bond but not from a lone pair.

The doubly bridged butterfly structure **A** has three electron-pair bonding components for E–E bonding, i.e. one electron-sharing σ bond and two E–H donor–acceptor bonds. It may thus be considered to contain a triple bond where the electron-sharing π bonds of a classical triple bond are replaced by dative bonds. Note that the E–E distances in **A** (Si = 2.23 Å, Ge = 2.39 Å, Sn = 2.78 Å, Pb = 2.93 Å) are clearly longer than in the linear structure **E** (Si = 1.98 Å, Ge = 2.05 Å, Sn = 2.41 Å, Pb = 2.48 Å), and yet, the former species are much lower in energy than the latter. This is because the dative bonds in **A** do not require the large excitation energy $^2\Pi \rightarrow {}^4\Sigma^-$ of the fragments EH. We want to point out that the charge donation from the E–H bonds in **A** is stronger than the hypothetical sideward donation of the electron lone pairs, because the lone-pair orbitals of the heavy atoms E have mainly s-character and because hydrogen is more electronegative than Si–Pb.

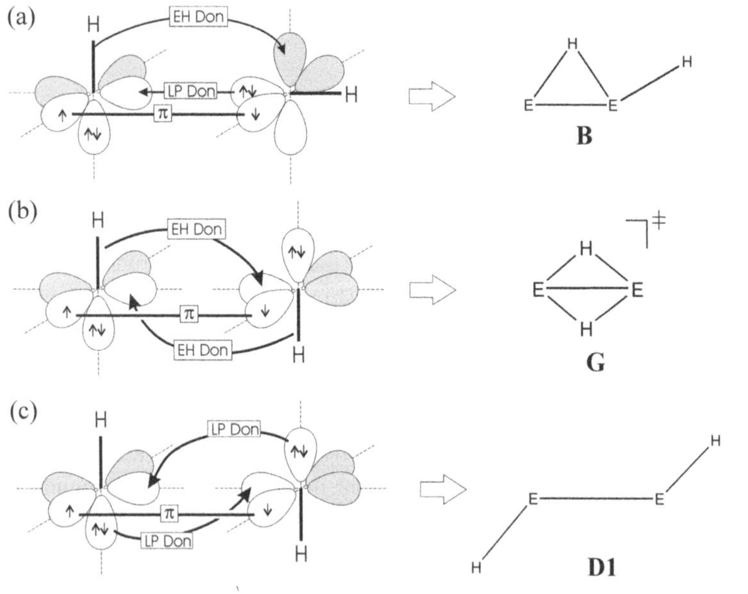

Fig. 14 Qualitative model for the orbital interactions between two EH molecules in different orientations where the unpaired electrons yield a π bond

The unpaired electrons of the EH fragments in the $^2\Pi$ ground state may also couple and form an electron-sharing π bond. Figure 14 displays three different orientations of the fragments with a HE-EH π bond. The arrangement in Fig. 14a shows that the E–H bond of the left fragments may donate into the vacant p(π) AO of the right EH species. In order to maximize the donor–acceptor interactions, the E–H bond and the vacant p(π) AO are tilting which induces a concomitant uplift of the E–H bond of the right fragment. This explains nicely the very unusual geometry of structure **B** where the terminal E–H bond is at the same side as the bridging hydrogen atom. The donation of the electron lone pair from the right to the left moiety complements the octet shell of this atom E. The octet rule is thus fulfilled. Isomer **B** also possesses three electron-pair bonds between atoms E which consist of one electron-sharing π bond, one E–H dative bond and one lone-pair dative bond. Rotation of the right E–H fragment by 90° leads to structure **G** which enables two E–H dative bonds besides the electron-sharing π bond (Fig. 14b). Structure **G** is therefore lower in energy than isomer **B,** but it is a transition state for the degenerate wing-flapping motion of the global energy minimum structure **A** [128]. The opposite tilting of the E–H fragments leads to structure **D1** (Fig. 14c) where the dative bonds come from the lone-pair orbitals. The electron donation is therefore weaker in **D1** which is higher in energy structures **B** and **G**. Steric repulsion enforces a trans-arrangement of bulky substituents in compounds REER which therefore adopt the structures **D1** for E=Si–Sn and **D2** or E=Pb in agreement with the relative energies of the parent systems (Fig. 10). **D1** has three electron-pair bonds which consist of one electron-sharing π bond and two lone-pair dative bonds while **D2** has one electron-sharing σ bond. (A similar explanation for the *trans*-bent geometry of HEEH (E = Si – Pb) has been given in [127]. The *trans*-bent structure of valence isoelectronic digallium compounds $[RGaGaR]^{2-}$ was discussed in terms of HOMO–LUMO mixing by [141].) For a more detailed discussion, we refer to the original literature [96].

Is it possible to sketch the bonding situation of the energy minimum structures **A**, **B**, **D1** and **D2** with simple Lewis formulas and possibly resonance forms? The answer is yes if the difference between an electron-sharing bond and a dative bond is depicted. Figure 15 shows a viable way how this can be done. The arrows indicate

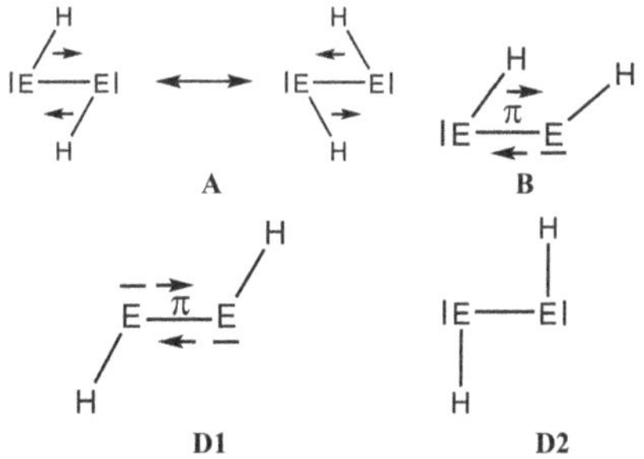

Fig. 15 Suggested Lewis structures for isomers **A**, **B**, **D1** and **D2** of E_2H_2 isomers

whether the dative bond comes from a lone-pair orbital or from an E–H bond. The sketches are slightly more complicate than standard Lewis structures, but they convey information about the different type of electron-pair bonding.

5 Concluding Remarks

The electron-pair model of Gilbert Lewis remains 100 years after its introduction a powerful tool for creative chemical research where it continues to serve as a guideline for finding new molecules and for understanding molecular structures *if* it is combined with quantum chemical calculations. The understanding of chemical bonding which originates through a quantum theoretical resonance phenomenon in terms of electron pairs requires the knowledge of the electronic structure which can be analysed with a variety of modern theoretical methods [142, 143]. The legacy of Gilbert Lewis includes both the electron-pair bonding model with its associated rules and the permanent willingness for "maintaining an opening of mind; so that, when the solution of these problems, which now seem so baffling, is ultimately offered, its acceptance will not be retarded by the conventions and the inadequate mental abstractions of the past" [15]. We do not find better words than the original statement by this pioneer of chemistry.

Acknowledgements The manuscript was written while one of the authors (GF) was guest professor at the Indian Institute of Technology Bombay. He wants to express his gratitude to Prof. Raghavan B. Sunoj and his colleagues for the warm hospitality. It is a pleasure to acknowledge very helpful comments by both referees and by Prof. Mike Mingos which improved our work. This work was supported by the Deutsche Forschungsgemeinschaft.

References

1. Lewis GN (1916) J Am Chem Soc 38:762
2. Lewis GN (1916) J Am Chem Soc 38:768
3. Heitler W, London F (1927) Z Physik 44:455
4. Hellmann H (1937) Einführung in die Quantenchemie. Deuticke, Leipzig and Wien
5. Andrae D (ed) (2015) Hans Hellmann: Einführung in die Quantenchemie. Springer Spektrum, Heidelberg
6. Lewis GN (1916) J Am Chem Soc 38:773
7. Lewis GN (1923) Valence and the structure of atoms and molecules. American Chemical Society Monograph Series, New York. http://babel.hathitrust.org/cgi/pt?id=uc1.$b35072;view=1up;seq=1
8. Langmuir I (1919) J Am Chem Soc 41:868
9. Langmuir I (1919) J Am Chem Soc 41:1543
10. Langmuir I (1920) J Am Chem Soc 42:274
11. Kohler R (1974) Irving Langmuir and the 'Octet' theory of valence. Hist Stud Phys Sci 4:39–87

12. Kohler R (1975) The Lewis–Langmuir theory of valence and the chemical community. Hist Stud Phys Sci 6:431–438
13. Lewis GN (1923) Valence and the structure of atoms and molecules. American Chemical Society Monograph Series, New York, p 42. http://babel.hathitrust.org/cgi/pt?id=uc1.$b35072;view=1up;seq=1
14. Lewis GN (1923) Valence and the structure of atoms and molecules. American Chemical Society Monograph Series, New York, p 163. http://babel.hathitrust.org/cgi/pt?id=uc1.$b35072;view=1up;seq=1
15. Lewis GN (1923) Valence and the structure of atoms and molecules. American Chemical Society Monograph Series, New York, p 165. http://babel.hathitrust.org/cgi/pt?id=uc1.$b35072;view=1up;seq=1
16. Lewis GN (1923) Valence and the structure of atoms and molecules. American Chemical Society Monograph Series, New York, p 142. http://babel.hathitrust.org/cgi/pt?id=uc1.$b35072;view=1up;seq=1
17. Pauling L (1939) The nature of the chemical bond. Cornell University Press, New York
18. Pauling L (1960) The nature of the chemical bond, 3rd edn. Cornell University Press, New York
19. Lewis GN (1938) J Frankl Inst 226:293
20. Haaland A (1989) Angew Chem Int Ed Engl 28:992
21. Dewar MJS (1951) Bull Soc Chim Fr 18:C79
22. Chatt J, Duncanson LA (1953) J Chem Soc 2939
23. Leigh GJ, Winterton N (eds) (2002) Modern coordination chemistry: the legacy of Joseph Chatt. The Royal Society, London
24. Frenking G (2002) In: Leigh GJ, Winterton N (eds) Modern coordination chemistry: the legacy of Joseph Chatt. The Royal Society, London, p 111
25. Frenking G, Fröhlich N (2000) Chem Rev 100:717
26. Hieber W (1942) Die Chemie 55:25
27. Orgel LE (1955) J Chem Phys 23:1004
28. Tonner R, Öxler F, Neumüller B, Petz W, Frenking G (2006) Angew Chem Int Ed 45:8038
29. Tonner R, Frenking G (2007) Angew Chem Int Ed 46:8695
30. Tonner R, Frenking G (2008) Chem Eur J 14:3260
31. Tonner R, Frenking G (2008) Chem Eur J 14:3273
32. Tonner R, Frenking G (2009) Pure Appl Chem 81:597
33. Takagi N, Shimizu T, Frenking G (2009) Chem Eur J 15:3448
34. Takagi N, Shimizu T, Frenking G (2009) Chem Eur J 15:8593
35. Klein S, Tonner R, Frenking G (2010) Chem Eur J 16:10160
36. Esterhuysen C, Frenking G (2011) Chem Eur J 17:9944
37. Takagi N, Frenking G (2011) Theor Chem Acc 129:615
38. Barua SR, Allen WD, Kraka E, Jerabek P, Sure R, Frenking G (2013) Chem Eur J 19:15941
39. Celik MA, Sure R, Klein S, Kinjo R, Bertrand G, Frenking G (2012) Chem Eur J 18:5676
40. Kinjo R, Donnadieu B, Celik MA, Frenking G, Bertrand G (2011) Science 333:610
41. Mondal KC, Roesky HW, Klinke F, Schwarzer MC, Frenking G, Niepötter B, Wolf H, Herbst-Irmer R, Stalke D (2013) Angew Chem Int Ed 52:2963
42. Xiong Y, Yao S, Inoue S, Epping JD, Driess M (2013) Angew Chem Int Ed 52:7147
43. Xiong Y, Yao S, Tan G, Inoue S, Driess M (2013) J Am Chem Soc 135:5004
44. Li Y, Mondal KC, Roesky HW, Zhu H, Stollberg P, Herbst-Irmer R, Stalke D, Andrade DM (2013) J Am Chem Soc 135:12422
45. Wang Y, Xie Y, Wei P, King RB, Schaefer HF III, Schleyer PVR, Robinson GH (2008) Science 321:1069
46. Sidiropoulos A, Jones C, Stasch A, Klein S, Frenking G (2009) Angew Chem Int Ed 48:9701
47. Jones C, Sidiropoulos A, Holzmann N, Frenking G, Stasch A (2012) J Chem Soc Chem Commun 48:9855

48. Wang Y, Xie Y, Wie P, King RB, Schaefer HF III, Schleyer PVR, Robinson G (2008) J Am Chem Soc 130:14970
49. Back O, Kuchenbeiser G, Donnadieu B, Bertrand G (2009) Angew Chem Int Ed 48:5530
50. Russell CA (2010) Angew Chem In Ed 49:9572
51. Abraham MY, Wang Y, Xie Y, Wei P, Schaefer HF III, Schleyer PVR, Robinson GH (2010) Chem Eur J 16:432
52. Brauschweig H, Dewhurst RD, Hammond K, Mies J, Radacki K, Vargas A (2012) Science 336:1420
53. Abraham MY, Wang Y, Xie Y, Wei P, Schaefer HF, Schleyer PVR, Robinson GH (2011) J Am Chem Soc 133:8874
54. Ghadwal RS, Azhakar R, Roesky HW, Pröpper K, Dittrich B, Goedecke C, Frenking G (2012) J Chem Soc Chem Commun 48:8186
55. Ghadwal RS, Azhakar R, Roesky HW, Pröpper K, Dittrich B, Klein S, Frenking G (2011) J Am Chem Soc 133:17552
56. Mondal KC, Samuel PP, Roesky HW, Aysin R, Leites L, Neudeck S, Lübben J, Dittrich B, Holzmann N, Hermann M, Frenking G (2014) J Am Chem Soc 136:8919
57. Himmel D, Krossing I, Schnepf A (2014) Angew Chem Int Ed 53:370
58. Frenking G (2014) Angew Chem Int Ed 53:6040
59. Himmel D, Krossing I, Schnepf A (2014) Angew Chem Int Ed 53:6047
60. Diels O, Wolf B (1906) Ber Dt Chem Ges 39:689
61. Jensen P, Johns JWC (1964) J Mol Spectrosc 118:248
62. Koput J (2000) Chem Phys Lett 320:237
63. Ellern A, Drews T, Seppelt K (2001) Z Anorg Allg Chem 627:73
64. Moore CE (1971) Atomic energy levels; NSDRS-NBS 35. U.S. National Bureau of Standards, Washington, DC
65. Huber KP, Herzberg G (1979) Molecular spectra and molecular structure IV. Constants of diatomic molecules. Van Nostrand-Reinhold, New York
66. Dixon AR, Xue T, Sanov A (2015) Angew Chem Int Ed 54:8764
67. Schröder D, Heinemann C, Schwarz H, Harvey JN, Dua S, Blanksby SJ, Bowie JH (1998) Chem Eur J 4:2550
68. Zhang Q, Li W-L, Xu C, Chen M, Zhou M, Li J, Andrada DM, Frenking G (2015) Angew Chem Int Ed 54:11078
69. Fröhlich N, Frenking G (2000) Chem Rev 100:717
70. Ramirez F, Desai NB, Hansen B, McKelvie N (1961) J Am Chem Soc 83:3539
71. Schmidbaur H (1983) Angew Chem Int Ed 22:907
72. Kaska WC (1983) Coord Chem Rev 48:1
73. Kolodiazhnyi OI (1996) Tetrahedron 52:1855
74. Petz W, Frenking G (2010) Top Organomet Chem 30:49
75. Walker JD, Poli R (1989) Polyhedron 8:1293
76. Jensen WP, Gehrke H, Jones DR, Suh I-H, Jacobson RA (1996) Z Kristallogr 211:829
77. Herrmann WA (2002) Angew Chem Int Ed Engl 41:1290
78. Dyker CA, Lavallo V, Donnadieu B, Bertrand G (2008) Angew Chem 120:3250
79. Dyker CA, Lavallo V, Donnadieu B, Bertrand G (2008) Angew Chem Int Ed 47:3206
80. Chen W-C, Shen J-S, Jurca T, Peng C-J, Lin Y-H, Wang Y-P, Shih W-C, Yap GPA, Ong T-G (2015) Angew Chem Int Ed 54:15207
81. Hsu Y-C, Shen J-S, Lin B-C, Chen W-C, Chan Y-T, Ching W-M, Yap GPA, Hsu C-P, Ong T-G (2015) Angew Chem Int Ed 54:2420
82. Chen W-C, Lee C-Y, Lin B-C, Hsu Y-C, Shen J-S, Hsu C-P, Yap GPA, Ong T-G (2014) J Am Chem Soc 136:914
83. Chen W-C, Hsu Y-C, Lee C-Y, Yap GPA, Ong T-G (2013) Organometallics 32:2435
84. Petz W, Öxler F, Neumüller B, Tonner R, Frenking G (2009) Eur J Inorg Chem 4507
85. Inés B, Patil M, Carreras J, Goddard R, Thiel W, Alcarazo M (2011) Angew Chem Int Ed 50:8400

86. Frenking G, Tonner R (2009) Pure Appl Chem 81:597
87. Frenking G, Tonner R, Klein S, Takagi N, Shimizu T, Krapp A, Pandey KK, Parameswaran P (2014) Chem Soc Rev 43:5106
88. Petz W (2015) Coord Chem Rev 291:1
89. Viehe HG, Janousek Z, Gompper R, Lach D (1973) Angew Chem Int Ed Engl 12:566
90. Petz W, Kutschera C, Heitbaum M, Frenking G, Tonner R, Neumüller B (2005) Inorg Chem 44:1263
91. Tonner R, Heydenrych G, Frenking G (2008) ChemPhysChem 9:1474
92. Braunschweig H, Dewhurst RD, Hupp F, Nutz M, Radacki K, Tate CW, Vargas A, Ye Y (2015) Nature 522:327
93. Frenking G (2015) Nature 522:297
94. Ishida S, Iwamoto T, Kabuto C, Kira M (2003) Nature 421:725
95. Iwamoto T, Masuda H, Kabuto C, Kira M (2005) Organometallics 24:197
96. Lein M, Krapp A, Frenking G (2005) J Am Chem Soc 127:6290
97. Ghadval RS, Roesky HW, Merkel S, Henn J, Stalke D (2009) Angew Chem Int Ed 48:5683
98. Filippou AC, Chernov O, Schnakenburg G (2009) Angew Chem Int Ed 48:5687
99. Soleilhavoup M, Bertrand G (2015) Acc Chem Res 48:256
100. Andrada DA, Frenking G (2015) Angew Chem Int Ed 54:12319
101. Mondal KC, Roesky HW, Schwarzer MC, Frenking G, Neudeck S, Tkach I, Wolf H, Kratzert D, Herbst-Irmer R, Niepötter B, Stalke D (2013) Angew Chem Int Ed 52:1801
102. Lewis GN (1923) Valence and the structure of atoms and molecules. American Chemical Society Monograph Series, New York, p 94. http://babel.hathitrust.org/cgi/pt?id=uc1.$b35072;view=1up;seq=1
103. Goldberg DE, Harris DH, Lappert MF, Thomas KM (1976) J Chem Soc Chem Commun 261
104. West R, Fink MJ, Michl J (1981) Science 214:1343
105. Masamune S, Hanzawa Y, Williams DJ (1982) J Am Chem Soc 104:6163
106. Davidson PJ, Harris DH, Lappert MF (1976) J Chem Soc Dalton Trans 2268
107. Stürmann M, Saak W, Marsmann H, Weidenbruch M (1999) Angew Chem Int Ed 38:187
108. Weidenbruch M (2002) J Organomet Chem 646:39
109. Grützmacher H, Fässler TF (2000) Chem Eur J 6:2317
110. Power PP (1999) Chem Rev 99:3463
111. Weidenbruch M (1999) Eur J Inorg Chem 373
112. Power PP (1998) J Chem Soc Dalton Trans 2939
113. Binkley JS (1984) J Am Chem Soc 106:603
114. Kalcher J, Sax A, Olbrich G (1984) Int J Quantum Chem 25:543
115. Köhler H-J, Lischka H (1984) Chem Phys Lett 112:33
116. Clabo DA, Schaefer HF (1986) J Chem Phys 84:1664
117. Thies BS, Grev RS, Schaefer HF (1987) Chem Phys Lett 140:355
118. Koseki S, Gordon MS (1988) J Phys Chem 92:364
119. Koseki S, Gordon MS (1989) J Phys Chem 93:118
120. Colegrove BT, Schaefer HF (1990) J Phys Chem 94:5593
121. Colegrove BT, Schaefer HF (1991) J Am Chem Soc 113:1557
122. Grev RS, Schaefer HF (1992) J Chem Phys 97:7990
123. Grev RS, De Leeuw BJ, Schaefer HF (1990) Chem Phys Lett 165:257
124. Grev RS (1991) Adv Organomet Chem 33:125
125. Palagyi Z, Schaefer HF, Kapuy E (1993) J Am Chem Soc 115:6901
126. Li Q-S, Lü R-H, Xie Y, Schaefer HF (2002) J Comput Chem 23:1642
127. Nagase S, Kobayashi K, Takagi N (2000) J Organomet Chem 611:264
128. Bogey M, Bolvin H, Demuyneck C, Destombes J-L (1991) Phys Rev Lett 66:413
129. Cordonnier M, Bogey M, Demuynck C, Destombes J-L (1992) J Chem Phys 97:7984
130. Wang X, Andrews L, Kushto G (2002) J Phys Chem A 106:5809
131. Wang X, Andrews L, Chertihin GV, Souer PF (2002) J Phys Chem A 106:6302
132. Andrews L, Wang X (2002) J Phys Chem A 106:7697

133. Wang X, Andrews L (2003) J Am Chem Soc 125:6581
134. Chen Y, Hartmann M, Diedenhofen M, Frenking G (2001) Angew Chem Int Ed 40:2052
135. Pu L, Twamley B, Power PP (2000) J Am Chem Soc 122:3524
136. Trinquier G, Malrieu J-P (1987) J Am Chem Soc 109:5303
137. Trinquier G, Malrieu J-P (1989) J Am Chem Soc 111:5916
138. Luo Y-R (2007) Comprehensive handbook of chemical bond energies. CRC, Boca Raton
139. Kaupp M (2014) Chemical bonding of main-group elements. In: Frenking G, Shaik S (eds) The chemical bond. Chemical bonding across the periodic table. Wiley-VCH, Weinheim, pp 1–24
140. Lewis GN (1923) Valence and the structure of atoms and molecules. American Chemical Society Monograph Series, New York, p 96. http://babel.hathitrust.org/cgi/pt?id=uc1.$b35072;view=1up;seq=1
141. Bytheway I, Lin Z (1998) J Am Chem Soc 120:12133
142. Frenking G, Shaik S (eds) (2014) The chemical bond. 1. Fundamental aspects of chemical bonding. Wiley-VCH, Weinheim
143. Frenking G, Shaik S (eds) (2014) The chemical bond. 2. Chemical bonding across the periodic table. Wiley-VCH, Weinheim

Struct Bond (2016) 169: 157–180
DOI: 10.1007/430_2015_210

Published online: 23 February 2016

Structure and Bonding Patterns in Large Molecular Ligated Metal Clusters

Jean-Yves Saillard and Jean-François Halet

Abstract Although there will always be an Edisonian component to a search for new cluster compounds, the greater the understanding of the underlying chemistry, the more focused and efficient the search. It is why the rapid expansion of the synthesis and characterization of ligated transition-metal clusters over the last decades has been accompanied by theories about their bonding and electronic properties with the aid of conceptual ideas and theoretical models such as the development of electron-counting rules which govern the relationship between the structure and the electron count. This review summarizes these theoretical models, their historical development, their limits, and using a selection of specific examples among the extensive panoply of large ligated metal clusters available in the literature, shows how they can help in understanding their structural and electronic properties.

Keywords Aluminum · Capping principle · Closed-shell principle · Clusters · Condensed clusters · Copper · Electron-counting rules · Gallium · Gold · Jellium model · Nanoclusters · Palladium · Polyhedral skeletal electron pair theory · Silver · Tensor surface harmonic

Contents

J.-Y. Saillard (✉) and J.-F. Halet (✉)
Institut des Sciences Chimiques de Rennes, UMR 6226 CNRS-Université de Rennes 1, 35042 Rennes cedex, France
e-mail: saillard@univ-rennes1.fr; halet@univ-rennes1.fr

Abbreviations

AO	Atomic orbital
ccp	Cubic close packed
Cp*	Pentamethylcyclopentadienyl
CVE	Cluster valence electron
DFT	Density functional theory
EAN	Effective atomic number
Et	Ethyl
fcc	Face-centered cubic
FO	Frontier orbital
hcp	Hexagonal close packed
HOMO	Highest occupied molecular orbital
i-Pr	Isopropyl
LUMO	Lowest unoccupied molecular orbital
Me	Methyl
MO	Molecular orbital
n-Pr	n-propyl
NR_2	Organoamino
Ph	Phenyl
PR_3	Organophosphine
PSEPT	Polyhedral skeletal electron pair theory
p-Tol	4-Methylphenyl
SEP	Skeletal electron pair
SMO	Skeletal molecular orbital
SR	Organothiolato
TSH	Tensor surface harmonic

1 Introduction

For nearly a century, the Lewis' valence theory [1] and the subsequent development of the effective atomic number (EAN) rule [2, 3] as well as the valence bond theory [4] have constituted the fundamental basis concepts used for rationalizing the structure and bonding in a tremendously large area of covalent chemistry [5]. However, there are families of compounds, which have been, at least in part, reluctant to stick to this conventional two-center/two-electron approach, in particular those in which hypervalency and/or hypercoordination are present. This is the case, of course, in cluster chemistry where hypercoordination, associated with electron delocalization and/or electron deficiency, is common and renders impotent the use of EAN rules. Nevertheless, for a closed-shell cluster, as for any chemically stable (i.e., viable [6]) molecule, there is a relationship between its number of valence electrons and its structure and stability. This relationship is based on the cluster molecular shape and, as for the EAN rules, the closed-shell requirement principle [7, 8]. This principle requires a significant energy gap separating the highest occupied molecular orbital (HOMO) from the lowest unoccupied molecular orbital (LUMO), which is usually a necessary condition for thermodynamic (Jahn–Teller) and kinetic stability. This situation is generally satisfied when the bonding and nonbonding molecular orbitals (MOs) are occupied and the antibonding MOs are unoccupied.[1] Formally, changing the number of valence electrons in a stable molecule, which initially satisfied the closed-shell requirements, results in this principle no longer being satisfied. As a consequence, the molecular structure should change in order to adapt the number of bonding, nonbonding, and antibonding MOs to the new electron count. It follows that a given structure is associated with a given electron count and vice versa. Of course, there are exceptions to this structure/electron count bijection (e.g., think of isomerism), but this is a very general principle in covalent chemistry. The rules which govern the structure/electron count relationships are called electron-counting rules.

2 Polyhedral Skeletal Electron Pair Theory

Electron-counting rules have a scope of application, outside of which, they cannot apply satisfyingly. Organic chemistry, for instance, is largely governed by the octet rule (in fact a connectivity/electron count relationship), whereas the 18-electron rule dominates transition-metal organometallic chemistry [8]. In the case of cluster chemistry, the first electron-counting rules to be set up concerned borane $[B_nH_m]^{x-}$ and related carborane clusters and were later extended to other main-group and transition-metal clusters, thanks to the isolobal analogy [9, 10]. These rules emerged empirically from the seminal works of Williams, Rudolph, Wade, and Mingos [11–20], which were built on some earlier theoretical MO analysis [21–30]. This set of consistent rules, which links the shape and nature of the cluster skeletal polyhedron to the number of electron pairs associated with, is named the polyhedral skeletal electron pair theory (PSEPT) [7, 31–33]. In organometallic chemistry, they are often called the Wade–Mingos rules.

[1] In some cases, nonbonding orbitals can be unoccupied because lying too high in energy for being accessible.

3 Tensor Surface Harmonic (TSH) Theory

Interesting attempts to provide a rationalization of these rules from topological approaches have been successively realized [34, 35]. However, the major breakthrough was made by Stone and colleagues, who provided a very elegant theoretical demonstration of the fundamental PSEPT rules, the basis of the tensor surface harmonic (TSH) theory [36–42]. This approach is based on the idea initiated by Lipscomb and coworkers [43] and Ruedenberg and coworkers [44] that the bonding onto the skeletal envelope of a pseudo-spherical cluster can be understood by looking first at the simple model of a free electron constrained to move on a sphere. The scalar solutions of the corresponding angular Schrödinger equation are the familiar $Y_{L,M}(\theta, \phi)$ spherical harmonics (called surface harmonics by Stone), where θ and ϕ are the spherical angular coordinates. In this model, the $Y_{L,M}$ energy is $-L(L+1)$ in atomic units. Now, going from the sphere to a particular n-vertex polyhedral spherical cluster of which the angular coordinates θ_i and ϕ_i of each individual vertex ($i = 1, n$) are perfectly defined, Stone proposes to identify each $Y_{L,M}(\theta, \phi)$ with a skeletal MO (SMO), i.e., with a linear combination of atomic orbitals centered on the atoms occupying the polyhedron vertices, the individual $Y_{L,M}(\theta_i, \phi_i)$ values being the coefficients in the linear combination [36, 37].[2] This step implicitly incorporates the potentials of the nuclei at the polyhedron vertices (not taken into account so far) so that the energy of the $\Psi_{L,M}$ SMO is different from that of the $Y_{L,M}$ it is associated with, but, as for the particle-on-a-sphere problem, it increases with increasing L. Stone has shown that within the Hückel approximation, this energy can be written as $E_l = \alpha + (2e/n) \times \beta \times P_L(\cos \omega)$, where α and β are the standard Hückel parameters, e is the number of polyhedron edges, P_L is a Legendre polynomial of degree L, and ω is the angular separation between two neighboring vertices (considered as a unique value for a given polyhedron) [36, 37].

Of course, this approach works only when all molecular fragments constituting the polyhedral cluster participate in the bonding with one single frontier orbital (FO), being of σ-type. In the case where each fragment participates with several FOs of different types (e.g., radial σ- and tangential π-type), interactions between $\Psi_{L,M}$ SMOs made of different FO types should be considered. But overall, this model works only when the molecular fragments possess only radial (σ-type) FOs, such as an *ns* AO or an *sp* hybrid which points toward the center of the sphere. Indeed, the $Y_{L,M}(\theta, \phi)$ spherical harmonics are unable to reproduce the intrinsic local nodes that the π- and δ-type tangential fragment FOs (and their combinations) display at the nuclei, i.e., at the polyhedron vertices. The presence of these local nodes on the skeleton surface infers on the bonding/antibonding nature of the SMOs which are made of such FOs. For example, the BH fragments constituting a pseudo-spherical borane cluster $[B_nH_n]^{2-}$ possess 3 FOs each, one of σ-type and two of π-type (Fig. 1). As discussed above, the scalar $Y_{L,M}$ solutions of the particle-on-a-sphere Schrödinger equation can be used to derive the SMOs constructed on the

[2] This process is nicely detailed for the octahedron example in Ref. [7].

Fig. 1 Comparison of the frontier orbitals of E, ER, and ML_3 isolobal fragments (E = main-group; M = transition-metal). The electron occupation corresponds to C, BH, and $Fe(CO)_3$

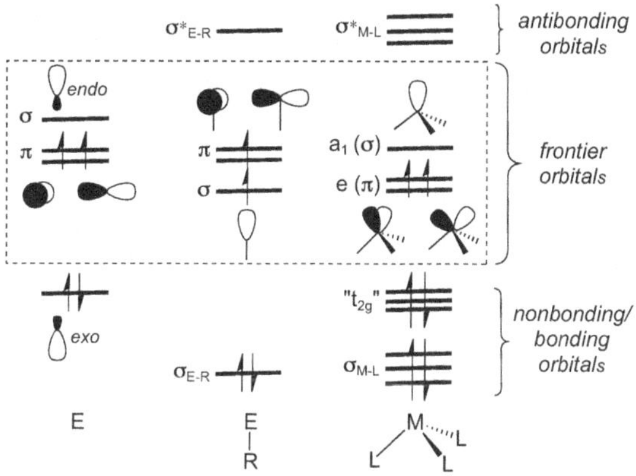

σ-type BH FOs. To derive the SMOs constructed on the π-type FOs, Stone introduced the vector surface harmonics (tensors of order one) $V_{L,M}$ and $\overline{V}_{L,M}$ which are also solutions of the considered two-dimensional Schrödinger equation and constructed from each $Y_{L,M}$ as follows:

$$V^{\pi}_{L,M} = \nabla Y_{L,M} \text{ and } \overline{V}^{\pi}_{L,M} = r \wedge V^{\pi}_{L,M}$$

where the effect of the vector product $r \wedge$ is a local 90° rotation of all the p^{π} components in $V^{\pi}_{L,M}$ about the radial vector. These orthogonal "spherical" vectors have two components (one in θ and one in ϕ), which can be derived from the $Y_{L,M}$ they are associated with. Then n $\Psi^{\pi}_{L,M}$ and n $\overline{\psi}^{\pi}_{L,M}$ π-type SMOs can be written as linear combinations of the local tangential $2p_{\theta i}$ and $2p_{\phi i}$ boron AOs, the associated coefficients being $V^{\theta}_{L,M}(\theta_i, \phi_i)$ and $V^{\phi}_{L,M}(\theta_i, \phi_i)$ for $\Psi^{\pi}_{L,M}$ and $\overline{V}^{\theta}_{L,M}(\theta_i, \phi_i)$ and $\overline{V}^{\phi}_{L,M}(\theta_i, \phi_i)$ for $\overline{\psi}^{\pi}_{L,M}$ [36, 37]. There is no $\Psi^{\pi}_{0,M}$ and $\overline{\psi}^{\pi}_{0,M}$ and $\Psi^{\pi}_{L,M}$ and $\overline{\psi}^{\pi}_{L,M}$ differ by their symmetry properties with respect to the inversion center; they are called even and odd functions, respectively. As for the σ SMOs, their Hückel energies depend linearly on $2e/n$. Thus, assuming no antibonding orbitals occupied, the cluster bonding energy also depends on $2e/n$. From this relationship and the Euler theorem on polyhedra [45], it can be shown that the bonding energy is maximized when the considered polyhedron has triangular faces, thus accounting for the deltahedral nature of $[B_nH_n]^{2-}$ [36, 37]. It can also be shown that the $\Psi^{\pi}_{L,M}$ even functions are bonding, whereas the $\overline{\psi}^{\pi}_{L,M}$ odd ones are antibonding and that the combinations of σ-type FOs $\Psi^{\sigma}_{L,M}$ for which $L > 0$ will interact with their $\Psi^{\pi}_{L,M}$ counterparts in such a way that they will be destabilized. Thus, one is left with only one (pure) σ-type SMO, namely, $\Psi^{\sigma}_{L,M}$ and n $\Psi^{\pi}_{L,M}$ bonding SMOs, all the other SMOs being antibonding, so that the closed-shell principle requires $n+1$ skeletal electron pairs (SEPs) for a pseudo-spherical n-vertex cluster made of fragments possessing 3 FOs (one radial (σ-type) and two tangential (π-type)). This is the fundamental electron-counting rule for the *closo* (pseudo-spherical deltahedral

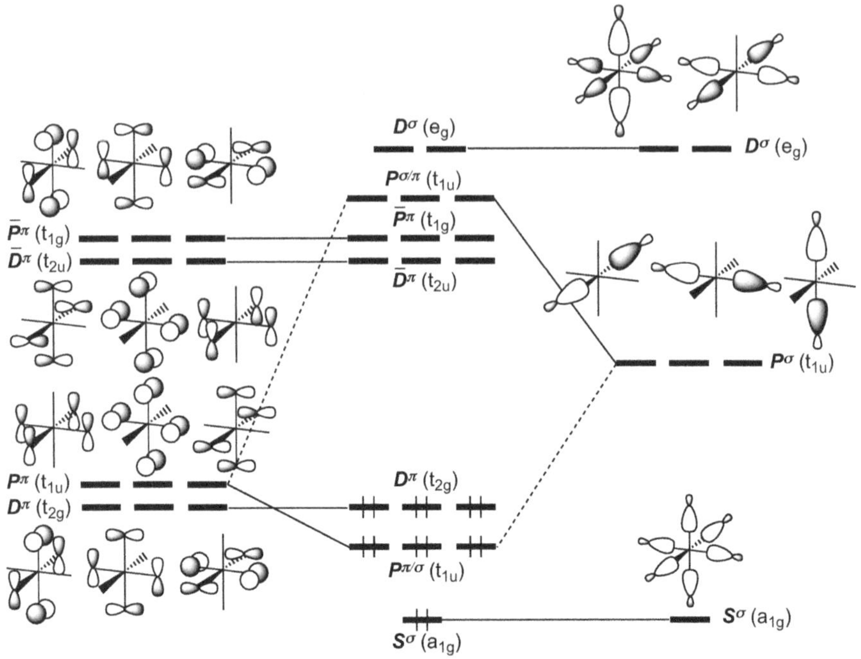

Fig. 2 Stone molecular orbitals for the octahedral $[B_6H_6]^{2-}$ cluster with σ/π mixing (*middle*). The L^π and L^σ orbitals are given on the *left* and on the *right*, respectively

structure) $[B_nH_n]^{2-}$ clusters. This is illustrated in Fig. 2 for the textbook octahedral $[B_6H_6]^{2-}$ cluster.

The isolobal analogy [9, 10] tells us that this $(n+1)$-SEP rule can be in principle extended to any type of *closo* skeleton made of fragments possessing a similar set of 3 FOs (see Fig. 1), irrespectively of their energy and precise shape or composition. This is indeed the case of many transition-metal carbonyl or phosphine organometallic clusters, most of them built from conical ML_3 fragments. For example, the three octahedral clusters $[B_6H_6]^{2-}$, $[Ru_6(CO)_{18}]^{2-}$ [46], and $[Ru_4(CO)_{12}Bi_2]$ [47] are 7-SEP species.[3]

The TSH theory can also handle SMOs made of δ-type orbitals, i.e., tangential FOs having two local nodal planes at the skeleton vertices. Such orbitals are indeed involved in skeletal bonding in some clusters of early transition metals with π-donor ligands [7, 48]. Following the approach used for treating the π-type orbitals, two tensor surface harmonics of order 2 are constructed from each $Y_{L,M}$ [36, 37]:

[3] Organometallic chemists often count electrons in adding to the skeletal electrons all the other electrons lying in the metal coordination sphere, but not participating significantly to the skeletal bonding. An ML_3 fragment will thus contribute to this count with 12 additional electrons (6 nonbonding "t_{2g}" d-type electrons and the 6 electrons coming from the ligands) (see Fig. 1). Within this electron-counting scheme, $[Ru_6(CO)_{18}]^{2-}$ and $[Ru_4(CO)_{12}Bi_2]$ are 86 (14 + 6 x 12) and 62 (14 + 4 x 12) electron species, respectively.

$$T^{\delta}{}_{L,M} = \nabla\nabla Y_{L,M}(\text{even}) \text{ and}$$

$$\overline{T}^{\delta}{}_{L,M} = r \wedge T^{\delta}{}_{L,M} \quad (\text{odd, } 45^{\circ} \text{ rotation about the radial vector})$$

The expansion coefficients in the SMOs are calculated from these two tensors in a similar way as described above for the π-type orbitals. A detailed procedure can be found in [7]. Thus, from the TSH theory results it is also possible to build electron-counting rules for clusters in which δ-type orbitals participate in the skeletal bonding.

Not only does the TSH theory provide a rationalization of the favored number of skeletal electrons in a given polyhedron, but also it offers a reasonable approximation of the $\Psi_{L,M}$ SMOs. In particular, going from the ideal spherical symmetry to the real symmetry group of the current polyhedron, it is generally easy to derive their irreducible representations from their L and M indices and thus getting a fairly good idea of their shapes and of the possible interactions between them. Symmetry considerations can also be used for extending the electron-counting rules to clusters having incomplete spherical shapes, i.e., *nido*, *arachno*, *hypho*, etc., species which can be described as pseudo-spherical deltahedra of which one, two, three, etc., vertices are missing (unoccupied), respectively. In the case of boranes and other clusters made of 3-FO fragments, the $(n+1)$ rule is maintained, n being the total number of deltahedron vertices, including the missing (unoccupied) ones.

4 Application and Extension of the Polyhedral Skeletal Electron Pair Theory

The power of the PSEPT originates from the fact that it covers almost the entire field of the structural chemistry of clusters of sub-nanometer size and a significant part of the nanosized ones. Indeed, it has been extended to an incredibly large number of specific types of clusters [49]. This includes lowering of symmetry due to oblate or prolate distortions away from the pseudo-spherical shape [7] or due to the heterogeneity in the fragments constituting the cluster cage (e.g., mixed main-group/transition-metal clusters) [50–56]. This includes also clusters with capped faces (the capping principle [57] states that capping a face should not in general modify the favored number of skeletal electrons and, in a more general way, clusters made of fused polyhedra).[4] Electron-counting rules for condensed polyhedra were originally set up by Mingos [58, 59] who established that the total electron count of a cluster made of two polyhedra sharing a vertex, edge, or face is equal to the sum of the total electron count of the parent polyhedra minus the number of electrons characteristic to the shared vertex, edge, or face, respectively. A

[4] This simple rule may not apply in the case of multicapped clusters, but symmetry considerations may be used to evaluate the number of additional bonding MOs present (see Ref [7]).

somewhat related approach developed later by Jemmis et al. is known as the *mno* rules [60–63]. These rules, established on the basis of simple MO arguments, state that the SEP requirement of condensed polyhedral clusters is $m+n+o+p-q$, where $m=$ number of polyhedra, $n=$ number of vertices, $o=$ number of single-vertex condensation, $p=$ number of missing vertices, and $q=$ number of caps [60–63].

Of particular interest for nanosize clusters, which are generally filled compact structures, is the multispherical approach. Indeed, the TSH theory is based on the concept of electrons restrained to lie onto a sphere, not allowed to roam inside a spherical volume. Therefore, the trick resides in considering the cluster as made of concentric successive shells of atoms. Each sphere can be first treated independently within the TSH formalism and in a second step the SMOs of each individual sphere are let to interact with those of the neighboring spheres, symmetry permitting. Such an approach can also be used when encapsulated atoms or di-, tri-, or polyatomic units are encapsulated within an outer sphere. Mingos has shown that when the outer sphere is made of n_s fragments involved principally in radial (σ-type) bonding with the encapsulated unit and when the interactions within this outer sphere are negligible, the total cluster valence electron (CVE) count of such multispherical species can be expressed as $CVE = 12n_s + \Delta_i$, where Δ_i is the number of electrons associated with the interstitial moiety. This is the so-called inclusion principle [64–68]. For example, considering that the contributions to CVE of Au, Cl, and a phosphine in $[Au_{13}Cl_2(PMePh_2)_{10}]^{3+}$ [69] (Fig. 3) are 11, 1, and 2, respectively, and taking account the positive charge, one is left with $CVE = 13 \times 11 + 2 \times 1 + 10 \times 2 - 3 = 162$. Considering that its Au_{13} skeletal structure is a centered icosahedron ($n_s = 12$) and that the electron count associated with a single Au atom is 18, the $12n_s + \Delta_i$ law is satisfied ($12 \times 12 + 18 = 162$). Not only gold clusters but also a significant number of group 10 (and some group 9) transition-metal carbonyl clusters satisfy also this rule. This means that in these species which exhibit compact metal packing, the n_s outer fragments interact principally through radial (σ-type) bonding. Another typical example is $[Ni_{38}Pt_6(CO)_{48}]^{6-}$ [70] (Fig. 4), where a Pt_6 octahedron is encapsulated within a Ni_{38} outer sphere (fcc packing) and for which $CVE = 38 \times 10 + 6 \times 10 + 48 \times 2 + 6 = 542$. Assuming that the favored total electron count for a metal octahedron is 86 (i.e., 14 skeletal electrons (7 SEPs) $+ 6 \times 12$ (see above)), then $CVE = 12 \times 38 + 86 = 542$. Note that the 86-CVE (7-SEP) count of the encapsulated octahedron means that tangential (π-type) orbitals on the Pt atoms participate in the bonding.

In the above examples, the fragments constituting the outer sphere are supposed noninteracting between them and are involved only in radial bonding with the encapsulated moiety. In the case where tangential (π-type) bonding is fully operating and outer fragments are close to each other and therefore interact, the favored total valence electron count is dominated by the outer sphere and for a deltahedral framework is given by $CVE = 14n_s + 2$. Intermediate situations with significant but weaker involvement of tangential orbitals result in $CVE = 12n_s + 24$ [7, 64–67].

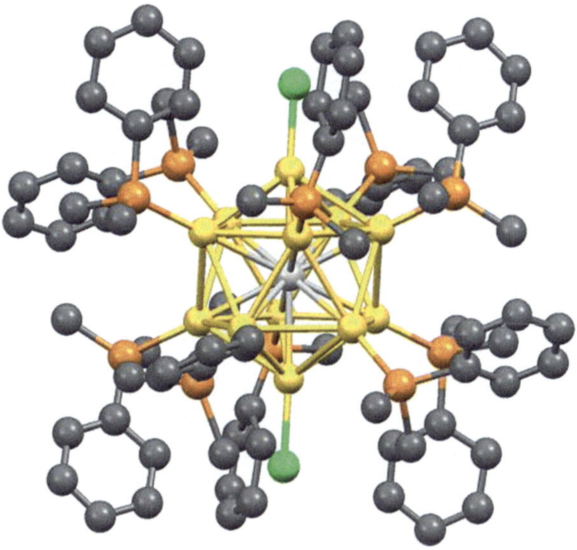

Fig. 3 Structural arrangement of $[Au_{13}Cl_2(PMePh_2)_{10}]^{3+}$ [69]. *Yellow*, *gray*, *green*, *orange*, and *black spheres* are Au (*outer*), Au (*inner*), Cl, P, and C, respectively. Hydrogen atoms are not shown for clarity

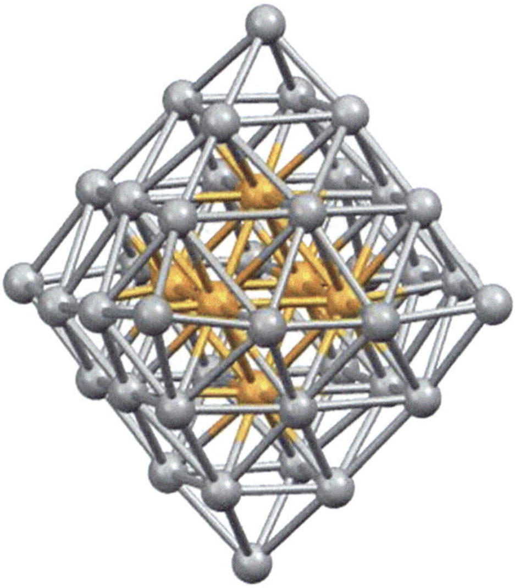

Fig. 4 Structural arrangement of the metallic core of $[Ni_{38}Pt_6(CO)_{48}]^{6-}$ [70]. *Gray* and *orange spheres* are Ni (*outer*) and Pt, respectively. Surrounding carbonyl groups are not shown for clarity

These principles can be used to good effect to rationalize the electron count of the nanosized group 9 and 10 carbonyl/phosphine homo- and hetero-metal transition-metal clusters [71–74], where metal–metal bonding is substantial. Let us illustrate it for the large ellipsoidal-shaped icosahedral-based cluster $Pd_{59}(CO)_{32}(PMe_3)_{21}$, for instance, characterized by Dahl and coworkers [75]. This cluster possesses 696 CVEs (59 x 10 (Pd) + 32 × 2 (CO) + 21 × 2 (PMe_3) = 696). Among several ways to dissect its complex geometric structure, one is

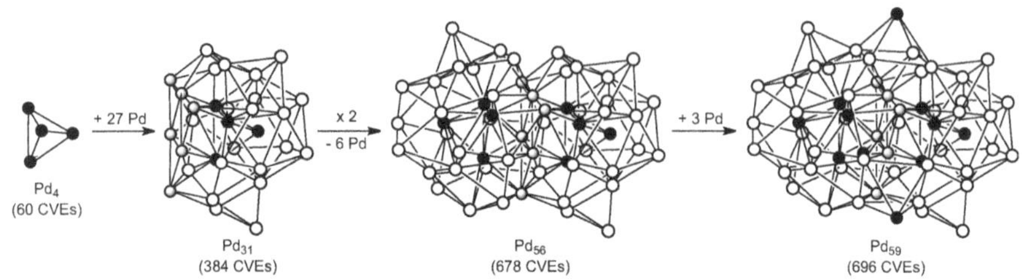

Fig. 5 "Retrosynthesis" of the ellipsoidal-shaped Pd_{59} core in $Pd_{59}(CO)_{32}(PMe_3)_{21}$

proposed in Fig. 5, using successfully two of the cluster principles mentioned above: (a) cluster inclusion considering a large cluster with a central polyhedral core with radial bonding alone to an outer shell of metals (CVE count of $12n_s + \Delta_i$) and (b) cluster fusion where the electron count of the common fragment is subtracted from that of the two clusters fused. Thus, in its "retrosynthesis" (see Fig. 5), this cluster can be considered as a fused metal dimer of two 31-atom clusters containing a Pd_4 tetrahedral cluster core ($12n_s + \Delta_i = 12 \times 27 + 60 = 384$ CVEs). The fusion is somewhat complex and consists of a shared 6-atom triangular raft (90 CVEs) between the two 31-atom clusters yielding a 56-atom dimer. The remaining three atoms are the apical atoms of three additional square pyramidal clusters (74 CVEs) in which both basal pairs of atoms are shared (2 x 34), one with each 31-atom cluster, thereby bridging the two fused clusters [49, 55]. This gives a cluster electron count of $2 \times 384 - 90 + 3 \times 74 - 3 \times 68 = 696$ in line with the chemical composition and actual electron count.

As elegantly described by Dahl and coworkers, over 20 different geometries have been crystallographically identified in the family of ligated homometal clusters $Pd_n(CO)_x(PR_3)_y$ ranging in numbers of metal atoms per cluster from 4 to 165 (see Table 1 for a few representatives with $n \geq 30$) – some large hetero-metal species of the same kind have also been reported [73]. They exhibit closed-packed Pd_n frameworks which can be viewed as pieces of cubic closed-packed (ccp) or mixed cubic closed-packed/hexagonal closed-packed (ccp/hcp) bulk stackings or alternatively icosahedral-based (single, interpenetrating, face-fused, or multishell icosahedral) structures. It turns out that most of these compounds obey (or nearly obey) one way or the other the electron-counting rules established for high-nuclearity closed-packed clusters.

Among these large clusters, some with icosahedral-based structures can be described as "clusters of clusters" $s_n(N)$ (where n is the number of icosahedra and N is the nuclearity) based on vertex-sharing icosahedra as building blocks, as initially proposed by Teo and Zhang in the mid-1980s for high-nuclearity Au/Ag clusters [35, 86]. Indeed, a growth sequence, termed "polyicosahedricity," was tentatively proposed for the formation of these "clusters of clusters" starting from a single 13-atom icosahedron and adding successive vertex-sharing icosahedral units [87].

Table 1 Representative examples of $Pd_n(CO)_x(PR_3)_y$ clusters, $n \geq 30$

Cluster	Structure[a]	CVE[b]	ΔE_{H-L} (eV)[c]	References
$Pd_{30}(CO)_{26}(PEt_3)_{10}$	ccp	372	0.18	[76]
$Pd_{34}(CO)_{24}(PEt_3)_{12}$	ico	412		[77]
$Pd_{35}(CO)_{23}(PMe_3)_{15}$	ico	426	0.12	[78]
$Pd_{37}(CO)_{28}[P(p\text{-Tol})_3]_{12}$	ico	450	0.23	[79]
$Pd_{38}(CO)_{28}(PEt_3)_{12}$	irr	460	0.32	[80]
$Pd_{39}(CO)_{23}(PEt_3)_{16}$	ico	468	0.03	[78]
$Pd_{52}(CO)_{36}(PEt_3)_{14}$	hcp/ccp	620	0.01	[81]
$Pd_{54}(CO)_{40}(PEt_3)_{14}$	hcp/ccp	648	0.01	[76]
$Pd_{59}(CO)_{32}(PMe_3)_{21}$	ico	696	0.12	[78]
$Pd_{66}(CO)_{45}(PEt_3)_{16}$	hcp/ccp	782	0.03	[81]
$Pd_{69}(CO)_{36}(PEt_3)_{18}$	ico	798	0.23	[82]
$Pd_{145}(CO)_{60}(PEt_3)_{30}$	ico	1,630	0.00	[83]
$Pd_{164-x}Pt_{x+1}(CO)_{72}(PPh_3)_{20}$[d]	ico	1,834		[84]

[a]tc-octa, tetracapped octahedral; hc-octa, hexacapped octahedral; hcp, hexagonal close packed; ccp, cubic close packed; ico, icosahedral; irr, irregular core geometry
[b]CVE count
[c]HOMO–LUMO gap (when available, computed at the DFT/BP86 level of theory) [85]
[d]$x \sim 7$

5 Jellium Model

As discussed above, the conventional PSEPT reaches its limits in the rationalization of large, compact, multispherical clusters. Obviously, the underlying TSH theory which is based on the electron-on-a-sphere model is not the best appropriate way to provide a sufficiently good description of the electronic structure of such species. The spherical jellium model, which is related to the problem of an electron inside a sphere, was first proposed by Knight and coworkers to rationalize mass spectrometry experiments on sodium clusters which showed particularly large peaks for Na_n clusters with $n = 8$, 20, 40, 58, and 92 [88]. In this homogeneous electron gas model [89, 90], each nearly free electron is supposed not to interact with the individual nuclei but with a smoothened square potential, the "square" width being related to the diameter of the considered spherical cluster. To be short, in a perfect (non-smoothened) "square" potential, the potential would have a certain constant value (attractive) inside the sphere and another one outside. Thus, the corresponding Schrödinger equation is somewhat similar to that of a polyatomic atom, except that the radial potential is not Coulombic (i.e., not infinite at $r = 0$), but rounded square [91–93].[5] As for the atom case, the corresponding jellium orbitals can be expressed as $\Psi_{N,L,M} = f_{N,L}(r) \times Y_{L,M}(\theta, \phi)$. The radial part $f_{N,L}(r)$ depends on the analytical form of the central field potential. A difference with the atom case is than N can be associated with any positive L numbers. The energy level ordering for the one-electron spherical jellium model of Knight is:

[5] Variations of this model can be found in Refs. [89, 90].

1S < 1P < 1D < 2S < 1 F < 2P < 1G < 2D < 1H < 3S < 2 F < 3P < 1I < 2G, etc. This level ordering appears to be stable with respect to the shape of the potential. It follows that the closed-shell requirement principle is satisfied when the 1S, 1P, 1D, 2S, 1 F, etc., shells are successively filled up, giving rise to the so-called "magic" numbers of electrons: 2, 8, 18, 20, 34, 40, 58, 68, 90, etc., which can be compared to the noble-gas valence electron numbers 2, 8, 18, and 32. This simple model has been extended to nonspherical oblate and prolate cluster shapes [94]. One might expect that this model should apply only to systems in which the atoms contribute to the bonding with only one orbital of *ns* (or at least a σ-type) character. Indeed, as for the scalar $Y_{L,M}$ harmonics in the case of the hollow clusters described above, the $\Psi_{N,L,M}$ jellium orbitals are not able to reproduce the intrinsic local nodes that the π- and δ-type atomic orbitals display at the various nuclei positions, except for that lying at the cluster center (if any). A tensor harmonic approach, similar to that used in the TSH theory described above, should be used. However, to our knowledge, no such development has been explored so far. Nevertheless, it appears that many aluminum and gallium clusters follow the jellium counting rules, despite the fact they participate in the bonding with their *p* AOs, in addition to their *s* AOs (see below).

6 Application of the Jellium Model to Group 11 Clusters

As shown by Mingos and Teo a long time ago [95–97], the jellium model can be used for understanding the structure and diamagnetism of a large class of inorganic gold clusters. Indeed, in such species, the participation to metal–metal bonding of the occupied *5d* and vacant *6p* gold AOs can be neglected. For example, $[Au_{13}Cl_2(PMePh_2)_{10}]^{3+}$ [69], which depicts a centered icosahedral metal core, as already mentioned above, has a jellium 8-electron count (jellium configuration $1S^2$ $1P^6$). To obtain this number, one has to consider the jellium metallic core in its actual oxidation state, i.e., $[Au_{13}]^{5+}$ (more precisely $[Au@Au_{12}]^{5+}$), and count only the electrons providing from the active AOs, namely, the 6s atomic orbitals. One should note that despite the "mixed-valent" character of gold (averaged oxidation state = +0.38), this cluster is fairly stable and exhibits a closed-shell electron configuration. This is a common situation for this type of gold clusters. Thus, $[Au_{13}Cl_2(PMePh_2)_{10}]^{3+}$ satisfies the electron-counting rules of both the PSEPT (CVE = 162; see above) and the spherical jellium approach (8 electrons). The development of the chemistry of gold clusters with closed-shell jellium configuration, in particular gold–thiolato species, has been blossomed during the last decade, giving rise to nanosized species of increasing nuclearity. More recently, a similar silver chemistry has started to develop [98–100]. At least one related copper cluster is also known [101]. Mixed Au/Ag [98, 99] and Au/Cu species [102] have also been characterized. The exceptional stability of these clusters has been described by the "noble-gas superatom" analogy [103–105]. This concept has been particularly developed by Häkkinen and coworkers who have extensively investigated the electronic structure of gold clusters with the aid of DFT computations

[106, 107]. They were able to illustrate the spherical jellium configurations in projecting the cluster density of states on the spherical harmonic angular momenta. They also introduced the "divide and protect" concept [108], which applies to metal–thiolato clusters and makes a clear distinction between the "mixed-valent" cluster core and peripheral "oxidized" Au^I (or Ag^I) metal centers which are covalently bonded to thiolate ligands but interact only weakly with the "mixed-valent" core, through metallophilic interactions [109–112]. These outer Au^I atoms form with the thiolate ligands $Au_n(SR)_m$ "staples" which protect (passivate) the surface of the "mixed-valent" core and are anchored to it by Au (core)–S bonds. For example, the 8-electron cluster $[Au_{25}(SR)_{18}]^-$ is made of a centered icosahedral core surrounded by six $Au_2(SR)_3$ staples and can be reformulated as $[Au_{13}]^{5+}\left\{[Au^I{}_2(SR)_3]_6\right\}^{6-}$ [113]. Similarly, $Au_{102}(SR)_{44}$ [114] can be reformulated as $[Au_{79}]^{21+}[Au^I{}_{23}(SR)_{44}]^{21-}$; the $[Au_{79}]^{21+}$ core depicts a 49-atom Marks decahedron surrounded by a 20-atom shell, i.e., $[Au_{49}@Au_{20}]^{21+}$, and possesses the "magic" electron count of 58 (jellium configuration $1S^2$ $1P^6$ $1D^{10}$ $2S^2$ $1F^{14}$ $2P^6$ $1G^{18}$) [103]. Assuming a closed-shell superatom core and various structural constraints on the nature and number of possible $Au_n(SR)_m$ "staples," it should be in principle possible to make predictions for new thiolato–gold nanoclusters. This is a goal that theoreticians are presently addressing [115, 116].

At this stage of the discussion, it is important to emphasize on the fact that the jellium electron count, which determines the spherical shape of the cluster core, is independent of the number of electrons provided by the surrounding ligands. For example, compounds $[Au_{13}Cl_2(PMePh_2)_{10}]^{3+}$ [69] and $[Ag_{21}\{S_2P(Oi\text{-}Pr)_2\}_{12}]^+$ [100], which have both an 8-electron centered icosahedral core and can be reformulated as $[Au_{13}]^{5+}[Cl_2(PMePh_2)_{10}]^{2-}$ and $[Ag_{13}]^{5+}[Ag^I{}_8\{S_2P(Oi\text{-}Pr)_2\}_{12}]^{4-}$, respectively, differ importantly by their core-ligand bonds (12 and 18, respectively). Nevertheless, they have the same metal core electronic structure, independently from that associated with the core-ligand bonds.

The reason lies in the fact that the ligand lone pairs interact with high-lying accepting valence AOs on the core surface metal atoms which are of p_π or exo sp_σ nature and are different from that (endo sp_σ) involved in the jellium MOs (Fig. 6) and which may or may not participate in core-ligand bonds. In the latter case, they remain unoccupied (high-lying) metal-ligand nonbonding orbitals. This is why the jellium electronic structure is fairly independent from the presence and nature of the passivating shell. This is exemplified by the similar plots of the 1S and 1P Kohn–Sham orbitals of $[Ag_{21}\{S_2P(Oi\text{-}Pr)_2\}_{12}]^+$ and its bare $[Ag_{13}]^{5+}$ core (Fig. 7) [100].

Another important remark concerns the number of electrons, which can be accommodated within a particular M_n core. Not only shell closure is required, but also the electrons have to occupy bonding (possibly nonbonding) orbitals. The number of accessible orbitals depends on the number of atoms constituting the core (one valence *s* orbital by atom). A simple Hückel calculation on an icosahedron or a centered icosahedron (I_h symmetry), for example, indicates that only four bonding orbitals (of a_{1g} and t_{1u} symmetry) are present, all the other ones being antibonding. Thus, such architecture can accommodate only 2 ($1S^2$) or 8 ($1S^2$ $1P^6$)

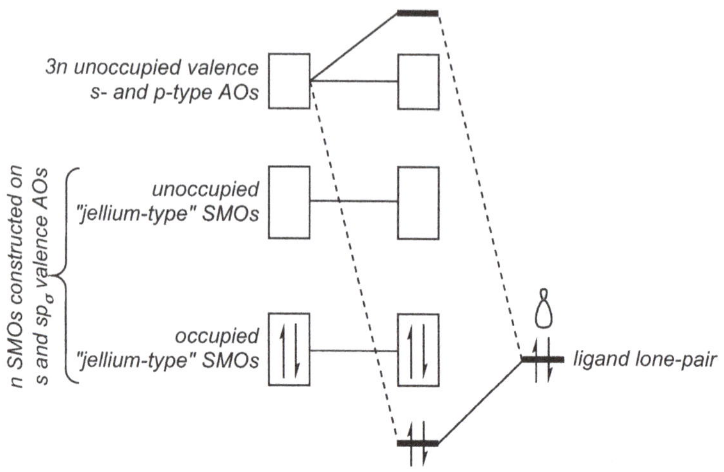

Fig. 6 Schematic description of the interaction of a two-electron ligand with the *n*-atom mixed-valent core of a closed-shell superatom

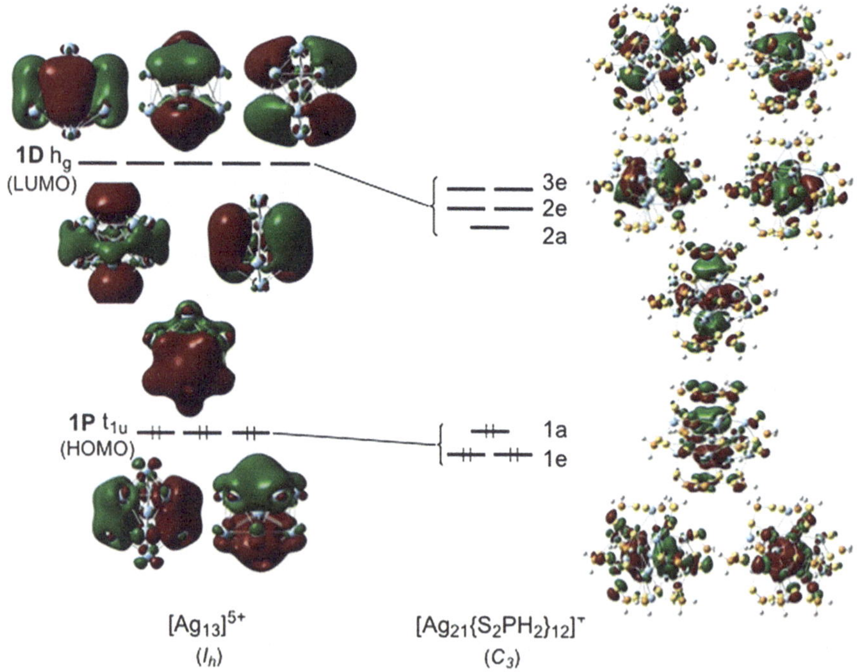

Fig. 7 The frontier Kohn–Sham SMOs of $[Ag_{21}]^{5+}$ and $[Ag_{21}\{S_2P(Oi\text{-}Pr)_2\}_{12}]^+$ [100]

electrons. Examples of centered icosahedral 8-electron clusters are not uncommon (see above) and 2-electron species exist also, as for instance in $[Cu_{25}H_{22}(PPh_3)_{12}]^+$ [101] which can be reformulated $[Cu_{13}]^{11+}[Cu_{12}H_{22}(PPh_3)_{12}]^{10-}$ [117].[6] To accommodate 18 electrons, a supplementary metal shell is necessary, as in $[Ag_{44}(SR)_{30}]^{4-}$, i.e., $[Ag_{12}@Ag_{20}]^{14+}[\{Ag^I_2(SR)_5\}_6]^{14-}$ [98, 99]. Indeed, the Ag_{12} I_h icosahedron embedded in an Ag_{20} pentagonal dodecahedron provides nine accessible (all bonding) Hückel orbitals of a_g, $t_{1u,}$ and h_g symmetry, allowing the $1S^2$ $1P^6$ $1D^{10}$ spherical jellium configuration.

As already mentioned, the jellium model can be applied to prolate or oblate clusters [94], the peculiar electron count of which being simply explained as resulting from of the degeneracy splitting of a partly occupied HOMO, associated with a Jahn–Teller distortion away from spherical symmetry. This approach has been successfully exploited in the past by Mingos et al. and resulted in the prediction of structures which were later experimentally confirmed [96, 118]. Mingos has also developed simple electron-counting rules for cluster frameworks which can be viewed as resulting from the condensation of several spherical skeletons. This elegant approach, which uses an analogy with Mulliken's united atom concept for diatomic molecule, is called the united cluster model [96]. For example, interaction between the jellium SMOs of two fused icosahedra generates a level ordering depending on its strength, i.e., on the number of shared vertices: one (single-vertex sharing), two (edge sharing), three (face sharing), or even stronger interpenetration. The occupation of the resulting accessible orbitals results in successive favored jellium electron counts of 16, 14, 12, 10, and 8, depending on the degree of the icosahedra interpenetration. The model nicely fits with the observed electron counts of the known polyspherical gold clusters. For example, the 14-electron count of $Au_{38}(SR)_{24}$ is associated with an $[Au_{23}]^{9+}$ core made of two face-sharing icosahedra and passivated by an $[Au^I_{15}(SR)_{24}]^{9-}$-covering shell. More complex structures, such as $[Au_{30}(SR)_{18}S]$ (interpenetrated cuboctahedra), $[Au_{36}(SR)_{24}]$ (tetrahedron of cuboctahedra), or $[Au_{14}(PR_3)_{10}]^{4+}$ (linked tricapped tetrahedra), have also been rationalized within the united cluster model [96].

As said above, the extreme scarcity of superatom-type copper clusters [101] contrasts with the abundance of gold (and to some extent silver) species. In fact, when applied to copper salts, the general synthetic process for such Group 11 nanoclusters (reduction by BH_4^-) leads to the formation of Cu^I polyhydrides [119, 120]. Assuming H and Cu as being isolobal (one electron in a unique *ns* FO), it is tempting to look at such species as being H/Cu "polymetallic" clusters. For example, the Cu^I octanuclear cluster $[Cu_8(H)\{Se_2P(O\textit{i}\text{-}Pr)_2\}_6]^{2+}$ [121], the skeleton of which exhibiting a hydride-centered Cu_4 tetrahedron tetracapped by four external metals ($[H@Cu_4@Cu_4]^{7+}$ (see left side of Fig. 8)), can be viewed as made of a centered tetrahedral 2-electron superatomic core ($1S^2$ configuration). However, such an electron counting does not apply for larger nanosized copper hydride clusters. To fill the gap with such species, an interesting extension of the jellium

[6] The tetracapped tetrahedron can also accommodate 8 electrons (see ref. [117]).

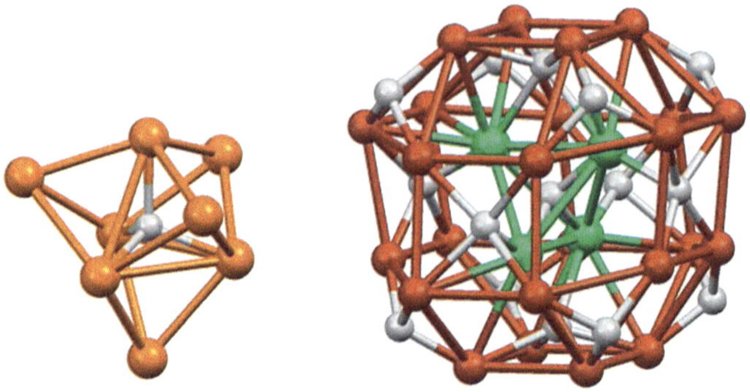

Fig. 8 Cluster core of $[Cu_8(H)\{Se_2P(Oi\text{-}Pr)_2\}_6]^{2+}$ *(left)* [121] and $[Cu_{28}(H)_{15}(S_2CNn\text{-}Pr_2)_{12}]^+$ *(right)* [123]. *Orange, green,* and *gray spheres* are Cu *(outer)*, Cu *(inner)*, and H, respectively. Surrounding ligands are not shown for clarity

model, called "jelliumatic shell model" was recently proposed by Teo and Yang [122]. The idea is based on a multishell approach, each concentric shell being supposed to obey the jellium model with an associated characteristic electron count. Each shell is also supposed to interact weakly with its neighboring shells, so that the total "jelliumatic" electron count is the sum of that of the individual shells. For example, the spherical 30-electron $[Cu_{28}(H)_{15}(S_2CNn\text{-}Pr_2)_{12}]^+$ clusters [123] (right side of Fig 8) can be divided into three concentric "jelliumatic" shells, $[H]^-$@$[Cu_4H_6]^{2+}$@$[Cu_{24}(H_8)(S_2CNn\text{-}Pr_2)_{12}]$ of 2, 8, and 20 electrons, respectively. The model of Teo and Yang appears to be applicable to other known polyhydrides, as well as other multispherical species such as the well-known matryoshka cluster $[As@Ni_{12}@As_{20}]^{3-}$ [124] which is described as $[As]^{3-}$@$[Ni_{12}]^{8-}$@$[As_{20}]^{8+}$ ($8+8+92=108$ electrons) [122]. Other multispherical approaches have been also used to rationalize the stability of this latter structure [125–127]. So far, more experimental structures and compositions are needed for validating the "jelliumatic shell model." On the other hand, a theoretical justification based on a more complex radial potential describing the average electron–nuclei interaction (not square-like but somewhat step-like) is still missing.

7 Application of the Jellium Model to Metalloid Al and Ga Clusters

The noble-gas superatom model covers not only a large field of viable ligated clusters of gold, silver, and other late transition metals but also that of the so-called protected metalloid aluminum and gallium closed-shell clusters, the chemistry of which has been largely developed by Schnöckel and coworkers [128–132]. A typical example is the pseudo-octahedral organometallic cluster $[Al_{50}Cp^*_{12}]$ ($Cp^* =$ pentamethylcyclopentadienyl) which can be written as $[Al_8$@Al_{30}@

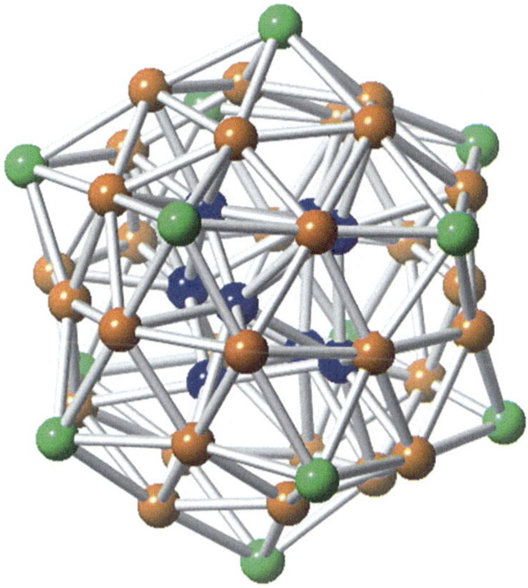

Fig. 9 Structural arrangement of the metallic core of $[Al_{50}Cp^*_{12}]$ described as $[Al_8@Al_{30}@(AlCp^*)_{12}]$ [133]. *Blue*, *orange*, and *green spheres* are Al of the Al_8, Al_{30}, and Al_{12} shells, respectively. Surrounding Cp* groups are not shown for clarity

$(AlCp^*)_{12}]$ (Fig. 9). This cluster is made of a central distorted square antiprism surrounded by an icosidodecahedron, the 12 pentagonal faces of which being capped by an AlCp* unit [133]. DFT calculations have established that this cluster is a 138-electron closed-shell superatom with an $[Al_{50}]^{12+}$ jellium core of $1S^2$ $1P^6$ $1D^{10}$ $2S^2$ $1F^{14}$ $2P^6$ $1G^{18}$ $2D^{10}$ $3S^2$ $1H^{22}$ $3S^2$ $2F^{14}$ $3P^6$ $1I^{26}$ configuration [134, 135]. In this view, the twelve Cp* anions constitute the passivating shell of the cluster. The same group has also investigated the electronic structure of $[Ga_{23}(NR_2)_{11}]$ ($R = SiMe_3$) [135] also made by Schnöckel and colleagues [136]. This less symmetrical but still pseudo-spherical Ga-centered species can be formulated as $[Ga@Ga_{11}@\{Ga(NR_2)_{11}\}_{12}]$ and was found to be a 58-electron superatom ($1S^2$ $1P^6$ $1D^{10}$ $2S^2$ $1F^{14}$ $2P^6$ $1G^{18}$) with a $[Ga_{23}]^{11+}$ jellium core [135].

As discussed above, the closed-shell electron counts predicted by the spherical jellium model are generally supposed to be observed when each atom belonging to the cluster core participates to the bonding with only one orbital, this latter being of σ type (basically, one *s* AO, or one *sp* endo hybrid for peripheral atoms). It is noteworthy that in the case of the last examples, the atomic average valence electron number is larger than 2, meaning that the *3p* (Al) or *4p* (Ga) orbitals participate importantly to cluster bonding. This situation is nicely illustrated by the mass spectrometry characterized $[Al_{13}]^-$ anion [137]. This 40-electron centered icosahedral species [138] and its neutral parent have been extensively investigated theoretically (see, e.g., [139, 140] and references therein) [141]. As noted above, if only the σ-type AOs would contribute to skeletal bonding, an 8-electron jellium count with the 1S 1P-filled shell would result. The $[Al_{13}]^-$ jellium configuration being $1S^2$ $1P^6$ $1D^{10}$ $2S^2$ $1F^{14}$ $2P^6$, there are 16 occupied SMOs which are derived from bonding combinations of *3p* (Al) AOs, of both p_π or p_σ (or exo sp_σ) types. Thus, before applying the jellium model, one should check that the atoms constituting the cluster cage provide to the bonding with enough atomic orbitals to satisfy the occupation of low-lying SMOs. An interesting counterexample is $Al_4Cp^*_4$ [142]. In this small tetrahedral cluster, each Al^ICp^* fragment provides to the bonding with one single endo sp_σ hybrid and two electrons. DFT calculations support the view of an 8-electron jellium count ($1S^2$ $1P^6$) with an a_1^2 t_2^6 configuration in T_d symmetry [143]. Simple Hückel considerations lead to the expectation that the t_2 σ-type (1P) combinations should be antibonding (thus unavailable). Consequently, there is no way to involve participation of the $3p_\pi$ (Al) AOs, already participating in Al–Cp* bonding, unless aluminum hypercoordination is considered, i.e., delocalization is assumed in both Al–Al and Al–Cp* moieties [142, 144].

8 Large Ligated Metal Clusters vs. Nanoparticles

It is largely established that the HOMO–LUMO gap of a polynuclear cluster is expected to decrease with respect to increasing its nuclearity and eventually vanishes when the cluster reaches a certain size to resemble more and more to bulk metal. The question of when this vanishing occurs is not simple, since it depends not only on the nuclearity but also on the shape and ligand coverage of the particle and of course on the nature of the metal. The disappearance of the HOMO–LUMO gap should coincide with the loss of the domain where the electron-counting rules apply, because they are based on a closed-shell principle. Nevertheless, one might expect that the smallest "metallic" particles should exhibit a hole or a pseudogap in their density of states near the Fermi level. This is a situation which allows that the structural stability is not associated with a single electron count as for molecules (see Introduction), but with a range of allowed electron counts, as found in solid-state chemistry [49, 145]. Concomitantly, non-atomically precise structures should arise and compact packing (hcp or fcc) should largely dominate. In the case of gold, a rough estimation based on a simple electron gas model suggests that the transition should occur around 150 atoms [146]. This is confirmed by various experimental and/or theoretical investigations [147, 148]. In the case of palladium, DFT calculations performed at the BP86 level of theory show a small but substantial HOMO–LUMO gap of 0.12 eV for $Pd_{59}(CO)_{32}(PH_3)_{21}$, for instance, used to mimic $Pd_{59}(CO)_{32}(PMe_3)_{21}$, indicating electronic molecular behavior. Must we conclude that large ligated palladium clusters are molecules rather than pieces of bulk metals? The answer may not be straightforward, but it is worth mentioning that DFT calculations carried on a series of $Pd_n(CO)_x(PR_3)_y$ clusters ($n = 30$–145) reveal HOMO–LUMO gaps for many of them (see Table 1), explaining why many of these species seem to be electron-precise with respect to the electron-counting rules recalled above [85]. However, there are also some examples for which no (or virtually no) HOMO–LUMO gap is computed (see Table 1) and which may favor open-shell rather than closed-shell electronic configurations. This situation might also favor the possibility of several electron counts for (roughly) the same cluster geometry – some large transition-metal clusters show some reversible redox aptitude [72] (see, e.g., [149]).

9 Concluding Remarks

Structure and bonding are intimately linked to the number of valence electrons. In the case of chemically stable (viable) species, the relationship between the structure and the electron count is governed by electron-counting rules. It is important to keep in mind that such rules assume the existence of a substantial HOMO–LUMO gap. Applying them, for example, to open-shell molecules or to metallic nanoparticles may lead to partially or completely wrong conclusions. Whereas the Lewis valence theory and the subsequently developed EAN rules apply fairly well to systems where localized bonding prevails, in the case of cluster species, delocalized approaches based on nearly free-electron spherical models have been used to develop the electron-counting rules of the PSEPT and the superatom theory. The degree of sophistication of these rules, which as a whole covers the entire field of cluster chemistry, is noteworthy. Theoretical investigations have also contributed significantly to rationalize the structural chemistry of clusters. One example among many of such contributions coming out of DFT calculations is the "divide and protect" concept. Such investigations are also necessary to understand bonding and properties in open-shell systems [52, 53, 149], as well as physical and chemical properties in general. However, there are size limits in the possibility of calculations on very large ligated or bare metal and nanoparticles up to several hundreds of atoms [146, 148, 150]. For viable predictions, an unbiased structure search using algorithms that are able to systematically scan the potential surface of n-atomic clusters would then be needed. Several techniques have been proposed and tested for this purpose, but they require a large number of calculations and geometry optimizations, which limit their application [151, 152] (see, e.g., [153]). Their improvements should enable more detailed studies of the effect of cluster size and composition on all static properties, e.g., structural, electronic, magnetic, optical, thermal, and catalytic, as well as dynamic properties.

Acknowledgments We would like to thank Dr Samia Kahlal (Rennes) for providing helpful comments.

References

1. Lewis GN (1916) J Am Chem Soc 38:762–785
2. Langmuir I (1921) Science 54:59–67
3. Sidgwick NV (1927) The Electronic Theory of Valency. Clarendon, Oxford
4. Heitler W, London F (1927) Z Phys 44:455–472
5. Pauling L (1960) The nature of the chemical bond, 3rd edn. Cornell University Press, Ithaca/New York
6. Hoffmann R, Schleyer PVR, Schaefer HF III (2008) Angew Chem Int Ed 47:7164–7167
7. Mingos DMP, Wales DJ (1990) Introduction to cluster chemistry. Prentice-Hall, Englewood Cliffs

8. Albright TA, Burdett JK, Whangbo M-H (2013) Orbital Interactions in Chemistry, 2nd edn. Wiley, New York
9. Elian M, Chen MM-L, Mingos DMP, Hoffmann R (1976) Inorg Chem 15:1148–1155
10. Hoffmann R (1982) Angew Chem Int Ed 21:711–724
11. Williams RE (1971) Inorg Chem 10:210–214
12. Williams RE (1992) Chem Rev 92:177–207
13. Rudolph RW (1976) Acc Chem Res 9:446–452
14. Wade K (1971) J Chem Soc D Chem Commun 792–793
15. Wade K (1972) Inorg Nucl Chem Lett 8:559–562
16. Wade K (1972) Inorg Nucl Chem Lett 8:563–566
17. Wade K (1972) Inorg Nucl Chem Lett 8:823–827
18. Wade K (1976) Adv Inorg Chem Radiochem 18:1–66
19. Wade K (1980) Some bonding considerations. In: Johnson BFG (ed) Transition metal clusters. Wiley, Chichester, pp 193–264, chapter 3
20. Mingos DMP (1972) Nat Phys Sci 236:99–102
21. Longuet-Higgins HC, de V Roberts M (1955) Proc R Soc A 230:110–119
22. Longuet-Higgins HC (1957) Quart Rev Chem Soc 11:121–133
23. Hoffmann R, Lipscomb WN (1962) J Chem Phys 36:2179–2189
24. Hoffmann R, Lipscomb WN (1962) J Chem Phys 36:3489–2493
25. Hoffmann R, Lipscomb WN (1962) J Chem Phys 37:2872–2883
26. Lipscomb WN (1963) Boron Hydrides. Benjamin, New York
27. Klanberg F, Muetterties EL (1966) Inorg Chem 5:1955–1960
28. Eatin DR, Guggenberger LJ, Muetterties EL (1967) Inorg Chem 6:1271–1281
29. Guggenberger LJ (1968) Inorg Chem 7:2260–2264
30. Guggenberger LJ (1969) Inorg Chem 8:2771–2774
31. Mason R, Thomas KM, Mingos DMP (1973) J Am Chem Soc 95:3802–3804
32. Mingos DMP, Johnston RL (1987) Struct Bond 68:29–87
33. Mingos DMP (1991) Pure Appl Chem 63:807–812
34. King RB, Rouvray DH (1977) J Am Chem Soc 99:7834–7840
35. Teo BK, Zhang H (1995) Coord Chem Rev 143:611–636
36. Stone AJ (1980) Mol Phys 41:1339–1354
37. Stone AJ (1981) Inorg Chem 20:563–571
38. Stone AJ, Alderton MJ (1982) Inorg Chem 21:2297–2302
39. Stone AJ (1984) Polyhedron 3:2051–2068
40. Wales DJ, Mingos DMP (1989) Inorg Chem 28:2148–2154
41. Wales DJ, Mingos DMP, Slee T, Lin Z (1989) Acc Chem Res 23:17–22
42. Johnston RL (1997) Struct Bond 87:1–34
43. Eberhardt WH, Crawford B, Lipscomb WN (1954) J Chem Phys 22:989–1001
44. Hoffman DK, Ruedenberg K, Verkade JG (1977) Struct Bond 33:57–96
45. Lakatos I (1976) Proofs and Refutations. Cambridge University Press, London
46. Jackson PF, Johnson BFG, Lewis J, McPartlin M, Nelson WJH (1979) J Chem Soc Chem Commun 735–736
47. Hay CM, Johnson BFG, Lewis J, Raithby PR, Whitton AJ (1988) J Chem Soc Dalton Trans 2091–2097
48. Johnston RL, Mingos DMP (1985) Inorg Chem 25:1661–1671
49. Fehlner TP, Halet J-F, Saillard J-Y (2007) Molecular clusters. A bridge to solid state chemistry. Cambridge University Press, Cambridge
50. Halet J-F, Hoffmann R, Saillard J-Y (1985) Inorg Chem 24:1695–1700
51. Halet J-F (1995) Coord Chem Rev 143:637–678
52. Halet J-F, Saillard J-Y (1997) Struct Bond 87:81–109
53. Gautier R, Halet J-F, Saillard J-Y (1999) Hexacapped cubic transition metal clusters and derivatives: a theoretical approach. In: Braunstein P, Oro L, Raithby PR (eds) Metal clusters in chemistry, vol 3. Wiley, Weinheim, pp 1643–1663

54. Gautier R, Halet J-F, Saillard J-Y (2009) Computational methods: transition metals clusters. In: Solomon EI, Scott RA, King RB (eds) Computational inorganic and bioinorganic chemistry. Wiley, Chichester, pp 539–550
55. Halet J-F, Saillard J-Y (2013) Theoretical treatment of ligated clusters containing transition metals. In: Reedijk J, Poeppelmeier K (eds) Comprehensive inorganic chemistry II, vol 9, Theory and methods. Elsevier, Oxford, pp 869–885, Alvarez S volume ed
56. Roy DK, Ghosh S, Halet J-F (2014) J Cluster Sci 25:225–237
57. Mingos DMP, Forsyth MI (1977) J Chem Soc Dalton Trans 610–616
58. Mingos DMP (1983) J Chem Soc Chem Commun 706–708
59. Mingos DMP (1984) Acc Chem Res 17:311–319
60. Jemmis ED, Balakrishnarajan MM, Pancharatna PD (2001) J Am Chem Soc 123:4313–4323
61. Jemmis ED, Balakrishnarajan MM, Pancharatna PD (2002) Chem Rev 102:93–144
62. Shameema O, Jemmis ED (2009) Computational studies: boranes. In: Solomon EI, Scott RA, King RB (eds) Computational inorganic and bioinorganic chemistry. Wiley, Chichester, pp 539–550
63. Priyakumari CP, Jemmis ED (2014) Electron-counting rules in cluster bonding – polyhedral boranes, elemental boron, and boron-rich solids. In: Frenking G, Shaik S (eds) The chemical bonding, vol 2, Chemical bonding across the periodic table. Wiley, Weinheim, pp 113–147, chapter 5
64. Mingos DMP (1985) J Chem Soc Chem Commun 1352–1354
65. Mingos DMP (1986) Chem Soc Rev 15:31–61
66. Mingos DMP, Lin Z (1988) J Organomet Chem 341:523–534
67. Mingos DMP, Lin Z (1988) J Chem Soc Dalton Trans 1657–1664
68. Halet J-F (1992) Theoretical aspects of the bonding in transition metal carbonyl clusters with interstitial main group atoms. In: Gielen M (ed) Topics in physical organometallic chemistry, vol 4. Freund Publishing House, London, pp 221–288
69. Briant CE, Theobald BRC, White JW, Bell LK, Mingos DMP, Welch AJ (1981) J Chem Soc Chem Commun 201–202
70. Ceriotti A, Demartin F, Longoni G, Manassero M, Marchionna M, Piva G, Sansoni M (1985) Angew Chem Int Ed Engl 24:697–698
71. Belyakova OA, Slovokhotov YL (2003) Russ Chem Bull Int Ed 52:229–2327
72. Femoni C, Iapalucci MC, Kaswalder F, Longoni G, Zacchini S (2006) Coord Chem Rev 250:1580–1604
73. Mednikov EG, Dahl LF (2010) Philos Trans R Soc A (London) 368:1301–1332
74. Ciabatti I, Femoni C, Iapalucci MC, Longoni G, Zacchini S (2014) J Cluster Sci 25:115–146
75. Tran NT, Kawano M, Powell DR, Dahl LF (1998) J Am Chem Soc 120:10986–10987
76. Mednikov EG, Ivanov SA, Dahl LF (2003) Angew Chem Int Ed 42:323–327
77. Mednikov EG, Kanteeva NI (1995) Russ Chem Bull 44:163–166
78. Tran NT, Kawano M, Dahl LF (2001) Dalton Trans 19:2731–2748
79. Mednikov EG, Dahl LF (2008) J Am Chem Soc 130:14813–14821
80. Mednikov EG, Eremenko NK, Slovokhotov YL, Struchkov YT (1987) J Chem Soc Chem Commun 218–219
81. Mednikov EG, Ivanov SA, Slovokhotova I, Dahl LF (2005) Angew Chem Int Ed 44:6848–6854
82. Tran NT, Dahl LF (2003) Angew Chem Int Ed 42:3533–3537
83. Tran NT, Powell DR, Dahl LF (2000) Angew Chem Int Ed 39:4121–4125
84. Mednikov EG, Jewell MC, Dahl LF (2005) J Am Chem Soc 129:11619–11630
85. Marchal R, Manca G, Furet E, Kahlal S, Saillard J-Y, Halet J-F (2014) J Cluster Sci 26:41–51
86. Teo BK, Zhang H (2002) In: Feldheim DL, Foss CA Jr (eds) Metal Nanoparticles: Synthesis, Characterization, and Applications. Marcel Dekker, New York, pp 55–88, chapter 3
87. Teo BK, Zhang H (1991) Proc Natl Acad Sci U S A 88:5067–5071
88. Knight WD, Clemenger K, de Heer WA, Saunders W, Chou MY, Cohen ML (1984) Phys Rev Lett 52:2141–2143

89. Ashcroft NW, Mermin ND (1976) Solid state physics. Holt/Rinehart and Winston, New York, pp 285–288
90. Louie SG, Froyen S, Cohen ML (1982) Phys Rev B 26:1738–1742
91. de Heer WA (1983) Rev Mod Phys 65:611–678
92. Lin Z, Slee T, Mingos DMP (1990) Chem Phys 142:321–334
93. Mingos DMP, Slee T, Lin Z (1990) Chem Rev 90:385–402
94. Clemenger K (1985) Phys Rev B 32:1359–1362
95. Mingos DMP (2014) Struct Bond 162:1–65
96. Mingos DMP (2015) Dalton Trans 44:6680–6695
97. Teo BK (2014) J Cluster Sci 25:5–28
98. Desireddy A, Conn BC, Guo J, Yoon B, Barnett RN, Monahan BN, Kirschbaum K, Griffith WP, Whetten RL, Landman U, Bigioni TP (2013) Nature 501:399–402
99. Yang H, Wang Y, Huang H, Gell L, Lehtovaara L, Malola S, Häkkinen H, Zheng N (2013) Nature Commun 4:2422
100. Dhayal R-S, Liao J-H, Liu Y-C, Chiang M-H, Kahlal S, Saillard J-Y, Liu C-W (2015) Angew Chem Int Ed Engl 54:3702–3706
101. Nguyen T-AD, Jones ZR, Goldsmith BR, Buratto WR, Wu G, Scott SL, Hayton TW (2015) J Am Chem Soc 137:13319–13324
102. Yang H, Wang Y, Yan J, Chen X, Zhang X, Häkkinen H, Zheng N (2014) J Am Chem Soc 136:7197–7200
103. Walter M, Akola J, Lopez-Acevedo O, Jadzinsky PD, Calero G, Ackerson CJ, Whetten RL, Grönbeck H, Häkkinen H (2008) Proc Natl Acad Sci U S A 105:9157–9162
104. Castleman AW Jr, Khanna SN (2009) J Phys Chem C 113:2664–2675
105. Häkkinen H (2015) Electronic structure: shell structure and the superatom concept. In: Tsukuda T, Häkkinen H (eds) Protected metal clusters: from fundamentals to applications, vol 9. Elsevier, Amsterdam, pp 189–222
106. Häkkinen H (2008) Chem Soc Rev 37:1847–1859
107. Häkkinen H (2012) Theoretical studies of gold nanoclusters in various chemical environments: when the size matters. In: Louis C, Pluchery O (eds) From gold nanoparticles for physics, chemistry and biology, vol 9. Imperial College Press, London, pp 233–272, chapter 9
108. Häkkinen H, Walter M, Grönbeck H (2006) J Phys Chem B 110:9927–9931
109. Sculfort S, Braunstein P (2011) Chem Soc Rev 40:2741–2760
110. Schmidbaur H, Schier A (2012) Chem Soc Rev 41:370–412
111. Schmidbaur H, Schier A (2015) Angew Chem Int Ed 54:746–784
112. Schmidbaur H, Schier A (2015) Organometallics 34:2048–2066
113. Akola J, Walter M, Whetten RL, Häkkinen H, Grönbeck H (2008) J Am Chem Soc 130:3756–3757
114. Jadzinsky PD, Calero G, Ackerson CJ, Bushnell DA, Kornberg RD (2007) Science 318:430–433
115. Pei Y, Zeng XC (2012) Nanoscale 4:4054–4072
116. Jiang DE (2013) Nanoscale 5:7149–7160
117. Freitag K, Banh H, Gemel C, Seidel RW, Kahlal S, Saillard J-Y, Fischer RA (2014) Chem Commun 50:8681–8684
118. Mingos DMP (1976) J Chem Soc Dalton Trans 1163–1169
119. Latouche C, Liu CW, Saillard J-Y (2014) J Cluster Sci 25:147–171
120. Dayal RS, Liu C-W (2015) Acc Chem Res. doi:10.1021/acs.accounts.5b00375
121. Edwards AJ, Dhayal RS, Liao P-K, Liao J-H, Chiang M-H, Piltz RO, Kahlal S, Saillard J-Y, Liu CW (2014) Angew Chem Int Ed 53:7214–7218
122. Teo BK, Yang S-Y (2015) J Cluster Sci 26:1923–1941
123. Dhayal RS, Liao J-H, Lin Y-R, Liao P-K, Kahlal S, Saillard J-Y, Liu CW (2013) J Am Chem Soc 135:4704–4707
124. Moses MJ, Fettinger JC, Eichhorn BW (2003) Science 300:778–780
125. King RB, Zhao J (2006) Chem Commun 4204–4205

126. Sheong FK, Chen W-J, Lin Z (2015) Dalton Trans 44:7251–7257
127. Rauhalahtib M, Muñoz-Castro A (2015) RSC Adv 5:18782–18787
128. Schnöckel H (2005) Dalton Trans 3131–3136
129. Schnöckel H (2008) Dalton Trans 4344–4362
130. Burgert R, Schnöckel H (2008) Chem Commun 2075–2089
131. Schnöckel H (2010) Chem Rev 110:4125–4163
132. Schnöckel H, Schnepf A (2011) Aluminium and gallium clusters: metalloid clusters and their relationship to the bulk phases, to naked clusters and to nanoscaled materials. In: Aldridge A, Downs AJ (eds) The group 13 metals aluminium, gallium, indium and thallium: chemical patterns and peculiarities. Wiley, Chichester, pp 402–487, Chapter 6
133. Vollet J, Hartig JR, Schnöckel H (2004) Angew Chem Int Ed 44:3186–3189
134. Clayborne RA, Lopez-Acevedo O, Whetten RL, Grönbeck H, Häkkinen H (2011) Eur J Inorg Chem 2649–2652
135. Lopez-Acevedo O, Clayborne PA, Häkkinen H (2011) Phys Rev B 84:035434/1–035434/6
136. Hartig J, Stosser A, Hauser P, Schnöckel H (2007) Angew Chem Int Ed 46:1658–1662
137. Taylor KJ, Pettiette CL, Craycraft MJ, Chesnovsky O, Smalley RE (1988) Chem Phys Lett 152:347–351
138. Cheng HP, Berry RS, Whetten RL (1991) Phys Rev B 43:10647–10653
139. Bergeron DE, Roach PJ, Castelman AW Jr, Jones NO, Khanna SN (2005) Science 307:231–235
140. Varns R, Strange P (2012) Phys Status Solidi B 249:2179–2189
141. King RB, Silaghi-Dumitrescu I (2008) Dalton Trans 6083–6088
142. Dohmeier C, Robl C, Tacke M, Schnöckel H (1991) Angew Chem Int Ed 30:564–565
143. Clayborne PA, Olga Lopez-Acevedo O, Robert L, Whetten RL, Grönbeck H, Häkkinen H (2011) J Chem Phys 135:094701/1–094701/5
144. Purath A, Dohmeier C, Ecker A, Schnöckel H (1998) Organometallics 17:1894–1896
145. Burdett JK (1995) Chemical bonding in solids. Oxford University Press, Oxford
146. Malola S, Häkkinen H (2015) Europhys News 46:23–26
147. Negishi Y, Nakazaki T, Malola S, Takano S, Niihori Y, Kurashige W, Yamazoe S, Tsukuda T, Häkkinen H (2015) J Am Chem Soc 137:1206–1212
148. Tlahuice-Flores A (2015) Phys Chem Chem Phys 17:5551–5555
149. Bencharif M, Cador O, Cattey H, Ebner A, Halet J-F, Kahlal, S, Meier W, Mugnier Y, Saillard J-Y, Schwarz P, Trodi FZ, Wachter J, Zabel M (2008) Eur J Inorg Chem 1959–1968
150. Weigend F, Ahlrichs R (2010) Philos Trans R Soc A (London) 368:1245–1263
151. Oganov AR (ed) (2010) Modern methods of crystal structure prediction. Wiley-VCH, Berlin
152. Marchal R, Carbonnière P, Pouchan C (2009) J Chem Phys 131:114105–114113
153. Marchal R, Manca G, Kahlal S, Carbonnière P, Pouchan C, Halet J-F, Saillard J-Y (2012) Eur J Inorg Chem 4856–4866

Struct Bond (2016) 169: 181–198
DOI: 10.1007/430_2015_209

Published online: 29 March 2016

Electronic Properties of Endohedral Clusters of Group 14

Vaida Arcisauskaite, Xiao Jin, José M. Goicoechea, and John E. McGrady

Abstract The concept of stable "superatoms," molecular species which mimic the shell closures emphasised by Lewis and Kossel, has become an important paradigm of stability in cluster chemistry. In this review we discuss recent work, both experimental and theoretical, on the family of endohedral clusters $M@E_x$, where M is a transition metal ion and E is a member of group 14 (Si, Ge, Sn, Pb). The structural chemistry within this family is very varied, ranging from deltahedral motifs for the heavier tetrels to open 3-connected structures such as the hexagonal prism in $Cr@Si_{12}$. We explore the arguments that have been presented to rationalise these structural trends and their implications for chemical bonding.

Keywords Chemical bond · Density functional theory · Endohedral cluster · Germanium · Silicon

Contents

V. Arcisauskaite, X. Jin, J.M. Goicoechea, and J.E. McGrady (✉)
Department of Chemistry, University of Oxford, South Parks Road, Oxford OX1 3QZ, UK
e-mail: john.mcgrady@chem.ox.ac.uk

Abbreviations

HOMO Highest occupied molecular orbital
LUMO Lowest unoccupied molecular orbital

1 Introduction

One hundred years on from the seminal paper by Lewis [1], the octet rule remains one of the fundamental pillars of chemistry, one that is obeyed by the vast majority of main group compounds. The significance of the number 8 is now well understood in terms of the utilisation of four valence orbitals, *n*s and *n*p, and the logical extension of these ideas into the transition series leads naturally to the 18-electron rule (*n*s, *n*p, and (*n*−1)d). The energetic separation between the {*n*s, *n*p} and (*n*−1)d orbitals, along with their rather different radial extensions, means that exceptions to the 18-electron rule are far more common than exceptions to the 8-electron rule, but nevertheless, it underpins much of modern transition metal chemistry. Recognising that the shell structure of atomic orbitals is a direct consequence of the spherical symmetry of the atoms, we can extend the simple ideas of shell closure to more complex polyatomic clusters with approximate spherical symmetry. For example, the observation of intense peaks in the mass spectra of sodium clusters, Na_n, for $n = 8$, 18, and 20 [2] can be interpreted in terms of a spherical jellium model, wherein the *n* valence electrons are assumed to move in a spherical potential defined by averaging the positive charges of the Na^+ cores. The magic numbers 2, 8, 18, and 20 correspond to filling in the order $1s^2 1p^6 1d^{10} 2s^2$ – the delocalisation of the positive potential over a spherical surface rather than at a central point leads to the different ordering of shells compared to the normal atomic case ($1s^2 2s^2 2p^6$). Castleman's group has performed a number of detailed mass spectrometric investigations of this type and has shown, for example, that the $[Al_{13}]^-$ cluster, with 40 valence electrons ($1s^2 1p^6 1d^{10} 2s^2 2p^6 1f^{14}$, Fig. 1), is far more resistant to etching by dioxygen than its neighbours, $[Al_n]^-$ ($n \neq 13$) [3]. The analogy to the similarly inert noble gases led to the concept of "superatoms" and the so-called three-dimensional periodic table [4–11], which now features "super alkalis" such as K_3O [12], "super halogens" such as Al_{13} [13], and even a cluster mimic of carbon or silicon, Al_7^- [14], which can behave as either a di- or tetravalent fragment. The jellium model is less obviously applicable to clusters with more electronegative cores which tend to localise electrons, but even here advantage can still be taken of approximate spherical symmetry. A classic example comes from the work of Wade and Mingos who established the eponymous electron-counting rules for the *closo*-boranes and a host of closely related species [15–17]. The emphasis in this context is on how the removal of vertices from systems with a constant skeletal electron count leads to progressively less spherical geometries, as, for example, in the classic triad of *closo*-$[B_6H_6]^{2-}$, *nido*-B_5H_9, and *arachno*-B_4H_{10}.

In this review, we deal with a specific class of pseudo-spherical clusters, the endohedrally encapsulated $M@E_n$ family where M is a transition metal and E is one

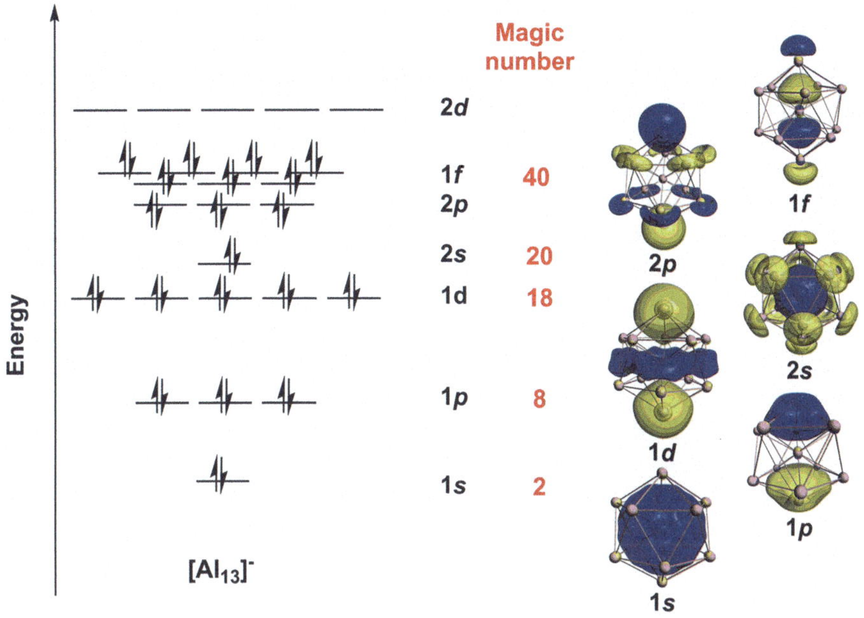

Fig. 1 Jellium model for the superatom $[Al_{13}]^-$

of the tetrel elements, Si, Ge, Sn, or Pb. Endohedral fullerenes, for example, $M@C_{60}$, have been reviewed extensively and are conceptually very different from the heavier analogues [18–23]. Experimental work in this area can broadly be divided into two distinct areas, gas-phase spectrometry/spectroscopy and condensed-phase structural studies, which are typically conducted by different research communities. For the silicon clusters, almost all the available data has come from the first of these sources: no X-ray crystallography has yet been reported on $M@Si_n$ clusters, but mass spectrometry and infrared spectroscopy have provided a rich source of information. In contrast, crystallographic data are relatively extensive in the clusters of the heavier tetrels, Ge, Sn, and Pb, and a number of remarkable structures have emerged in the last few years, primarily for the $n = 10$ and $n = 12$ families, and it is on these that we focus primarily in this review. Despite the very different experimental techniques applied to silicon clusters on the one hand and the heavier analogues on the other, common themes have emerged from the many studies of electronic structure that serve to connect these somewhat disparate fields. In comparison to the empty clusters such as $[Al_{13}]^-$, $[B_6H_6]^{2-}$, etc., the presence of an endohedral metal ion introduces a spherical electrostatic potential which will naturally tend to favour symmetrical arrangements of the atoms. Thus, we can anticipate that the structure of endohedral clusters will naturally align itself to models based on a spherical reference point [24–27], and indeed these clusters have featured prominently in the development of the "superatom" concept [4–8].

Amongst the $M@E_n$ family, the silicon species are the most extensively studied, primarily because they represent the simplest available models for point defects in the silicon wafers that are ubiquitous in modern electronic devices [28–31]. The high-temperature manufacturing process for bulk silicon wafers typically leaves a high concentration of transition metal impurities, particularly iron, copper, and nickel, in the silicon lattice. The solubility of these metals is, however, very limited at ambient temperatures, and the formation of metal silicides is instead thermodynamically favoured. The result is that diffusion of the metal ions through the silicon lattice leads to formation of metal silicide islands at isolated lattice irregularities, with detrimental impact on the performance of electronic devices [32]. The concentration of transition metal impurities, particularly copper, must therefore be maintained at the subparts per billion level for effective performance. A deeper understanding of the nucleation of metal ions into distinct clusters is therefore central to the "gettering" process, wherein defect sites are introduced into noncritical regions of the wafer to seed the growth of the silicide islands. The fact that M@Si_n clusters are typically more stable than their metal-free analogues, Si_n, also offers the possibility of rational design of silicon-based nanomaterials with tailored properties. Beyond any technological relevance, however, this family of clusters is of great interest because it challenges our understanding of chemical bonding and electron counting.

2 Endohedral Clusters of Silicon, M@Si_n

The study of small clusters containing both silicon and a transition element can be traced back to two papers published by Beck in the late 1980s [33, 34], where he described the formation of M@Si_n clusters in a supersonic beam. Laser vaporisation of a silicon wafer is used to generate a cluster vapour which is then quenched with helium carrier gas. If the carrier gas is seeded with metal carbonyl species (e.g. $Cr(CO)_6$), the CO ligands are stripped off in the laser vaporisation process, generating a series of cationic clusters which can be detected by time-of-flight mass spectrometry. The group VI hexacarbonyls, $Cr(CO)_6$ and $Mo(CO)_6$, generated a range of clusters with formula M@Si_n, with the mass envelopes strongly peaked around $n = 15$ and 16. There is a potential ambiguity in this methodology as the atomic mass of Si (28) is identical to the molecular mass of CO, but alternative protocols involving co-vaporisation of metal and silicon or laser vaporisation of the metal in the presence of SiH_4 produced very similar results. In 2001, Hiura, Miyazaki, and Kanayama used a quadrupole ion trap to isolate $[M@Si_nH_x]^+$ clusters from the reaction of SiH_4 with metal vapour generated by resistive heating [35]. The resulting time-resolved mass spectra show the stepwise growth of clusters with increasing n, but, in contrast to Beck's results, the envelope for $[W@Si_nH_x]^+$ terminates at $n = 12$, with no evidence for larger W@$Si_{15/16}$ clusters. The authors speculated that the discrepancy between the two data sets was a result of the higher reactivity of the vaporised silicon atoms used in Beck's experiments. Moreover, high-resolution spectra showed that the hydrogen content of the clusters, x, is reduced as n increases, and in the limit of $n = 12$, the most abundant tungsten cluster was completely devoid of hydrogens. The interpretation

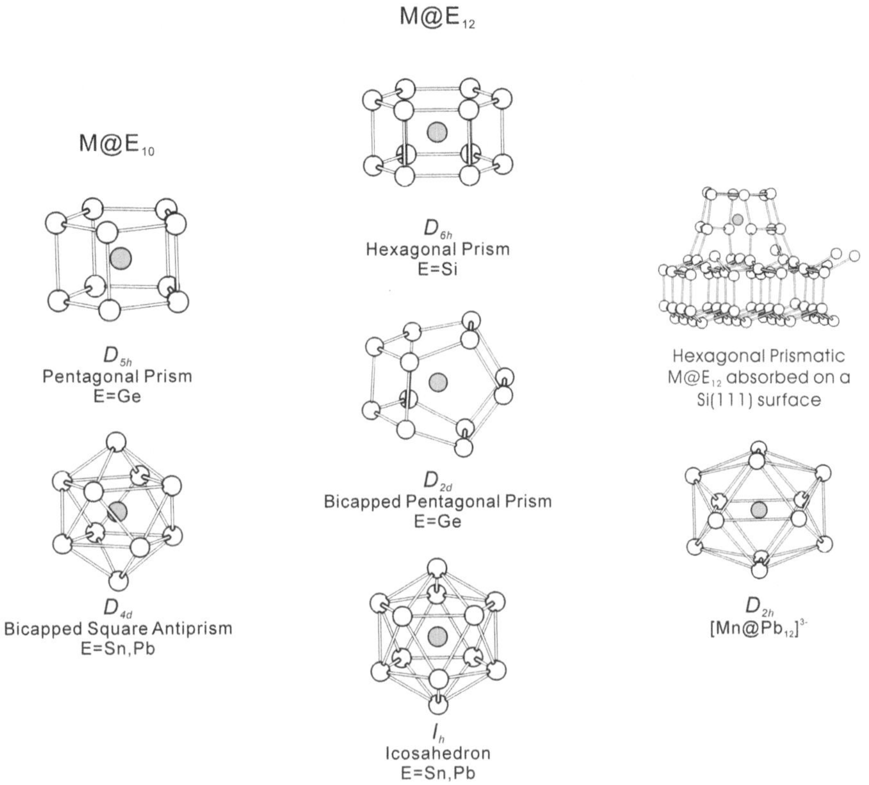

Fig. 2 Structural diversity of the $M@E_{10}$ and $M@E_{12}$ families, and a model of the latter absorbed on a Si (111) surface [36]

of this data is that the metal atom is completely encapsulated in $M@Si_{12}$, binding to all 12 silicon atoms and eliminating the need for additional hydrogens to saturate their valence. Hiura and co-workers also used density functional theory to establish the structure of $W@Si_{12}$, initially reporting a "basket-like" structure where the metal is encapsulated within the clusters and each silicon atom is bonded to three others. However, a note added in proof identified a more stable hexagonal prismatic structure (D_{6h}) (see Fig. 2) which is now generally recognised as the global minimum. Later work by Uchida and co-workers used scanning tunnelling microscopy to explore $[Ta@Si_{12}]^+$ clusters absorbed on reconstructed Si (111) surfaces and showed features consistent with intact hexagonal prismatic clusters [37]. XANES and EXAFS spectroscopies at the W L3 edge of $W@Si_x$ clusters deposited on silica are also consistent with the presence of substantial concentrations of hexagonal prismatic $W@Si_{12}$ [38]. Kanayama's group has also deposited amorphous films of $M@Si_{12}$ clusters on silica surfaces by reaction of the metal with SiH_4 followed by annealing at 500 C to remove the hydrogens [39]. The $Mo@Si_{12}$ films and $Nb@Si_{12}$ films were characterised as p-type and n-type semiconductors, respectively, both with mobilities much greater than that of a Mo/Si alloy where the arrangement of atoms is random. Thus, it seems that the structure of the metal/silicon aggregate can exert a dominant influence on the physical properties of the material. On this basis, the introduction of transition metal impurities into or onto bulk silicon in controlled geometries may prove a profitable avenue for future exploration.

One of the central questions that has emerged in the development of the chemistry of M@Si_n clusters is whether the metal is encapsulated within the cage or whether it simply replaces a silicon on the outside of the cluster. In 2002 Khanna, Rao, and Jena explored the potential energy surfaces for Cr@Si_n, $n = 11$–14 [40], showing in all cases that the metal is encapsulated by the silicon. For the $n = 12$ case, they identified the same hexagonal prismatic equilibrium structure described in Hiura's footnote and discussed its stability in the context of the 18-electron rule. The authors also noted that the corresponding anion, $[Cr@Si_{12}]^-$, should give two peaks in its photo-detachment spectrum, corresponding to singlet and triplet states of the neutral species, the predicted energies being 3.11 and 3.72 eV, respectively. The existence of experimental data supporting this prediction was noted in passing, and the data subsequently published by Bowen and co-workers in 2005 confirmed the presence of a narrow peak at 3.18 eV [41]. Khanna and co-workers also explored possible geometries for Cr@Si_{12} absorbed on a silicon surface, showing that in all cases, the Cr atom remains encapsulated within the cage and moreover that the most stable structures have a thirteenth silicon atom from the surface in the coordination sphere (see Fig. 2, right) [36]. The absence of hydrogen in the most abundant $[W@Si_{12}H_x]^+$ cluster reported by Hiura et al. [35] offers some experimental support for encapsulation, but more definitive evidence has come from the binding of small molecules. Ohara et al. showed that the abundance of $[Ti@Si_n]^+$ was reduced for $n < 11$ when water vapour was introduced, but the larger clusters ($n > 13$) were not affected [42, 43]. An endohedral metal is protected from the environment (i.e. water) by the silicon atoms, while an external metal is not, and so a change in reactivity with water is taken as evidence for a transition from exo- to endohedral structures around $n = 12$. Janssens and co-workers [44] developed this idea, using argon absorption as a probe of structure. Cationic clusters $[M_mSi_n]^+$ were generated by laser vaporisation, and addition of a small amount of argon leads to clusters with one or two additional bound atoms. The absence of peaks corresponding to cationic $[Si_n]^+$. Ar_x, suggests that binding to silicon is not favourable, and the detection of peaks due to $[M@Si_n]^+.Ar_x$ is then taken to be an indicative of an external exposed metal atom. The critical value of n, beyond which $[M@Si_n]^+.Ar_x$ is absent, depends on the metal atom, varying between 7 ($[Co@Si_7]^+$) and 12 ($[Ti@Si_{12}]^+$) across the first transition series. More recently, infrared multiple photon dissociation spectroscopy of clusters of vanadium and manganese bound to the heavier noble gas, Xe, has also been used by Ngan et al. to differentiate between the various possible isomeric forms [45, 46]. In this experiment, resonant absorption of photons by the cluster results in local heating and desorption of the Xe atom, allowing the infrared spectrum to be probed through variations in ion intensities with frequency. Unlike argon, the more polarisable xenon atom does bind to the external Si atoms of an endohedral cluster, and comparison of the measured vibrational modes with DFT-computed spectroscopic fingerprints for various candidate structures provides direct structural information. The spectra of both $[V@Si_{12}]^+$ and $[Mn@Si_{12}]^+$ are consistent with hexagonal prismatic architectures, albeit heavily distorted in the vanadium case. The structures of the 13-vertex analogues, $[M@Si_{13}]^+$, are derived from the hexagonal bipyramid by capping of a single hexagonal face, while the spectrum of the 14-vertex $[M@Si_{14}]^+$ is consistent with either a bicapped hexagonal prism (the two caps spanning a single square face) or a D_{3h}-symmetric cage with three-connected vertices.

Much of the interest in these clusters has been driven by the desire to assemble magnetic entities in ordered arrays for applications in electronics, although this goal has not yet been realised because the presence of an encapsulating Si_x cage tends to quench the moment, even for systems like $Cr@Si_{12}$ where the intrinsic moment of the isolated metal atom is high. The multiplicity of the ground state of isoelectronic $[Mn@Si_{12}]^+$ has, however, been controversial: in reference [45, 46] discussed in the previous paragraph, Ngan et al. could only reach acceptable agreement with the experimental infrared spectrum if the cluster was assumed to be in a triplet state. In contrast, X-ray MCD spectroscopy studies by Zamudio-Bayer et al. [47] concluded that the $[Mn@Si_{12}]^+$ cluster is in fact diamagnetic. We will return to this apparent paradox in the subsequent discussion of theoretical models. Khanna and co-workers in their study of possible geometries for $Cr@Si_{12}$ absorbed on a silicon surface noted that binding to the surface can restore the magnetic moment of the chromium atom, albeit not in the most stable structures located by density functional theory [36].

The role of theory in the development of the chemistry of the silicon clusters, both as a complement to experiment and as an interpretative tool, has already been alluded to in the previous paragraphs, and Hiura's identification of $[W@Si_{12}]^+$ prompted a number of papers dealing with the electronic structure of this and closely related species [48–67]. Although many flavours of density functional theory have been applied, Hiura's proposal of a hexagonal prismatic structure appears to be undisputed, at least for elements of group VI (Cr, Mo, W), but the interpretation of this result in qualitative electronic structure terms has provoked intense debate. In their initial report, Hiura et al. suggested that its stability could be rationalised in terms of an 18-electron rule: donation of 12 electrons from the Si_{12} cage to the metal raises the electron count to 18, leaving three electrons per Si centre to form the three Si–Si bonds [35]. This model, however, implies that all valence electrons of silicon lie in one hemisphere of the atom (i.e. directed towards the metal), which would represent a significant departure from the borane clusters, for example, where an external pair of B–H bonding electrons is assumed to be present at each vertex in all electron-counting schemes. It would also run counter to Lewis' ideas of an octet symmetrically distributed around the atom. Jellium-type models can also be applied in this context, although the extent to which the electrons can truly be regarded as moving in a spherically averaged potential is open to debate. In the jellium framework, configurations of $1s^2 1p^6 1d^{10}$ (18 electrons) and $1s^2 1p^6 1d^{10} 2s^2$ (20 electrons) have been invoked to account for the relatively high dissociation energies computed for hexagonal prismatic Cr@Si_{12} and Fe@Si_{12} [56–58]. An analysis of the molecular orbital array of Cr@Si_{12} (Fig. 3), however, shows that the $3d_{z^2}$ orbital, oriented along the sixfold axis of the hexagonal prism, is almost entirely nonbonding and vacant, suggesting that a 16-electron count is more appropriate in this case [56–58]. We note here that in classical 16-electron complexes of the transition series (e.g. square-planar Pt(II)), it is the np_z orbital rather than $(n-1)d_{z^2}$ that is unused (in the sense that it is neither occupied by a lone pair nor involved in bonding/antibonding interactions with the ligands). The fact that it is the relatively low-lying $3d_{z^2}$ orbital that is unused in Cr@Si_{12} renders the hexagonal prismatic clusters more prone to nucleophilic attack than typical 16-electron species, leading, for example, to the 13-coordinate surface-absorbed species shown in Fig. 2. In the controversial case of $[Mn@Si_{12}]^+$, the $3d_{z^2}$ orbital is further stabilised by the positive charge, raising the possibility of facile intramolecular charge transfer wherein a single electron is promoted from the cluster-based $2a_{2u}$ orbital to $4a_{1g}$ (primarily $3d_{z^2}$). Amongst these detailed discussions of electronic structure, the significance of the size of the cluster should not be underestimated: the 12 silicon atoms can enclose a cavity of finite volume, and this volume is much greater in the rather open hexagonal prism than in an alternative icosahedron, for example. In later sections, we will see that the intrinsically larger cavities in Ge, Sn, and Pb clusters allow the system to adopt very different geometries (as shown in Fig. 2).

The discussion above has established the presence of a significant HOMO–LUMO gap for the hexagonal prism, and also that it can be considered as a 16-electron species, but this does in itself not resolve the key question: **why** is the

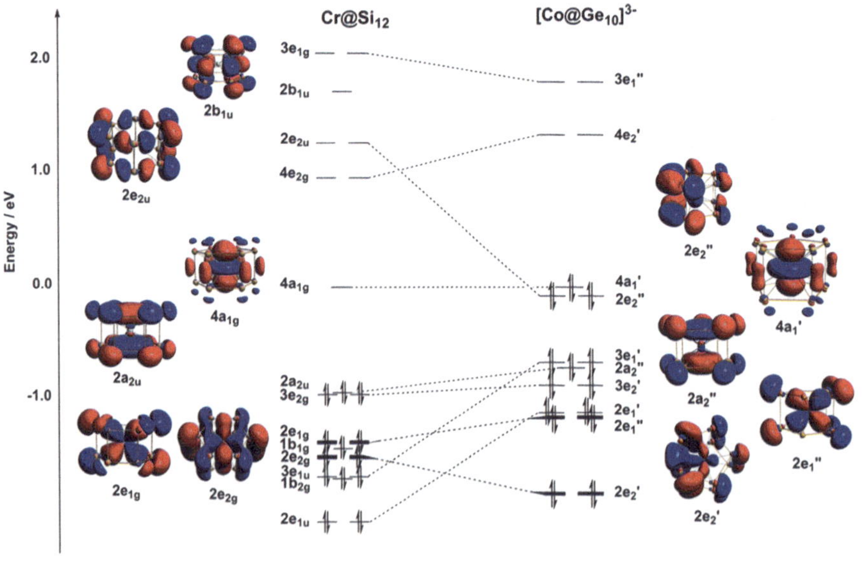

Fig. 3 Comparison of the electronic structures of Cr@Si_{12} (*left*) and $[Co@Ge_{10}]^{3-}$ (*right*)

hexagonal prismatic structure adopted in $Cr@Si_{12}$ in preference to the alternatives identified in Fig. 2 or indeed any other structure? If we ignore for a moment the central metal ion, three-connected architectures such as the hexagonal prism are characteristic of electron-precise electron counts ($5n$ valence electrons, where n is the number of vertices) – tetrahedral P_4 (20 electrons) is a classic example. In the case of $Cr@Si_{12}$, this would demand 60 valence electrons, while only 54 are available in total and 16 of these have been assigned to the metal. How can we reconcile the adoption of a structure that is typical of an electron precise count with the apparently much lower number of electrons available to the cage? The answer is that four pairs of electrons ($2e_{1g}$ and $2e_{2g}$) in Fig. 3 play a dual role, stabilising **both** the metal and the cage simultaneously. This idea is familiar enough in the context of classical electron counting: in $Cr(CO)_6$, for example, primarily ligand-based orbitals with a_{1g} and t_{1u} symmetry carry some metal 4s and 4p character and so are formally assigned to both the metal and the ligand, satisfying the valence electron demands of both. It is less common, however, to consider a dual role for the primarily metal 3d electrons: in the $Cr(CO)_6$ case, the six t_{2g} electrons are assigned solely to the metal, completing its 18-electron shell but not raising the count at the CO ligand. The difference in the case of $Cr@Si_{12}$ is that the neutral hexagonal prismatic Si_{12} is far from saturated, and so backbonding from the metal is essential to its stability. This synergy is apparent in the $2e_{1g}$ and $2e_{2g}$ orbitals of Fig. 3 which, while being dominantly Cr 3d, also have significant Si–Si bonding or Si lone pair character. Even when all available electrons are accounted for, the total electron count of 54 falls below the 60 required for a putative electron precise structure: three entirely cluster-based orbitals, $2b_{1u}$ and doubly degenerate $2e_{2u}$, remain vacant, and the cluster is therefore intrinsically electron deficient.

3 Endohedral Clusters of Germanium, M@Ge$_n$

Shortly after Hiura's report of the formation of $W@Si_{12}$, Gao and co-workers [68] produced a range of cobalt-containing germanium cluster anions by laser vaporisation of mixed cobalt/germanium powders. The resulting mass spectra are dominated by a cluster of formula $[Co@Ge_{10}]^-$, and the abrupt drop in signal intensity for $[Co@Ge_{11}]^-$ and all larger clusters was interpreted as evidence that the deca-germanium cluster had a cage structure, with the cobalt centre effectively shielded from further reaction. Subsequent studies by Nakajima and co-workers with earlier transition metals of group IV (Ti, Zr, Hf) showed the $M@Ge_{16}$ stoichiometry to be particularly abundant [69, 70], and their calculations suggested a T_d-symmetric Frank–Kasper-type polyhedron, similar to that identified previously by Kumar and Kawazoe for $Ti@Si_{16}$ [71]. Similarly, Neukermans et al. have shown through mass spectrometry that the $Cr@Ge_n$ and $Mn@Ge_n$ families show maximum abundances for $n = 15/16$, very much like the Si variants [72]. A number of theoretical papers have addressed the question of the structure and stability of these clusters [73–78], but the 10-vertex family, $M@Ge_{10}$, has been particularly well studied, and it is interesting to trace the development of bonding models used in this context. In Gao's original paper, the authors proposed a bicapped square antiprismatic structure for the very abundant $[Co@Ge_{10}]^-$ anion, rationalising it in terms of an 18-electron configuration (9 from the cobalt, 1 from the negative charge, and 1 from each of the 8 equatorial germanium atoms). The structure of this monoanion has subsequently been studied computationally by a number of groups, with alternative proposals for a bicapped antiprism [79], a C_{3v}-symmetric tetracapped trigonal prism [80, 81], and a C_s-symmetric "1-5-4" layered structure [82]. More definitive information emerged in 2009 with the simultaneous publication of the crystal structures of two isostructural 10-vertex species, 51-electron $[Fe@Ge_{10}]^{3-}$ and 52-electron $[Co@Ge_{10}]^{3-}$, by Goicoechea [24–26] and Fässler [83, 84], respectively. Both adopt a previously unprecedented D_{5h}-symmetric pentagonal prismatic geometry with each germanium vertex bonded to three others (Fig. 2) and in that sense are highly reminiscent of the hexagonal prismatic $Cr@Si_{12}$ cluster discussed above.

King and co-workers have discussed the origin of this unusual structure [80, 81], noting that if the cobalt centre accepts one electron to attain a d^{10} configuration and each vertex carries a radially oriented lone pair (20 electrons in total), the 22 remaining skeletal electrons lead to a prediction of a *closo*-deltahedral structure (the D_{4d}-symmetric square antiprism). They then argued that the preference instead for the pentagonal prism (D_{5h}) was driven by its larger internal volume compared to the bicapped square antiprism, making it more able to accommodate an endohedral metal. An alternative picture of bonding emerged from Fässler's original paper, where the analogy to the structure of the electron-precise (albeit unstable) hydrocarbon prismane, $C_{10}H_{10}$, was noted. Following the logic developed in the $Cr@Si_{12}$ case, we can view the three-connected pentagonal prism as the signature of an electron-precise (50-electron) count, which would imply, at least formally, the presence of a $[Ge_{10}]^{10-}$ cage (isoelectronic with $C_{10}H_{10}$) [83, 84]. From the perspective of the metal, the $3d_{z^2}$ orbital is now doubly occupied ($4a_1'$ in Fig. 3, right), suggesting an 18-electron count at the metal. These two observations can be reconciled by noting again the dual role of the $2e_1''$ and $2e_2'$ orbitals: they carry **both** Co–Ge and Ge–Ge bonding character and so contribute to the stability of both the metal and the cage in a synergic fashion. Of the 52 available valence electrons, a subset of 50 can then be assigned to the cage (the 10 lone pairs and 15 Ge–Ge bonding pairs of a putative electron-precise pentagonal prism span $3a_1' + 2a_2'' + 3e_1' + 2e_1'' + 3e_2' + 2e_2''$). Sixteen of these 50 can also be assigned to the metal, and the two additional electrons in $4a_1'$ raise its count to 18. The plots of the molecular orbitals of $Cr@Si_{12}$ and $[Co@Ge_{10}]^{3-}$ in Fig. 3 emphasise the close links between the two systems: both have prismatic cluster geometries where four of the d orbitals are involved in bonding, leaving the $3d_{z^2}$ orbital approximately nonbonding – in $Cr@Si_{12}$ it is vacant, while in $[Co@Ge_{10}]^{3-}$ it is occupied. The other significant difference between the two lies in the occupation of the cage-based orbitals: in $[Co@Ge_{10}]^{3-}$ all 25 orbitals required to make up an electron-precise count at the cluster are occupied, while in $Cr@Si_{12}$, three of the 30 cluster orbitals remain vacant.

The chemistry of the germanium clusters had one further surprise in store, in the form of the crystal structure of the $[Ru@Ge_{12}]^{3-}$ anion published in 2014 [85]. This cluster adopts a remarkable D_{2d}-symmetric structure based on square and pentagonal faces (variously described as a bicapped pentagonal prism (BPP) [86, 87] or "four pentagonal faces" (FPF) [88, 89] in the literature; see Fig. 2). This structural motif had been identified as the most stable isomer of $Au@Ge_{12}$, $[Au@Ge_{12}]^{-}$ [86, 90], and $Ni@Ge_{12}$ [91] and also $Ni@Si_{12}$ [92, 93] in different computational studies but had not previously been realised crystallographically. A second example of this structural motif, $[V@Ge_8As_4]^{3-}$, has subsequently been published by Dehnen and co-workers [94]. We can immediately identify a relationship between the BPP structure of $[Ru@Ge_{12}]^{3-}$ and the prismatic structures of $[Co@Ge_{10}]^{3-}$ and $Cr@Si_{12}$ (Fig. 2) all have the three-connected vertices that are the structural signature of precise electron counts. In the case of $[Ru@Ge_{12}]^{3-}$, an electron-precise count at the cluster demands 60 electrons in total, whereas only 59 are available. The SOMO is however rigorously localised on the cage (a point that is validated by the measured EPR spectrum) [85], and there are now precisely five doubly occupied orbitals with both Ru 4d *and* Ge lone pair or Ge–Ge bonding character (see Fig. 4). Developing the argument along the same lines as for

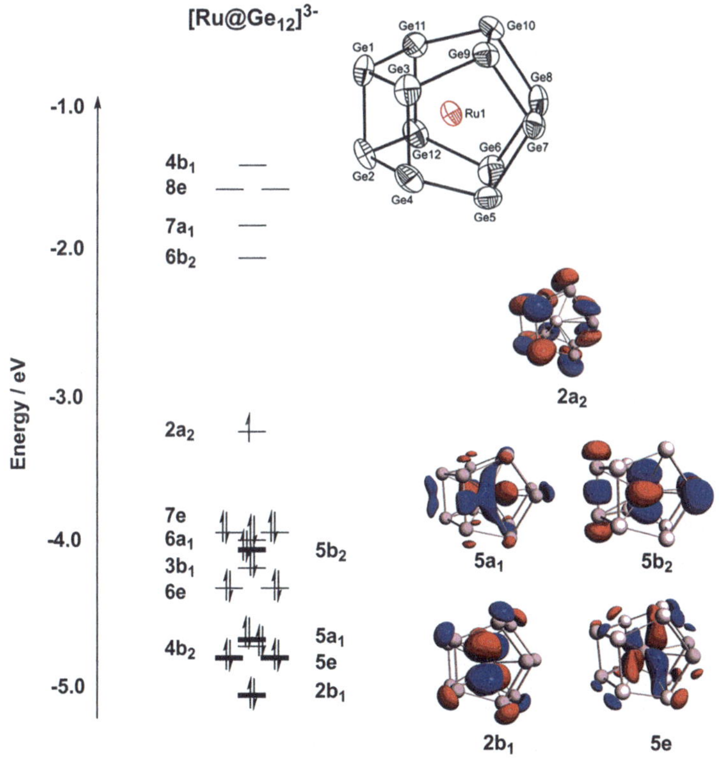

Fig. 4 Electronic structure of $[Ru@Ge_{12}]^{3-}$. Primarily metal-based orbitals are shown in *bold*

$[Co@Ge_{10}]^{3-}$, the 12 lone pairs and 18 Ge–Ge bonding orbitals transform as $6a_1 + 2a_2 + 3b_1 + 5b_2 + 7e$ in D_{2d} symmetry, precisely the ground-state configuration computed by DFT (notwithstanding the absence of one electron in the $2a_2$ SOMO). Thus, in this case all 59 electrons contribute to the electron-precise count at the cluster, while a subset of 18 of these contribute to the valence electron count at the metal. The electronic configuration is precisely that which simultaneously completes the valence shells of both the cluster and the encapsulated metal.

4 Endohedral Clusters of the Heavier Tetrels, M@Sn$_n$ and M@Pb$_n$

Like the germanium clusters described above, early progress in the chemistry of the endohedral tin and lead analogues can be traced back to the work of Gao in 2001 [95], where he also described the formation of anionic cobalt complexes of the form $[Co@Sn_n]^-$ and $[Co@Pb_n]^-$ via laser ablation of pressed tablets of Co/Sn/Pb powder mixtures. Prominent peaks in mass spectra are observed for both $[Co@Sn_{10}]^-$ and $[Co@Pb_{10}]^-$, along with one for the 12-vertex analogue $[Co@Pb_{12}]^-$. The authors proposed bicapped square antiprismatic and icosahedral structures for the 10- and 12-vertex clusters, respectively, noting the fact that the seven 3d electrons of cobalt could easily coordinate with lead atoms around it. A subsequent paper extended the series of $[M@Pb_n]^-$ clusters to include the entire first transition series from Ti to Cu, along with Pd and Ag from the second series, the results again suggesting high local stabilities for the 12-vertex clusters [96]. In 2004, Neukermans et al. reported the aluminium-centred species, $[Al@Pb_{10}]^+$ and $[Al@Pb_{12}]^+$ [97], and followed this in 2006 with a survey of the neutral and cationic tin and lead clusters of Cr, Mn, Cu, and Zn [72]. For tin, high abundances of $[M@Sn_n]^{0/+}$ are observed for $n = 13–16$ for the mid transition metals, Cr and Mn, although a particularly prominent peak for $[Mn@Sn_{12}]^+$ is also apparent. For the lead clusters, maximum abundances were observed for $M@Pb_{10}$ and $M@Pb_{12}$ for M = Cr, Mn. Gas-phase photoelectron spectroscopy by Wang and co-workers on the 12-vertex monoanionic family, $[M@Sn_{12}]^{1-}$ (M– Ti, V, Cr, Fe, Co, Ni, Cu, Nb, Pt, Au) [98], then confirmed that these clusters share a common icosahedral structure and are best formulated as $M^+@[Sn_{12}]^{2-}$. A subsequent comparison of the photoelectron spectroscopy of the ion pair, $K_2{}^+[Pt@Pb_{12}]^{2-}$, and the monoanion, $[PtPb_{12}]^{1-}$, also suggested an icosahedral geometry in both cases [99, 100].

The most conspicuous feature that distinguishes the cluster chemistry of Sn and Pb from that of the lighter elements is the availability of a large body of crystallographic data, primarily for highly anionic clusters that co-crystallise with large countercations. A survey of this data shows that deltahedral structures dominate rather than the three-connected architectures found for Si and Ge. Within the nine-vertex family, the series $[Co@Sn_9]^{5-}$, $[Ni@Sn_9]^{4-}$ [101], $[Cu@Sn_9]^{3-}$, and

$[Cu@Pb_9]^{3-}$ [102] have all been crystallographically characterised, and the structures vary between tricapped trigonal prismatic (D_{3h}) and monocapped square antiprismatic (C_{4v}) limits. The clusters are diamagnetic and can be straightforwardly formulated as $[E_9]^{4-}$ cages containing a closed-shell d^{10} metal ion (Co^{-1}, Ni^0, Cu^+). Given this formulation, it is interesting to contrast the structures with the empty $[E_9]^{4-}$ cages which adopt *nido*-monocapped square antiprismatic structures, consistent with Wade's rules for a 22-electron, 9-vertex polyhedron. The differences between the two structural types (tricapped trigonal prism and monocapped square antiprism) are, in fact, rather minor, but it seems that the presence of an endohedral metal favours the retention of a more symmetric, spherical geometry. The 10-vertex family is represented by $[Ni@Pb_{10}]^{2-}$ [103], a bicapped square antiprismatic structure (Fig. 2) characteristic of a *closo*-$[Pd_{10}]^{2-}$ cage, but it is the 12-vertex family, $M@E_{12}$, that has provided the richest and most diverse chemistry of the heavier elements. Eichhorn and co-workers have shown that the reaction of $[Pb_9]^{4-}$ with sources of low-valent group X metals Ni, Pd, and Pt generates the dianionic clusters $[M@Pb_{12}]^{2-}$ [99, 100]. Similar synthetic strategies using group IX metals have also recently yielded the isoelectronic and isostructural Rh analogue, $[Rh@Pb_{12}]^{3-}$ [104], and the iridium complex, $[Ir@Sn_{12}]^{3-}$ [105]. All of these species have 60 valence electron counts and almost perfect icosahedral symmetry (Fig. 2), with the M–E and E–E bond lengths dispersed over a narrow range around the mean. Of the 60 available valence electrons, 50 can be assigned to the cage (*closo*-$[E_{12}]^{2-}$) and the other 10 to the metal, completing the 18-electron count without any need for synergy. It is significant in this context that the empty $[Sn_{12}]^{2-}$ and $[Pb_{12}]^{2-}$ cages are in fact stable entities in their own right [106–108], and several authors have discussed the extent to which they, like $[B_{12}H_{12}]^{2-}$, can be considered to display three-dimensional aromaticity [109, 110]. This relative straightforward electron counting in $[M@Pb_{12}]^{2-}$ stands in stark contrast to $Cr@Si_{12}$, $[Fe/Co@Ge_{10}]^{3-}$, and $[Ru@Ge_{12}]^{3-}$, where four or more pairs of electrons must contribute simultaneously to the electron count of both metal and cluster in order to satisfy the valence electron demands of both. One recent structurally characterised example, a 58-electron species $[Mn@Pb_{12}]^{3-}$, however, hints at a bridge between the two classes [111]. This cluster is identifiably icosahedral, albeit very distinctly distorted into a D_{2h}-symmetric prolate geometry (see Fig. 2), and has a triplet ground state. In the context of the discussion in the introduction, it is therefore a structurally characterised example of a "magnetic superatom." The Mn–Pb distances vary over a much wider range (2.869(3)–3.308(4) Å) than seen in any of the other structurally characterised pseudo-icosahedral species. DFT calculations suggest that, just as in the $Cr@Si_{12}$, $[Fe/Co@Ge_{10}]^{3-}$, and $[Ru@Ge_{12}]^{3-}$ series, charge transfer from the endohedral Mn ion to the cage plays a central role in controlling stability. The key difference in the case of $[Mn@Pb_{12}]^{3-}$ is that the charge transfer is spin polarised: only minority spin density is delocalised onto the Pb_{12} cage (spin density $= -1.21$), with concomitant accumulation of spin-α density on the endohedral Mn ion (+3.21). The driving force is clearly the stability of the half-filled shell (d^5, Mn^{2+}), and the prolate distortion is precisely that required to stabilise a highly reduced $[Pb_{12}]^{5-}$ cage (although the degree of charge transfer clearly falls some way short of this highly polarised limit). The spatial distinction between spin-α and spin-β densities in the DFT solution is highly reminiscent of the open-shell $^1A_{2u}$ state of $[Mn@Si_{12}]^+$, and indeed the half-filled shell has emerged as a dominant feature in computational studies on M@Sn/Pb_{12} clusters [55, 112] and also from molecular beam experiments on Mn@Sn_{12} reported by Schäfer and co-workers [113].

5 Summary

In this review we have surveyed the development of the chemistry of endohedral clusters of silicon, germanium, tin, and lead over the past 25 years, with a particular emphasis on their structural characteristics. The chemistry has evolved from the early mass spectrometric studies which first hinted at the existence of these remarkable species, through spectroscopic work and ultimately to the characterisation of several examples in the solid state. The early experiments in the late 1980s coincided with the emergence of density functional theory as a viable tool in computational chemistry, and so from the outset, experiment has been accompanied by theory which has sought to predict structures and stabilities and to rationalise the wealth of data. At first glance the clusters of the light and heavy tetrels appear to be quite distinct, the former generally being characterised by open three-connected geometries, while the latter tend to adopt deltahedral structures. The three-connected clusters, exemplified by hexagonal prismatic Cr@Si_{12}, pentagonal prismatic $[Fe@Ge_{10}]^{3-}$, and D_{2d}-symmetric $[Ru@Ge_{12}]^{3-}$, present a paradox: their geometries are characteristic of high valence electron counts on the cage ($5n$), but this apparently leaves insufficient electrons to satisfy the valence requirements of the metal centres. We can reconcile this conflict by noting that some or all of the metal d electrons in fact play a dual role: they are simultaneously M–E and E–E bonding and therefore contribute to the global stability of both metal and cage. The heavier tin and lead clusters typically show less evidence for covalent mixing of the orbitals on the metal and the cage: the available valence electron count can usually be cleanly divided into subsets on the metal an on the cage. Thus, for example, the 52-electron count of $[Ni@Pb_{12}]^{2-}$ is simply formulated as a d^{10} Ni^0 atom inside a *closo*-($4n+2=42$-electron) cage. The distinction between these deltahedral and three-connected is not, however, always clear cut, as illustrated by the $[Mn@Pb_{12}]^{3-}$ ion where spin-polarised electron transfer from the metal drives a substantial distortion from perfect icosahedral symmetry but is not sufficiently strong to push the system into the D_{2d}-symmetric geometry of $[Ru@Ge_{12}]^{3-}$. The use of metals from the early and middle regions of the transition series therefore opens up the exciting possibility of isolating further examples of magnetic "superatoms," where the stability of the half-filled, rather than filled, d shell provides the driving force for stability.

Acknowledgments We acknowledge the financial support of the EPSRC (EP/K021435/1).

References

1. Lewis G (1916) J Am Chem Soc 38:762
2. Knight WD, Clemenger K, de Heer WA, Saunders WA, Chou MY, Cohen ML (1984) Phys Rev Lett 52:2141
3. Leuchtner RE, Harms AC, Castleman AW Jr (1989) J Chem Phys 91:2753
4. Reveles JU, Clayborne PA, Reber AC, Khanna SN, Pradhan K, Sen P, Pederson MR (2009) Nat Chem 1:310
5. Castleman AW Jr, Khanna SN (2009) J Phys Chem C 113:2664
6. Castleman AW Jr (2011) J Phys Chem Lett 2:1062
7. Medel VM, Reveles JU, Reber AC, Khanna SN, Castleman AW Jr (2011) Phys Rev B 84:075435
8. Luo Z, Castleman AW Jr (2014) Acc Chem Res 47:2931
9. Janssens E, Neukermans S, Lievens P (2004) Curr Opin Solid State Mater Sci 8:185
10. Khanna SN, Jena P (1995) Phys Rev B 51:13705
11. Jena P (2013) J Phys Chem Lett 4:1432
12. Reber AC, Khanna SN, Castleman AW Jr (2007) J Am Chem Soc 129:10189
13. Bergeron DE, Castleman AW Jr, Morisato T, Khanna SN (2004) Science 304:84
14. Reveles JU, Khanna SN, Roach PJ, Castleman AW Jr (2006) Proc Natl Acad Sci U S A 103:18405
15. Wade K (1971) J Chem Soc D0 Chem Commun 792
16. Wade K (1976) Adv Inorg Chem Radiochem 18:1
17. Mingos DMP (1972) Nature Phys Sci 236:99
18. Cong H, Yu B, Akasaka T, Lu X (2013) Coord Chem Rev 257:2880
19. Lu X, Bao L, Akasaka T, Nagase S (2014) Chem Commun 50:14701
20. Garcia-Borràs M, Osuna S, Luis JM, Swart M, Solà M (2014) Chem Soc Rev 43:5089
21. Lu X, Feng L, Akasaka T, Nagase S (2012) Chem Soc Rev 41:7723
22. Rodriguez-Fortea A, Balch AL, Poblet JM (2011) Chem Soc Rev 40:3551
23. Zhao J, Huang X, Jin P, Chen Z (2015) Coord Chem Rev 289–290:315
24. Zhou B, Denning MS, Kays DL, Goicoechea JM (2009) J Am Chem Soc 131:2802
25. Krämer T, Duckworth JCA, Ingram MD, Zhou B, McGrady JE, Gorcoechea JM (2013) Dalton Trans 42:12120
26. Goicoechea JM, McGrady JE (2015) Dalton Trans 44:6755
27. McGrady JE (2004) J Chem Educ 81:733
28. Bernardini F, Picozzi S, Continenza A (2004) Appl Phys Lett 84:2289
29. Dalpian GM, da Silva AJR, Fazzio A (2003) Phys Rev B 68:113310
30. da Silva AJR, Fazzio A, Antonelli A (2004) Phys Rev B 70:193205
31. Zhang ZZ, Partoens B, Chang K, Peeters FM (2008) Phys Rev B 77:155201
32. Graff K (2000) Metal impurities in silicon device fabrication. Springer, Berlin
33. Beck SM (1987) J Chem Phys 87:4233
34. Beck SM (1989) J Chem Phys 90:6306
35. Hiura H, Miyazaki T, Kanayama T (2001) Phys Rev Lett 86:1733
36. Robles R, Khanna SN (2009) Phys Rev B 80:115414
37. Uchida N, Bolotov L, Miyazaki T, Kanayama T (2003) J Phys D Appl Phys 36:L43
38. Sun Z, Oyanagi H, Uchida N, Miyazaki T, Kanayama T (2009) J Phys D Appl Phys 42:015412
39. Uchida N, Kintou H, Matsushita Y, Tada T, Kanayama T (2008) Appl Phys Exp 1:121502
40. Khanna SN, Rao BK, Jena P (2002) Phys Rev Lett 89:016803
41. Zheng W, Nilles JM, Radisic D, Bowen KH Jr (2005) J Chem Phys 122:071101
42. Ohara M, Koyasu K, Nakajima A, Kaya K (2003) Chem Phys Lett 371:490
43. Koyasu K, Akutsu M, Mitsui M, Nakajima A (2005) J Am Chem Soc 127:4998
44. Janssens E, Gruene P, Meijer G, Wöste L, Lievens P, Fielicke A (2007) Phys Rev Lett 99:063401

45. Ngan VT, Janssens E, Claes P, Lyon JT, Fielicke A, Nguyen MT, Lievens P (2012) Chem Eur J 18:15788
46. Claes P, Janssens E, Ngan VT, Gruene P, Lyon JT, Harding DJ, Fielicke A, Nguyen MT, Lievens P (2011) Phys Rev Lett 107:173401
47. Zamudio-Bayer V, Leppert L, Hirsch K, Langenberg A, Rittmann J, Kossick M, Vogel M, Richter R, Terasaki A, Möller T, Issendorf BV, Kűmmel S, Lau JT (2013) Phys Rev B 88:115425
48. Xiao C, Hagelberg F, Lester WA Jr (2002) Phys Rev B 66:075425
49. Hagelberg F, Xiao C, Lester WA Jr (2003) Phys Rev B 67:035426
50. Han JG, Xiao X, Hagelberg F (2002) Struct Chem 13:173
51. Han JG, Hagelberg F (2009) J Comput Theor Nanosci 6:257
52. Kumar V (2003) Eur Phys J D 24:227
53. Kawamura H, Kumar V, Kawazoe Y (2005) Phys Rev B 71:075423
54. Singh AK, Briere TM, Kumar V, Kawazoe Y (2003) Phys Rev Lett 91:146802
55. Kumar V, Kawazoe Y (2003) Appl Phys Lett 83:2677
56. Reveles JU, Khanna SN (2006) Phys Rev B 74:035435
57. Reveles JU, Khanna SN (2005) Phys Rev B 72:165413
58. Abreu MB, Reber AC, Khanna SN (2014) J Phys Chem Lett 5:3492
59. Lu J, Nagase S (2003) Phys Rev Lett 90:115506
60. Lu J, Nagase S (2003) Chem Phys Lett 372:394
61. Chen Z, Neukermans S, Wang X, Janssens W, Zhou Z, Silverans RE, King RB, von Ragué Schleyer P, Lievens P (2006) J Am Chem Soc 128:12829
62. Andriotis AN, Mpourmpakis G, Froudakis GE, Menon M (2002) New J Phys 4:78
63. Mpourmpakis G, Froudakis GE, Andriotis AN, Menon M (2003) J Chem Phys 119:7498
64. Pacheco JM, Gueorguiey GK, Martins JL (2002) Phys Rev B 66:033401
65. Oliveira MIA, Rivelino R, de Brito Mota F, Gueorguiev GK (2014) J Phys Chem C 118:5501
66. He J, Wu K, Liu C, Sa R (2009) Chem Phys Lett 483:30
67. Guo LJ, Zhao GF, Gu YZ, Liu X, Zeng Z (2008) Phys Rev B 77:195417
68. Zhang X, Li GL, Gao Z (2001) Rapid Commun Mass Spectrom 15:1573
69. Furuse S, Koyasu K, Atobe J, Nakajima A (2008) J Chem Phys 129:064311
70. Atobe J, Koyasu K, Furuse S, Nakajima A (2012) Phys Chem Chem Phys 14:9403
71. Kumar V, Kawazoe Y (2001) Phys Rev Lett 87:045503
72. Neukermans S, Wang X, Veldeman N, Janssens E, Silverans RE, Lievens P (2006) Int J Mass Spectrom 252:145
73. Tang C, Liu M, Zhu W, Deng K (2011) Comput Theor Chem 969:56
74. Wang J, Ma L, Zhao J, Wang G (2008) J Phys Condens Matter 20:335223
75. Kapila N, Jindhal VK, Sharma H (2011) Physica B 406:4612
76. Zhao WJ, Wang YX (2009) J Mol Struct Theochem 901:18
77. Wang J, Han JG (2006) J Phys Chem A 110:12670
78. Kumar V, Kawazoe Y (2002) Appl Phys Lett 80:859
79. Li GL, Zhang X, Tang ZC, Gao Z (2002) Chem Phys Lett 359:203
80. King RB (2010) Struct Bond 140:1
81. Uta MM, Cioloboc D, King RB (2012) Inorg Chem 51:3498
82. Deng XJ, Kong XY, Xu XL, Xu HG, Zheng WJ (2014) ChemPhysChem 15:3987
83. Wang JQ, Stegmaier S, Fässler TF (2009) Angew Chem Int Ed 48:1998
84. Scharfe S, Kraus F, Stegmaier S, Schier A, Fässler TF (2011) Angew Chem Int Ed 50:3630
85. Espinoza-Quintero G, Duckworth JCA, Myers WK, McGrady JE, Goicoechea JM (2014) J Am Chem Soc 136:1210
86. Li XJ, Su KH (2009) Theor Chem Acc 124:345
87. Li XJ, Ren HJ, Yang LM (2012) J Nanomat. http://www.hindawi.com/journals/jnm/2012/518593/. Article ID 518593
88. Uchida N, Miyazaki T, Kanayama T (2006) Phys Rev B 74:205427
89. Miyazaki T, Hiura H, Kanayama T (2003) Eur Phys J D 24:241

90. Li XJ, Su KH, Yang XH, Song LM, Yang LM (2013) Comput Theor Chem 1010:32
91. Bandyopadhyay D, Sen P (2010) J Phys Chem A 114:1835
92. Zdetsis AD, Koukaras EN, Garoufalis CS (2009) J Math Chem 46:971
93. Koukouras EN, Garoufalis CS, Zdetsis AD (2006) Phys Rev B 73:235417
94. Mitzinger S, Broeckaert L, Mass W, Weigend F, Dehnen S (2015) Chem Commun 51:3866
95. Zhang X, Li GL, Xing XP, Zhao X, Tang ZC, Gao Z (2001) Rapid Commun Mass Spectrom 15:2399
96. Zhang X, Tang Z, Gao Z (2003) Rapid Commun Mass Spectrom 17:621
97. Neukermans S, Janssens E, Chen ZF, Silverans RE, von Ragué Schleyer P, Lievens P (2004) Phys Rev Lett 92:163401
98. Cui LF, Huang X, Wang LM, Li J, Wang LS (2007) Angew Chem Int Ed 46:742
99. Grubisic A, Wang HP, Li X, Ko YJ, Kocak FS, Pederson MR, Bowen KH, Eichhorn BW (2011) Proc Natl Acad Sci U S A 108:14757
100. Esentur EN, Fettinger J, Eichhorn BW (2006) J Am Chem Soc 128:9178
101. Hlukhyy V, Stegmaier S, van Wullen L, Fässler TF (2014) Chem Eur J 20:12157
102. Scharfe S, Fässler TF, Stegmaier S, Hoffmann SD, Ruhland K (2008) Chem Eur J 14:4479
103. Esenturk EN, Fettinger J, Eichhorn BW (2005) Chem Commun 247
104. Wang Y, Wang LL, Ruan HP, Luo BL, Sang RL, Xu L (2015) Chin J Struct Chem 34:1253
105. Wang JQ, Stegmaier S, Wahl B, Fässler TF (2010) Chem Eur J 16:1793
106. Cui LF, Huang X, Wang LM, Li J, Wang LS (2006) J Am Chem Soc 128:8390
107. Cui LF, Huang X, Wang LM, Li J, Wang LS (2006) J Phys Chem A 110:10169
108. Wang LS, Cui LF (2008) Int Rev Phys Chem 27:139
109. Bai Tai T, Minh Tam N, Nguyen MT (2011) Chem Phys 388:1
110. Chen ZF, King RB (2005) Chem Rev 105:3613
111. Zhou B, Krämer T, Thompson AL, McGrady JE, Goicoechea JM (2011) Inorg Chem 50:8028
112. Kandalam AK, Chen G, Jena P (2008) Appl Phys Lett 92:143109
113. Rohrmann U, Schäfer S, Schäfer R (2009) J Phys Chem A 113:12115

Struct Bond (2016) 169: 199–248
DOI: 10.1007/430_2015_208

Published online: 10 March 2016

The Rich Structural Chemistry Displayed by the Carbon Monoxide as a Ligand to Metal Complexes

Shengda Ding and Michael B. Hall

Abstract The diatomic CO molecule is a very important ligand in organometallic chemistry. The bond between the carbonyl and a metal is moderately strong and consists of a sigma bond, formed by donation of electron density to the metal from the carbonyl's highest occupied molecular orbital (HOMO, the 5σ), and π bonds, formed by donation of electron density from the metal to the carbonyl's lowest unoccupied molecular orbital (LUMO, the 2π). The carbonyl may also serve as a bridging ligand connecting two or more metal atoms. Depending on the relative orientation between the carbonyl and metals, one may classify a bridging carbonyl as symmetric bridging, bent semibridging, linear semibridging, face bridging, and bridging isocarbonyls. The rich structural chemistry displayed arises from a complex interplay between the metal's electronic structure and the carbonyl's 5σ and 2π. In addition, the carbonyl's occupied 1π and 4σ orbitals may in certain cases donate electrons when it binds to electron-deficient metals, further complicating the electronic structure. Such complexity in the carbonyl–metal interaction raises challenges to the simple applications of Lewis bonding ideas and electron counting rules. Therefore, theoretical analyses have been applied, largely in a case-by-case pattern, to investigate the rationales behind the CO's rich structural chemistry.

Keywords Asymmetric · Backbonding · Bent · Bond order · Bridging carbonyl · Carbon monoxide · Computation · Isocarbonyl · Linear · Metal–metal bond · Metal–metal antibond · Molecular orbital diagram · Symmetric · Theoretical

---

S. Ding and M.B. Hall (✉)
Department of Chemistry, Texas A&M University, College Station, TX 77843, USA
e-mail: Hall@science.tamu.edu

Contents

1 Introduction

Lewis bonding theory, reviewed in the opening chapter, has been a good model to describe the interactions between atoms in a molecule before the dawn of quantum chemistry and molecular orbital theory. Bonds, made by atoms sharing electrons in pairs, are formed until all the atoms in the molecule possess the number of electrons that correspond to the next noble gas atom in the periodic table. This is the so-called effective atomic number (EAN) rule: the octet rule (for main group elements) and the 18-electron rule (for *d*-block transition metals). In the dative or coordinate covalent bond that was invoked to explain metal–ligand bond in Werner's transition metal complexes and Lewis acid–base pairs, the shared electrons originate from one of the atom in the ligand. It was not until much later that the concept of synergic bonding [1] that included the idea of backbonding with electrons on the metal shared with the ligand in the direction opposite to the usual dative bond developed and applied to the transition metal to carbonyl bonding (Chap. 1, Sect. 6.3). The rich structural chemistry displayed by polynuclear carbonyl complexes, especially those have bridging carbonyls, requires the application of molecular orbital theory to understand the details of the electron delocalization and backbonding interactions, before plausible conclusions can be drawn about how the system should be described in Lewis pair theory. In this review, an expansion to Lewis bonding theory concerning system containing bridging carbonyls and its derivatives is introduced, largely based on the contemporary theoretical and computational work.

1.1 Carbonyl as a Ligand

The carbonyl is classified as a strong ligand in the spectrochemical series with the capacity to split metal's *d*-orbitals to a large extent. The interaction between carbonyl and its bound metal is explained by the Dewar–Chatt–Duncanson model as presented in Fig. 1 [2, 3]. Due to C–O bond formation, the HOMO of CO (5σ), which serves as the σ-donor orbital to the metal, is primarily localized on the carbon as a mixture of C 2*s* and 2*p* orbitals such that its lobe points away from the oxygen. The LUMOs of CO are the degenerate 2π or π* constituted by the antibonding combinations of 2*p* orbitals of C and O with more C than O character. These two orbitals can withdraw electron density from occupied metal *d*-orbitals of appropriate symmetry through so-called π-backbonding or equivalently as π-acceptors. The large *d*-orbital splitting is created by stabilizing the *d*-π orbitals by π-backbonding while destabilizing the *d*-σ orbitals by σ-donation. Due to this twofold interaction between the metal (M) and CO, the M–C bond is usually strong and stable.

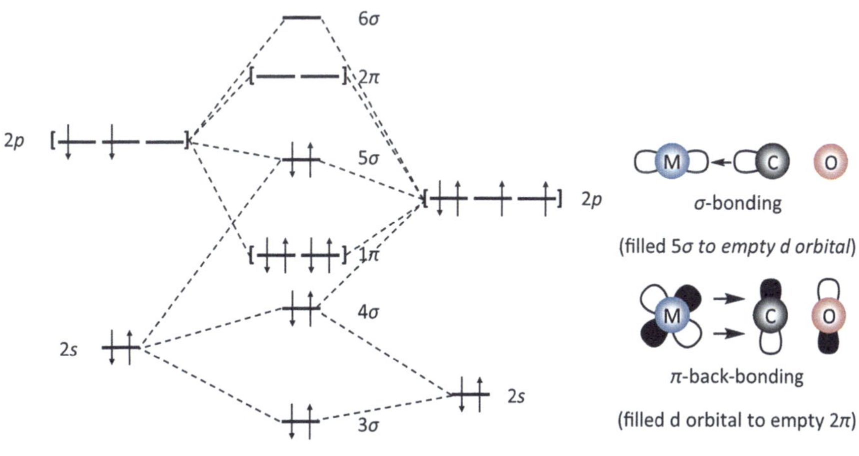

Fig. 1 The molecular orbital diagram of CO is on the *left* and the bonding between CO and metal (M) in the Dewar–Chatt–Duncanson model is on the *right*. The orbital degeneracy of the atomic orbitals is shown in *brackets*

1.2 Bridging Carbonyl and Its Classification

Carbonyl has a tendency to bridge metals in polynuclear complexes, especially for the complexes containing first-row transition metals. The general rule is the smaller the metals are, the more likely a bridging carbonyl can be found. The common scheme of classification for bridging carbonyls depends on the relative position between the carbonyl and the metals [4, 5]: terminal (I), bent semibridging (II), (symmetric) bridging (III), and linear semibridge (IV). The carbonyl may also bind to two metals through its carbon and oxygen separately (V) or bind to three or more metals (VI) at the same time. Types I–III carbonyls are typical scenarios reflecting the different extents that a CO is shared by two metals. The boundaries between types I and III carbonyls are not clearly defined; Crabtree proposed the use of the angle φ (defined as M_2–M_1–C, Fig. 2, where M_1 is defined to be the metal which is closer to C than M_2) to distinguish those carbonyls: terminal (I) has $\varphi > 75°$, bent semibridging (II) $50° > \varphi > 75°$, and (symmetric) bridging (III) $50° > \varphi$, while linear semibridging carbonyl (IV) has a wider range of φ between 40 and 75°. As a terminal carbonyl on M_1 gradually approaches M_2, the angle θ (defined by M_1–C–O, Fig. 2) gradually bends in the bent semibridging (II), but remains linear in the linear semibridging (IV). In the bent semibridging (II), the C–O vector is close to being perpendicular to the M_1–M_2 vector, i.e., $(\theta - \delta) \approx 90°$, while in the linear semibridging (IV), the C–O vector remains essentially linear with M_1, $\theta \approx 180°$ (>165° or 170°). Types V (bridging isocarbonyls) and VI (face bridging carbonyls) are substantially different from others in structure and usually little ambiguity arises. Rare cases of other bridging modes of carbonyls have been described in the previous publications [6, 7], but they receive very little theoretical coverage and will not be described in following sections.

Many reviews analyzed the structures [4, 5, 8–10], spectra [6, 11], and electron density topologies [12] of bridging carbonyls. However, no systematic review about theoretical and molecular orbital analysis of the bridging carbonyls is currently available. A selection of about ten systems covering different types of bridging carbonyls is presented here with full analysis of their molecular orbitals to help connect the theoretical descriptions of chemical bonds and electronic structures to the observable molecular structures.

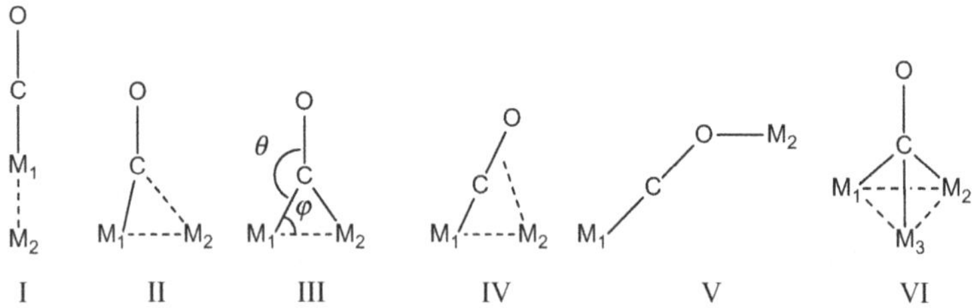

Fig. 2 The structures of different types of carbonyls. The lines connecting atoms do not necessarily indicate bonds between atoms

1.3 LXZ Ligand System and Notations

The ligand–metal σ bonds are commonly described as dative, or coordinate covalent, bonds, where the ligand donates two electrons from a lone pair to the metal. The 18-electron (18-*e*) rule, based on summing the electrons provided by dative bonds and the electrons remaining on the metal, is useful in predicting the stability of complexes and the direction of chemical reactions. It is extremely successful with mid and late transition metals bound with multiple ligands, especially π-acid ligands including the carbonyl. However, the two-electron dative bond description in the Lewis bonding structure may only cover a subset of the numerous types of ligand–metal interactions. It needs clarification and extension to include other types of interactions that occur in carbonyl complexes.

The first thing to do is to distinguish the dative bonds and common covalent bonds, the shared electron pair bond of Lewis. Upon the rupture of a covalent (Lewis) two-center, two-electron bond between the metal and the coordinating atom, the previously shared two electrons may either go to one atom (heterolytic cleavage) or be split between two atoms (homolytic cleavage). Usually, the bond that undergoes heterolytic cleavage is assigned as a dative bond, while that undergoing homolytic cleavage is assigned as a covalent bond [13, 14]. A ligand that forms a covalent bond with the metal donates one "net" electron to the count of 18 electrons and is recognized as an X ligand (one-electron ligand). In contrast, for the ligand–metal bond that undergoes heterolytic metal–ligand bond cleavage, the two electrons may follow either the ligand or the metal, indicating zero or two "net" electrons are contributed from the ligand for the 18-*e* count; therefore, such a ligand is classified as Z (zero-electron ligand) or L (two-electron ligand), respectively [15]. Conventionally, the dative bond is drawn as an arrowed line and the covalent bond is drawn as a plain line. In some cases, especially in mechanistic schemes, donations of one electron will be drawn as a half-arrow (Fig. 3).

CO is usually recognized simply as an L ligand, providing two electrons to the bound metal, though it also removes electron density from metal via its π-backbonding. However, if one considers each 2π orbital as a two-electron acceptor, CO can be rewritten as LZ even LZZ ligand depending on the strength of backbonding (Fig. 4). The representation of the carbonyl as an XX (equals to LZ) ligand, with two covalent bonds connecting the metal and the carbon, was actually used by Pauling in the 1940s and was summarized in his 1960 book to illustrate the electronic structure of $Ni(CO)_4$ [1], i.e., the ideas of synergic binding models, which later became the Dewar–Chatt–Duncanson model [2, 3]. However, extreme backbonding, represented in LZZ ligand, is not likely on monometallic complexes

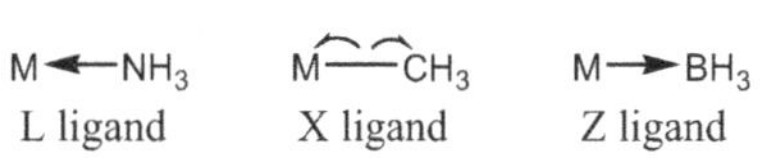

Fig. 3 Illustrative drawing of L/X/Z ligands and their interactions with the metal. The half-arrows for the "X ligand" show the electron movement for the homolytic bond cleavage

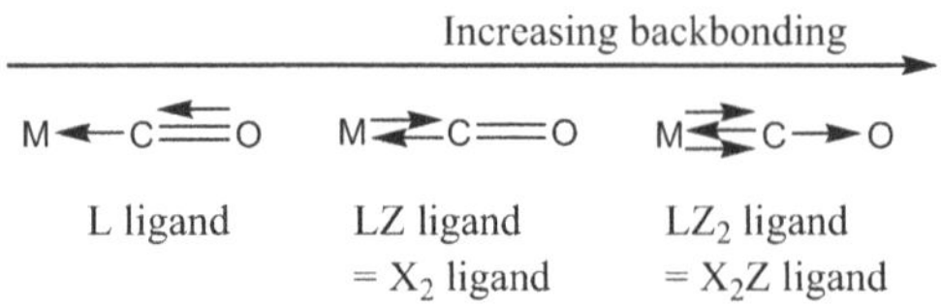

Fig. 4 Examples of backbonding and their relationship to L/X/Z ligand types. It is assumed dative bonds do not trigger formal electron transfers. The bonds were drawn to avoid formal charges, as the electrons available for a generalized metal (M) are unknown, and to satisfy the octet rule for C and O

but may occur to some degree in electron-rich complexes featuring bridging CO, where the chances are better that the 2π orbitals can be filled with electrons donated from multiple metals and serve as Z ligands.

1.4 Three-Centered Interaction Between Ligand and Metal

The situation is more complex for ligands interacting with multiple metal cores, as three-centered bonds are not completely compatible with simple Lewis structures. Here are two typical scenarios:

1. In an electron-deficient molecule, more than two atoms share a Lewis pair. For example, a hydrogen atom can bind to two metals as a bridging ligand, but the hydrogen only has one electron and one orbital (1*s*) to participate in bonding. Molecular orbital theory solves the issue by having a three-center, two-electron bond, such as the "banana-like" orbital often used to represent the three-center bridging BHB bond in B_2H_6. Similar issues exist for bridging H in transition metal systems, where in the LXZ system, the hydrogen is recognized as an X ligand for M_1 in forming an H–M_1 bond and then the H–M_1 bond is an L ligand when it donates two electrons to M_2 (Fig. 5a). In Lewis terminology, the bridging H and M_1 share two electrons (one from each) and these two electrons then make a coordinate covalent (dative) bond to M_2. Thus, the bridging hydride in a three-center, two-electron bond is an X(L) ligand. Of course, there is a contribution from the other resonance structure, in which M_1 and M_2 are reversed, an equal contribution if the metals are identical. Furthermore, additional resonance structures may contribute and the lines and arrows in Fig. 5a do not indicate a particular degree of electron flow. A similar situation applies to bridging carbonyls. Assuming that the CO π bonding is negligible and leaves the carbonyl with only two electrons on its 5σ orbitals for σ-donation to both metals. Here, the explanation is similar, the carbonyl establishes a dative bond with M_1 using its 5σ-electron pair, and this M_1–C dative bond further donates its electrons to M_2 (Fig. 5b). Such a carbonyl is counted as an L(L) ligand, and of course, related resonance structures would be involved to satisfy the molecule structure.

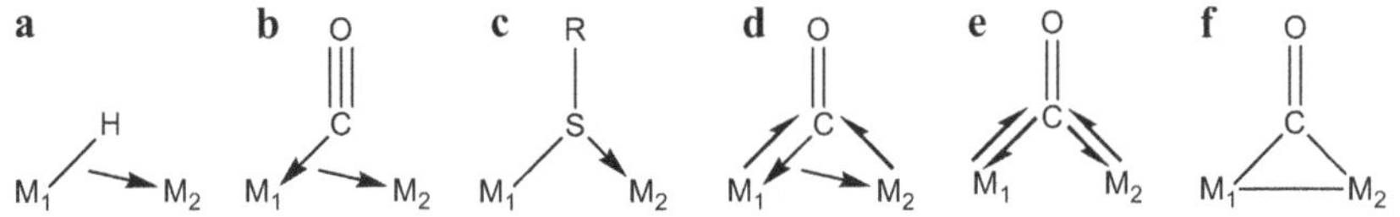

Fig. 5 Case studies of (**a**) bridging hydride, (**b**) bridging carbonyl with negligible π-backbonding, (**c**) bridging thiolate, (**d**) bridging carbonyl with strong π-backbonding, (**e**) bridging carbonyl with its σ-donation averaged over two metals, and (**f**) an alternative Lewis structural representation of a bridging CO with the addition of a metal–metal bond, an addition that is often invoked to satisfy the 18-*e* rule

2. Ligands with multiple lone pairs or orbitals can have more than one Lewis bond with one or more metals. Sometimes, the central metal(s) and the coordinating ligand may even establish interactions involving multiple Lewis bonds. Take thiolate as an example. It is usually recognized as an X ligand when it is bound to a single metal atom M_1; when a second metal M_2 coordinates to the same sulfur, the sulfur still has unused lone pairs available and donates one of them to establish a dative bond with M_2 (Fig. 5c). In such scenario, the thiolate may be classified as an XL ligand. The difference between XL ligand and previously mentioned X(L) ligand is the source of electrons. If the metal bound to the sulfur is electron deficient, then the sulfur may donate its *p* electrons to the central metal through a dative π bond, increasing the multiplicity of the M–S bond.

The bonding between bridging carbonyls and metals is actually more atypical. Although carbonyls have occupied 1π orbitals that could be candidates for further ligand donations, they are rarely used because they are not lone pairs and thus not as freely available. Conventionally, bridging carbonyls strengthen their bonds to multiple metals by backbonding to them in addition to the σ-donation already discussed. For the terminal carbonyl backbonding to a single metal, the two 2π orbitals from the CO are competing for the electrons from the same metal and have weaker π-overlap with metal *d*-orbitals. Thus, the total electron density transferred is modest. The situation is different for a bridging carbonyl as the 2π orbital in the M_1–C–M_2 plane now has σ directionality to bridged metals, hence higher overlap with the metal orbitals, and is accepting the electron density from two metals. In certain cases, so much electron density is transferred that the backbonding effect on the carbonyl is equivalent to a Z ligand; such backbonding in addition to the σ-donation, which was previously classified as L(L) ligand (Fig. 5b), results in that the bridging carbonyl ultimately is classified as L(L)Z ligand. Overall, the carbonyl donates two electrons and accepts one *d* electron from each metal (Fig. 5d). It seems tempting to average out the interaction utilizing the general rule that LZ ligand=XX ligand and to draw another Lewis structure (Fig. 5e), with a so-called "ketonic" carbonyl. This averaging process masks the importance of the three-center, two-electron (3c–2*e*) structure in Fig. 5b and sometimes causes defects in 18-*e* counts, for which metal–metal bond(s) must be added in turn to correct the defects. In fact, the argument that bridging carbonyls are "ketonic," along with 18-*e* rule, is often used to rationalize the metal–metal bond(s). However, such metal–metal bonds prove to be a myth in most cases. Examples are provided in the following sections to give detailed case-by-case analysis of the complex interaction between bridging carbonyl(s) and metals.

2 Symmetric Bridging Carbonyls

The symmetric bridging carbonyl, as its name suggests, is equally shared by the two metals to which it is bound. Important examples include well-known metal carbonyl complexes like $Fe_2(CO)_9$ and $Co_2(CO)_8$. Although these complexes were partly covered in a previous Structure and Bonding article [8], the theoretical and computational tools available then were too primitive to give proper descriptions of bridging carbonyls' effects on these systems. The results from modern theoretical tools are summarized here and in general provide a consensus view of their electronic structure.

2.1 *Bridging Carbonyl I: $Fe_2(CO)_9$*

2.1.1 A Historic Review and the Dilemma of the 18-Electron Rule

Soon after the synthesis of mononuclear transition metal carbonyls, polynuclear metal carbonyls featuring bridging carbonyls, such as $Fe_2(CO)_9$ and $Co_2(CO)_8$, were obtained and cyclopentadienyl anion containing complexes like $[(Cp)Fe(CO)_2]_2$ followed. The first example, $Fe_2(CO)_9$, features three identical bridging carbonyls and six terminal carbonyls in a high-symmetry (D_{3h}) structure that facilitated theoretical treatment and was extensively studied [16–22], though its X-ray structure was obtained quite late [4, 23]. These studies examined two fundamental questions: (1) what is the electronic nature of Fe–C(O)–Fe three-center bridge and (2) what role does the metal–metal interaction play in the complex, as the short distance between two irons and the 18-*e* rule imply a metal–metal bond?

At the first glance, there is strong temptation to assign the bridging carbonyls in the $Fe_2(CO)_9$ as X_2 ligands as the sharp Fe–C–Fe angle (77.6°) [4] reminds one of the ketonic carbonyl in organic compounds. In such a scenario, each bridging carbonyl forms a covalent bond with both irons which then have a formal oxidation number of +3. The coordinating environment of each iron is classified as FeL_3X_3 with a total electron count of 17, where the 18-*e* rule naturally pairs the odd electron on each iron into a metal–metal bond. This is the classical "textbook" description of the electronic structure of the $Fe_2(CO)_9$ (Fig. 6a). In our early understanding of these bridging carbonyls, it was concluded that the metal–metal bond is a prerequisite for symmetric bridging carbonyls; two metal atoms that are not directly

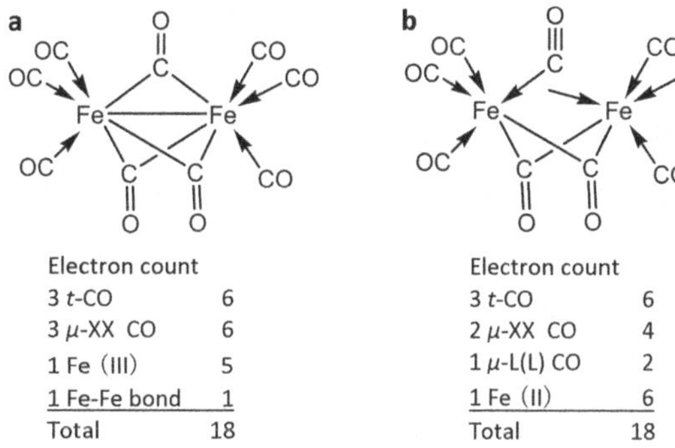

Fig. 6 The 18-*e* dilemma for $Fe_2(CO)_9$ and historic views of its electronic structures. The electronic structures featuring (**a**) all ketonic carbonyls and (**b**) a 3*c*–2*e* deficient Fe–CO–Fe bridging bond. Only one of six resonance structures is shown in **b**

bonded but bridged by a carbonyl would be too far apart and would have an excessively large M_1–C–M_2 bond angle [24].

There have long been opinions questioning the existence of the iron–iron bond in the $Fe_2(CO)_9$. Rather than an attractive interaction endorsed by a supposed iron–iron bond, Hoffmann et al. proposed that the interaction between two irons is repulsive with antibonding character while the short distance between two irons is the result of the three bridging carbonyls holding two irons closely together. Such an opinion is supported by numerous theoretical studies and is now consented to be correct by the chemical community though it puts the previous 18-*e* counting (Fig. 6a) to a vulnerable position. Analysis of the computational results can address this concern by recognizing the bridging carbonyls as two ketonic X_2 ligands and one μ-L(L) ligand. The μ-L(L) donor carbonyl, along with two irons, forms a 3c–2*e* electron-deficient bond, and the shared two electrons count toward into the 18-*e* requirement for both irons. Such an assignment (Fig. 6b) manages to fulfill the 18-*e* rule for both irons by using six resonance forms to create a high-symmetry electronic structure to match the geometric structure. This description provides a suitable solution, but molecular orbital theory provides an equivalent solution with a delocalized bonding.

2.1.2 Molecular Orbital Analysis of the $Fe_2(CO)_9$: Fe(*t*-CO)$_3$ and (μ-CO)$_3$ Fragments

In this and next subsection, the molecular orbital diagram of $Fe_2(CO)_9$ is going to be constructed from scratch. Such building procedures primarily follow Hoffmann's work of M_2L_9 [17] with references of Mealli [19] and Green [15]. The $Fe_2(CO)_9$ can be cleaved into three parts possessing high symmetries: two Fe(*t*-CO)$_3$ (C_{3v} point group) fragments and one (μ-CO)$_3$ (D_{3h} point group) fragment. This subsection assembles CO (whose molecular orbital diagram can be found in Fig. 1) and Fe atoms, while the next subsection assembles fragments into the full $Fe_2(CO)_9$ molecule.

Each iron is six coordinate in $Fe_2(CO)_9$ and the orientation of the six coordinating ligands on one iron forms a pseudo octahedral geometry, which requires six σ-acceptor orbitals oriented in the aforementioned orientation. To begin, imagine one Fe atom with six σ-acceptor orbitals in O_h symmetry (Fig. 7a). To further simplify the analysis, assume these six acceptor orbitals are generated by $3d'_{z2}3d'_{x2-y2}4s'^{1}4p'^{3}$ or d^2sp^3 hybridization (prime denotes orbitals and axes under imaginary O_h symmetry, while the counterparts without the prime refer to the actual C_{3v} point group; Fig. 7b), while the other three $3d'_{xy}$, $3d'_{xz}$, and $3d'_{yz}$ orbitals are essentially unchanged (Fig. 8a, b). It should be noted that d^2sp^3 hybridization is used to help visualize the orientations of the orbitals, but should not imply that the metal's 4*p* orbitals are going to contribute to the occupied molecular orbitals (MOs) to the degree indicated by the hybridization [25].

The real Fe(*t*-CO)$_3$ fragment taken from the $Fe_2(CO)_9$ can be envisioned as Fe(*t*-CO)$_3$(Va)$_3$ (Va is a vacant site; Fig. 7b); its symmetry, C_{3v}, is much lower than O_h; and critically, its principal threefold *z*-axis is not collinear with any of the three fourfold *x*', *y*', and *z*' axes in the O_h point group (Fig. 7c). Therefore, an analysis of changes of Cartesian coordinate systems is desired and was described by Orgel [26]. The direct consequence of the change of principal axis direction is that the coordinates need to be changed, and unfortunately, there is no *one-to-one* correlation between coordinates under O_h and C_{3v} (see Fig. 7c for coordinate change). In other words, linear recombination is required to convert $3d'_{xy}$, $3d'_{xz}$, *and* $3d'_{yz}$ in O_h into the correct irreducible representations in C_{3v}. The correlations given by character projection operations are, namely,

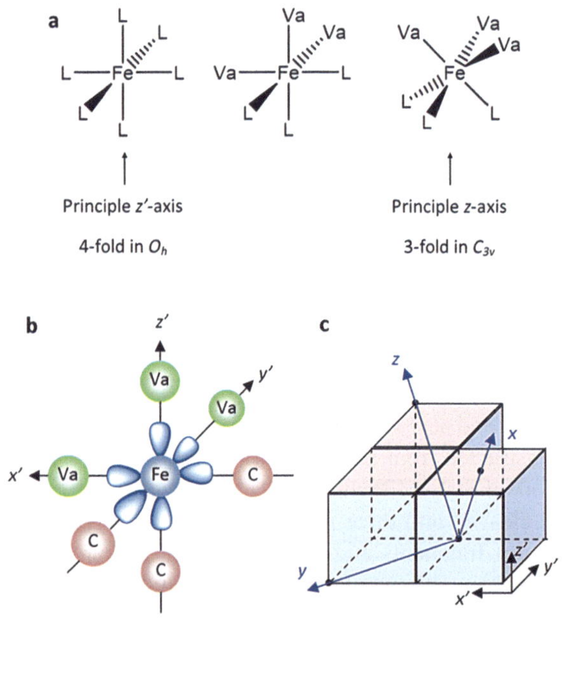

Fig. 7 (**a**) Illustration of principal axes in O_h- and C_{3v}-coordinating environment. (Va is the vacant cite). (**b**) Sketch of d^2sp^3 hybridization of Fe(*t*-CO)$_3$ fragment under pseudo-O_h symmetry. (**c**) The correlation between Cartesian coordinates under O_h and C_{3v} symmetry. (The unit vectors of *xyz* axes can be written as $\left(\sqrt{\tfrac{1}{3}}, \sqrt{\tfrac{1}{3}}, \sqrt{\tfrac{1}{3}}\right)$ $\left(\sqrt{\tfrac{1}{2}}, -\sqrt{\tfrac{1}{2}}, 0\right)$ and $\left(-\sqrt{\tfrac{1}{6}}, -\sqrt{\tfrac{1}{6}}, \sqrt{\tfrac{2}{3}}\right)$ under $x'y'z'$ coordinates

$$\varphi(a_1, d_{z^2}) = \sqrt{\frac{1}{3}}d'_{xy} + \sqrt{\frac{1}{3}}d'_{xz} + \sqrt{\frac{1}{3}}d'_{yz}$$
$$\varphi(e) = \left(\sqrt{\frac{2}{3}}d'_{xy} - \sqrt{\frac{1}{6}}d'_{xz} - \sqrt{\frac{1}{6}}d'_{yz}, \sqrt{\frac{1}{2}}d'_{xz} - \sqrt{\frac{1}{2}}d'_{yz}\right)$$

Due to the degeneracy of orbitals belonging to irreducible representation e in C_{3v}, their mathematical representations are not unique, and they may further mix with d'_{z2} and d'_{x2-z2} in O_h which also belong to irreducible representation e in C_{3v}. Such a change of coordinates does not look straightforward, but it is necessary as the $3d'_{z2}$ in O_h and $3d_{z2}$ in C_{3v} are actually different orbitals. (Note: d_{z2} orbital is indeed $d_{2z2\text{-}x2\text{-}y2}$ with a donut-shaped concentration of opposite sign in xy-plane.) The $3d'_{z2}$ used in the hybridization process is unrelated to $3d_{z2}$ in C_{3v}, since it originates from the linear combinations of the $3d'_{xy}$, $3d'_{xz}$, *and* $3d'_{yz}$ in O_h. The persistence of the $3d_{z2}$ in Fe(t-CO)$_3$ is theoretically helpful because its linear combination will form the metal–metal σ MO (bonding) and σ* MO (antibonding) between two Fe(t-CO)$_3$ fragments.

In addition to the change of coordinates, lowering the symmetry to C_{3v} lifts the threefold degeneracy present in O_h symmetry (Fig. 8c). Under the C_{3v} symmetry, the absence of a horizontal mirror plane (orthogonal to z-axis in C_{3v}) allows us to split the six hybrid (d^2sp^3) σ-acceptor orbitals into two orbital clusters, each

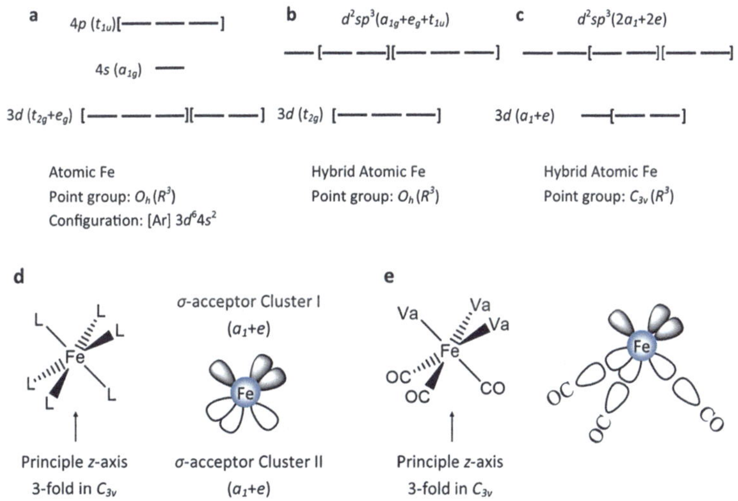

Fig. 8 The molecular orbital diagrams of atomic iron before (**a**) and after (**b**) d^2sp^3 hybridization under imaginary O_h symmetry. (*Square brackets* indicate degeneracy forced by symmetry; electrons were omitted for simplicity.) (**c**) Lowering (imaginarily) the symmetry from O_h to C_{3v} lifts certain degeneracy constrains, (**d**) splitting into two symmetry-forced degenerate σ-acceptor orbital clusters, under the (**e**) assumption σ-donation from three t-CO will go into σ-acceptor orbital cluster II

containing three σ-acceptor orbitals along the positive and negative directions of the *z*-axis, in a *fac* fashion (Fig. 8d). These six σ-acceptor orbitals still retain de facto degeneracy until ligands are added to the Fe atom to fulfill the C_{3v} symmetry by altering the chemical environment. Note that 3*d*, 4*s*, and 4*p* belong to different irreducible representations in O_h, but not in C_{3v}; thus, these orbital could mix with each other if necessary, but such mixing does not change the qualitative analysis.

The MO diagram of the three terminal carbonyls oriented in a *fac* fashion, *fac*-(t-CO)$_3$, in preparation to be attached to the Fe is presented in Fig. 9a. The frontier orbitals of three CO, 5σ, and 2π combine under C_{3v} symmetry, forming two orbital clusters: the σ-donor orbital cluster and π-acceptor orbital cluster. Similarly, the other carbonyl fragment consisting of the three coplanar bridging carbonyls that link two Fe(t-CO)$_3$ fragments in $Fe_2(CO)_9$ has a similar MO diagram but with higher D_{3h} symmetry (Fig. 9b).

The Fe(t-CO)$_3$ fragment can be assembled from the hybrid Fe atom in C_{3v} and *fac*-(t-CO)$_3$ as shown on Fig. 8e and 9c, where, because of the spatial orbital

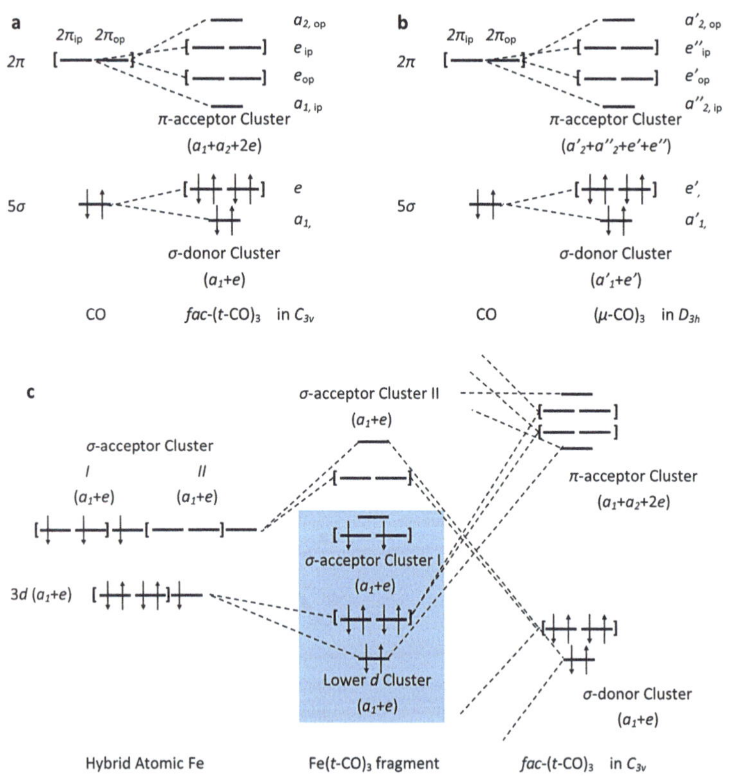

Fig. 9 The molecular orbital diagrams of (**a**) *fac*-(t-CO)$_3$, (**b**) (μ-CO)$_3$, and (**c**) Fe(t-CO)$_3$. Notations "op" and "ip" are used to distinguish out-of-plane and in-plane parts of degenerate π orbitals of CO; the reference plane is defined as it contains one CO under investigation and the threefold *z*-axis

overlap, the σ-donations from *fac*-(t-CO)$_3$ only interact with one of the two σ-acceptor orbital clusters (labeled as II in Fig. 9c), stabilizing the donor electron pairs that form the Fe–C σ bond and raising the energy of mainly σ-acceptor orbital cluster II that is now the Fe–C σ* MOs. Because of the energy ordering of the 3d, 4s, and 4p, the Fe 4s and 4p contribute more to the resulting antibonds (σ* MOs), while the Fe 3d contributes more to the, as yet nonbonding, σ-acceptor cluster I (d_{xz} and d_{yz} along with some 4s- and 4p-contribution). While the π-acceptor character of the COs lowers the energy of all three non-hybrid "pure" d-orbitals in the d-orbital cluster, it places d_{z2} orbital slightly lower than $d_{x2\text{-}y2}$ and d_{xy}. The final molecular orbital diagram of Fe(t-CO)$_3$ fragment is presented in Fig. 9c, where the highlighted MOs, d-orbital cluster and σ-acceptor cluster I, are recognized as the frontier orbitals and could actively interact with bridging (μ-CO)$_3$ while other orbitals, σ-acceptor cluster II and those mainly localized to CO, are either too high or too low in energy to participate in further orbital mixing.

2.1.3 Molecular Orbital Analysis of the $Fe_2(CO)_9$: The Metal–Metal and Metal-Bridging-CO Interactions

The full $Fe_2(CO)_9$ molecule consists of two Fe(t-CO)$_3$ fragments bridged by one (μ-CO)$_3$ fragment in D_{3h} symmetry. The non-hybrid d-orbitals (now with some 2π contributions from terminal carbonyls), namely, d_{z2}, d_{xy}, and $d_{x2\text{-}y2}$, on one Fe(t-CO)$_3$ fragment, can find exact counterparts on the other Fe(t-CO)$_3$ fragment and form three metal–metal bonding MOs (one σ and two primarily δ with respect to the Fe–Fe bond axis) [17] and three antibonding MOs as shown in the lower metal–metal bonding orbital cluster in Fig. 10a. All the bonds and antibonds in the lower bond orbital cluster are doubly occupied so that they contribute a net bond order of zero between two irons. The σ-acceptor orbital cluster I in [Fe(t-CO)$_3$]$_2$ reorganizes to give a pair of π and π* and a pair of σ and σ* MOs. The occupancies of these orbitals will be dependent on the interaction between the [Fe(t-CO)$_3$]$_2$ fragment and the (μ-CO)$_3$ fragment.

The final step is to insert the (μ-CO)$_3$ fragment into the center of the $D_{3\mathrm{h}}$ [Fe(t-CO)$_3$]$_2$ to fully assemble the $Fe_2(CO)_9$ molecule; consideration of its effects on orbital diagram from both σ-donation and π-acceptance will determine the final nature on the metal–metal interaction. As a bridging ligand, the μ-CO does not have the ideal orbital overlaps to inject or withdraw electron density from a single iron unit; in contrast, it is more likely to interact with the linear combinations of orbitals from both Fe(t-CO)$_3$ units. The σ-donation (Fig. 10b) transfers electron density from 5σ orbitals of the (μ-CO)$_3$ into symmetric combinations of the σ-acceptor cluster I. While π-accepting ability of the (μ-CO)$_3$ fragment is more complex due to orbital overlaps, the acceptance of Fe electron density by the out-of-plane (Fe_1-C-Fe_2 plane) $2\pi_{\mathrm{op}}$ orbital is weak due to its poor overlap, while that by the in-plane $2\pi_{\mathrm{ip}}$ orbital is strong. Examining just one CO, as shown in Fig. 10c, d, one can see that the $2\pi_{\mathrm{ip}}$ may interact with the antisymmetric combination (w.r.t. horizontal mirror plane) of d_{z2} orbitals (Fig. 10c) and d_{xz}(or d_{yz}) orbital (Fig. 10d). (Note: d_{xz},

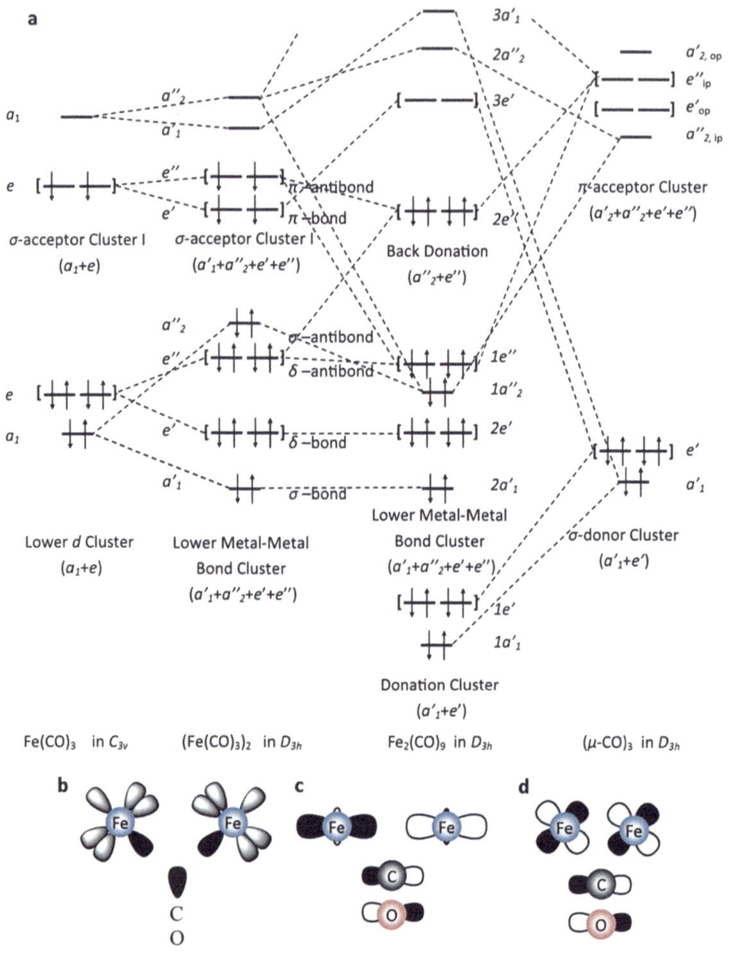

Fig. 10 (**a**) The full molecular orbital diagram of $Fe_2(CO)_9$. The bonding illustrations between irons and a bridging carbonyl: (**b**) the σ-donation and (**c** and **d**) two modes of orbital mixing of the π-acceptance

d_{yz}, and $d_{x2\text{-}y2}$, d_{xy} belong to the same irreducible representation so that it is inevitable they will mix with each other to certain extent.) Using all three COs in full D_{3h} symmetry, one can see that the a''_2 π-acceptor orbital from $(\mu\text{-CO})_3$ stabilizes the d_{z2}-d_{z2} σ* MO while its e'' π-acceptor orbitals stabilize the d_{xz}-d_{xz} and d_{yz}-d_{yz} π* MOs. These π-backbonding interactions from the bridging COs stabilize the antibonding combinations, while the σ-donations destabilize the bonding interactions. One should note that the d_{xz} and d_{yz}, originally in the d^2sp^3 hybrids, which are supposed to be the metal's σ-acceptor orbitals assume quite different roles when they are used to support bridging ligands rather than terminal ligands. This example is an excellent illustration of how the orientations of metal and ligand (bond angle in this case) may dramatically change the roles played by both counterparties in the interactions. The difference is ever more subtle in the cases of asymmetrically bridging carbonyls (vide infra).

From the preceding MO description, one can conclude that the occupancies of six MOs formed from the low-energy *d* cluster produce no net Fe–Fe bond as they all appear as bonding–antibonding pairs, making the configuration $(2a'_1)^2(2e')^2(1a''_2)^2(1e'')^2$ or $(\sigma)^2(\delta)^4(\sigma^*)^2(\delta^*)^4$. The highest occupied MO (HOMO), the $2e''$ orbitals, with four electrons partially delocalized onto the bridging COs is antibonding between the irons, which adds $(\pi^*)^4$ to the configuration. Alternatively, if the backbonding of the three bridging carbonyls was strong enough to deplete the *d* electrons on $2e''$ and $1a''_2$, the system would be treated as a de facto $d^{5\text{-}5}$ system like a hypothetic $M_2(L_3X_6)$ with a configuration of $(\sigma)^2(\delta)^4\,(\delta^*)^4$. Then there would be a single metal–metal bond [21], and the Lewis structure would have three ketonic carbonyls as bridges. However, the MO analysis only shows only four electrons ($2e''^4$) available for strong backbonding to the bridging COs; thus, the most appropriate Lewis structure has only two ketonic COs, a third bridging CO with a 3c–2*e* bond to the irons, and a nonbonding iron–iron interaction as in Fig. 6b. Ponec et al. was able to localize and visualize the 3c–2*e* bond along with four Fe–C bonds by domain average Fermi hole (DAFH) analysis of the Fe(μ-CO)$_3$Fe domain [27, 28]. Similar DAFH analysis on the imaginary D_{3h} $Cr_2(CO)_9$, which is expected to have a triple bond by 18-*e* rule, actually shows only three 3c–2*e* bonds evenly distributed, one with each bridging carbonyls [29]. In addition, the atoms-in-molecule (AIM) analysis based on the total electron density at Hartree–Fock level failed to detect a Fe–Fe (3,−1) bond critical point and adds further to the conclusion of no net bond between the two irons [30].

Mealli et al. [21] suggested that if the $1a''_2$ had large enough contributions from bridging carbonyls, it would not fully cancel the bonding contributions between two irons in $2a'_1$. A more recent DFT/electron density orbital partitioning study [22] by them inspected the orbital-by-orbital electron distribution and net partial σ bonding, along the Fe–Fe axis, was verified. Their analysis that the electron density in the $2e''$ π-antibonding orbitals surrounding the Fe–Fe axis like a cylinder does not directly cancel the σ bond electron density lying along the axis because these two interactions are spatially separated. The σ bonding in $2a'_1$ is only partially canceled by the σ-antibonding in $1a''_2$, and the remaining partial σ bonding coexists with the excess π-antibonding in $2e''$. Although the total net interaction may be antibonding, a partial σ bond can be claimed.

2.1.4 Closing Remarks About $Fe_2(CO)_9$

Though CO lies at the end of the spectrochemical series as a strong-field ligand, carbonyl's backbonding capacity is still relatively weak compared to what would be required to withdraw electrons from metals to form a covalent-style bond as it only has its antibonding π* to serve its backbonding function. A comprised presentation [15], using one μ-L(L) ligand and two X_2 ligands to build a conventional, easy-to-understand Lewis structure for the (μ-CO)$_3$ interaction, is adopted to reflect the backbonding strength in an average fashion. The story would be totally different if a bridging ligand has a lower-lying, unoccupied orbital for backbonding, like the *p* orbital in the singlet carbene like CH_2.

2.2 Bridging Carbonyl II: $Co_2(CO)_8$ and $Ni_2(CO)_7$

X-ray diffraction revealed that the solid structure of $Co_2(CO)_8$ features two bridging carbonyls [31]. It is similar to $Fe_2(CO)_9$, removing one bridging carbonyl and adding two electrons. Due to the loss of the third bridging carbonyl, the two Co (CO)$_3$ moieties slightly bend toward each other, using the two remaining bridging carbonyls as a hinge. Arguments have been made that a metal–metal bond, presumably bent [32, 33], may exist between two cobalt atoms. The early work even proposed that both linear and bent bonds were possible depending on the assignment of the unpaired electron of the Co(CO)$_3$ fragment. If the unpaired electron resides on the unused d^2sp^3 (due to the missing third CO) orbital, the two unpaired electron would couple to give a bent bond; alternatively, if the d^2sp^3 orbital is doubly occupied, the unpaired electron might reside on the *d*-orbital pointing to the other Co(CO)$_3$ fragment and they would complete a linear metal–metal bond [8]. Early extended Huckel MO calculations [33] contradicted such a bonding analysis, while CNDO calculations supported a bond, though it was claimed to be a *p–p* bond [16, 18].

Building the molecular orbitals of (Co(CO)$_3$)$_2$ begins exactly like that of the D_{3h} (Fe(CO)$_3$)$_2$, and then, the symmetry is lowered to C_{2v} to reflect the bending (Fig. 11). The interaction between (Co(CO)$_3$)$_2$ and two bridging carbonyls is fundamentally similar to that in $Fe_2(CO)_9$ except (Co(CO)$_3$)$_2$ has two more *d* electrons and they fill the HOMO, a_1 orbital (Fig. 11). Although the HOMO has significant bonding character, including primarily π-symmetry *d* contributions from both metals along with some *p*-contributions, and such an orbital gains some bent σ bond character due to the bending of Co(CO)$_3$ moieties, the Co–Co antibonding HOMO-1 and HOMO-2 cancel the net bonding so that no metal–metal bond exists between two cobalt atoms. Therefore, the configuration for upper *d*-orbitals is actually $(\pi)^2(\pi^*)^4$ with one excess antibonding MO. The bridging carbonyls alter [21] the order of bonding orbitals and antibonding orbitals in (Co (CO)$_3$)$_2$; thus, antibonding orbitals are filled first as in $Fe_2(CO)_9$ and the additional electrons go to Co–Co bonding orbitals that is destabilize by σ-donation from the bridging COs. The topology analysis of electron density at Hartree–Fock and DFT level cannot locate a $(3,-1)$ bond critical point indicative of a true Co–Co bond [34, 35]. Instead, a (3,1) ring critical point was located inside $Co_2C_{2,bridging}$ ring, suggesting that the only actual bonds are between two cobalt atoms and bridging carbonyls. Regardless of the absence of a $(3,-1)$ bond critical point, a bent Co–Co bond was claimed based on the fact that the most negative energy density is located in the “bent bond region” below the vector connecting two cobalt atoms and the

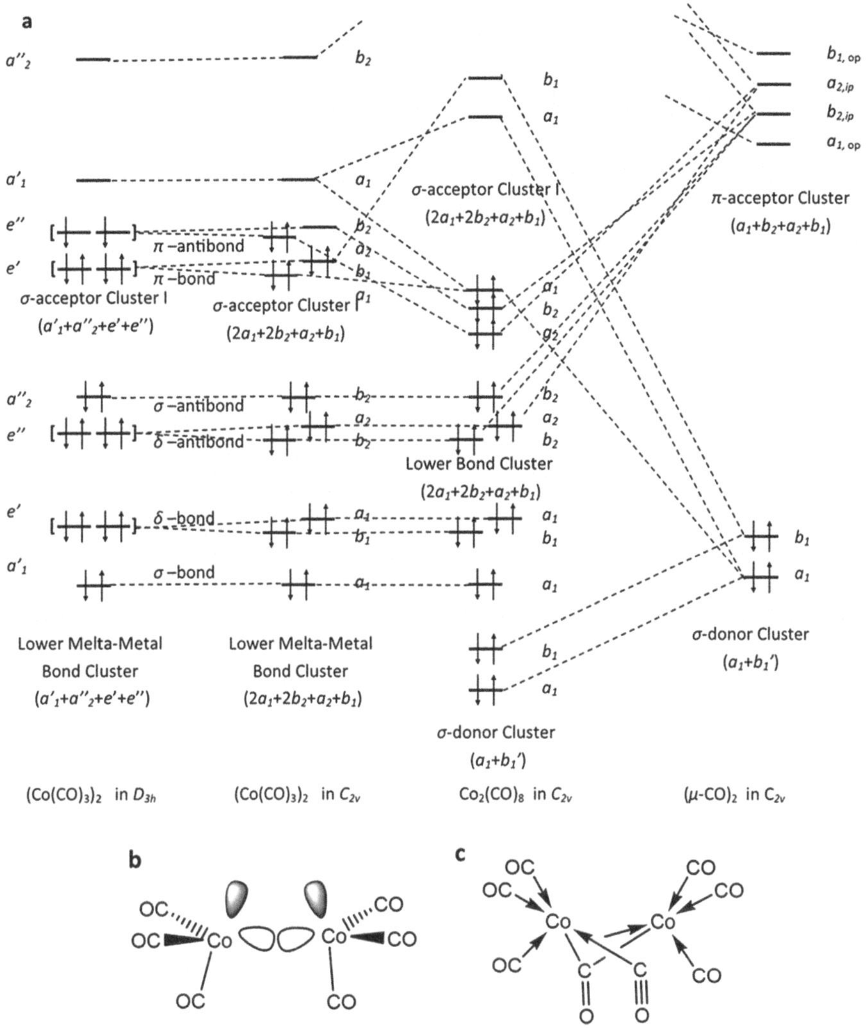

Fig. 11 (**a**) MO diagram of $Co_2(CO)_8$ and (**b**) the sketch of the HOMO; note that the bending of $Co(CO)_3$ moieties change the π-overlap into somewhat σ-overlap; (**c**) one of the four resonance structures featuring the electron-deficient forms

negative energy density usually means the stabilization by bonds [36]. However, such a conclusion was soon revised as the authors realized that the antibonds have a nodal plane which is perpendicular to the Co–Co vector and energy density and electron density are all zero on the nodal plane. The antibonds are "transparent" in this aspect of the topology analysis, and negative energy density leads one to a false conclusion that the bent Co–Co bond is not canceled [37]. However, in the argument for $Fe_2(CO)_9$, the cancelation may be incomplete as the metal–metal antibonding orbitals have more ligand contributions and less metal components but no formal bond is expected after such canceling is applied. The 3c–2*e* localized bonding scheme can also be applied here as presented in Fig. 11c, and it is verified by DAFH [27, 28].

These metal–metal antibonds actually help stabilize the whole molecule as they are bonding to the bridging COs. $Rh_2(CO)_8$ was calculated to have a structure similar to $Co_2(CO)_8$ featuring two bridging carbonyls and six terminal carbonyls. However, the ground-state configuration is slightly different as the HOMO of $Rh_2(CO)_8$ is a metal–metal antibonding orbital similar to the HOMO-1 in $Co_2(CO)_8$. The oxidation of $Rh_2(CO)_8$ removes one of the antibonding electrons and causes instability of $[Rh_2(CO)_8]^+$ in the original structure so it undergoes a reorganization to all-terminal-carbonyl isomer [38]. Similar structure also exists for the neutral $Co_2(CO)_8$ and these structures have direct Co–Co bonds and all-terminal COs.

The next species following the pattern $Fe_2(CO)_9$ and $Co_2(CO)_8$ is $Ni_2(CO)_7$, but it was only computationally investigated and features only one bridging carbonyl. Unlike $Fe_2(CO)_9$ and $Co_2(CO)_8$, the optimized $Ni_2(CO)_7$ has a relatively low symmetry (Fig. 12). Schaefer et al. [39] calculated $Ni_2(\mu\text{-CO})$ as a model to investigate the effects of the bridging carbonyl. Their calculations provided that two bare d^9s^1 Ni were bound together to yield the configuration $(\sigma_d)^2(\sigma_d{*})^2(\pi_d)^4(\pi_d{*})^4(\delta_d)^3(\delta_d{*})^3(\sigma_s)^2$, a triplet state with the Ni–Ni bond from the *s* orbitals. Upon the addition of the bridging carbonyl, the configuration became a mixture of $(\sigma_d)^2(\sigma_d{*})^2(\pi_d)^4(\pi_d{*})^4(\delta_d)^2(\delta_d{*})^4(\sigma_s)^2$ and $(\sigma_d)^2(\sigma_d{*})^2(\pi_d)^2(\pi_d{*})^4(\delta_d)^4(\delta_d{*})^4(\sigma_s)^2$ as in C_{2v} symmetry, among which one π_d and one δ_d belong to the same a_1 irreducible representation and mix with each other to accommodate a total of two bonding electrons. The net change after addition of the bridging carbonyl is that the antibonding $\delta_d{*}$ (a_2) orbital, which is well mixed with $\pi_d{*}$ (a_2) orbital, is stabilized and becomes doubly occupied, borrowing one electron from the bonding orbital δ_d (a_1) which is now vacant. Thus, the bridging CO in $Ni_2(\mu\text{-CO})$ likewise increases the occupancies of the metal–metal antibonding orbital. Although bridging COs increase the M–M antibonding and the M–C bonding at the same time, the latter is more important in stabilizing the overall structure. $Fe_2(CO)_9$ and $Co_2(CO)_8$ are indeed relatively stable while $Ni_2(CO)_7$ is not.

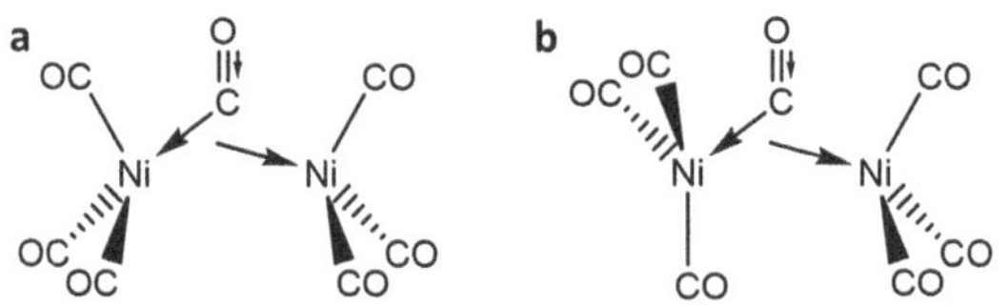

Fig. 12 (**a**) The proposed geometry of $Ni_2(CO)_7$ with a high symmetry but it was determined to be a transition state; (**b**) the optimized geometry of $Ni_2(CO)_7$ [39]. The carbonyl is drawn as an L (L) ligand as discussed previously

2.3 Bridging Carbonyl III: $(CpM)_2(\mu\text{-}AO)_2$ (M=Fe, Co, Ni; A=C, N)

The last examples for the discussion of symmetrically bridging diatomic π-acceptor ligands are a few related complexes: $(CpFe)_2(\mu\text{-}NO)_2$ [40, 41], $(CpCo)_2(\mu\text{-}CO)_2$ (though only its reduced form [42] and Cp* variant [43] were stable and subject to X-ray diffraction inspection), $(CpCo)_2(\mu\text{-}NO)_2$ [44, 45], and $(CpNi)_2(\mu\text{-}CO)_2$ [46], which are representative species in the $(CpM)_2(\mu\text{-}CO)_{2\text{-}x}(\mu\text{-}NO)_x$ family. Nitrosyls are even stronger π-acceptors than carbonyls, and their analyses can add to our general understanding of this general class on important ligands. The cyclopentadienyl anion is a six-electron donor, an analogue of $(t\text{-}CO)_3$, while NO^+ is isoelectronic to CO. Thus, $(Cp^-Co^0)_2(\mu\text{-}NO^+)_2$ is effectively (valence) "isoelectronic" to $Co_2(CO)_8$, but $(CpCo)_2(\mu\text{-}CO)_2$ has two fewer electrons. Unlike $Co_2(CO)_8$ with a puckered $Co_2(\mu\text{-}C)_2$ ring, many $(CpM)_2(\mu\text{-}CO)_{2\text{-}x}(\mu\text{-}NO)_x$ species appear to have a coplanar $M_2(\mu\text{-}N/C)_2$ ring. However, $(CpNi)_2(CO)_2$ is an apparent exception[46], and the structure of $(CpCo)_2(NO)_2$ is somewhat ambiguous, as the crystal structure appears planar[44] while IR spectra for solid state and solution show two NO vibrations indicative of a bent structure [45]. Thus, the species that are valence isoelectronic to $Co_2(CO)_8$ may, in fact, be puckered and the planar structure only appearing for species with fewer electrons.

This complex family was investigated by multiple computational tools, including extended Huckel [47], Fenske–Hall [48, 49], Hartree–Fock (HF) or self-consistent field (SCF) [50, 51], multi-configuration (MC) SCF [50] and generalized valence bond (GVB) [51], and density functional theory(DFT, B3LYP) [52]. As an illustrative example of this family, the construction of the MO diagram of $(CpCo)_2(\mu\text{-}NO)_2$ follows the example of Hall's work [49, 51, 52]. The $[CpCo]^-$ (d^9) fragment with its fivefold axis as the principal axis has a splitting pattern of the d-orbitals similar to that of the $Fe(CO)_3$ fragment: the d_{xz} and d_{yz} (e_1, orbitals of π-symmetry with respect to the Co–Co axis) orbitals are the highest-energy orbitals as they are destabilized by donation from the π HOMO of the Cp^- ring; the d_{xy} and $d_{x2\text{-}y2}$ (e_2, orbitals of δ-symmetry with respect to the Co–Co axis) are stabilized by the π LUMO of the Cp^- ring; and the d_{z2} (a_1), which points to the center of the ring, is only weakly destabilized by the totally symmetric π HOMO-1 of the Cp^- ring. In $[CpCo]^-$, the lower d-orbital cluster, i.e., d_{xy}, $d_{x2\text{-}y2}$, and d_{z2}, is fully occupied, and the higher d-orbital cluster, d_{xy} and $d_{x2\text{-}y2}$, has three electrons, exactly like the Co $(CO)_3$ fragment. When two $[CpCo]^-$ moieties are put together in D_{5h} symmetry, 12 electrons fill the Co–Co bonding and antibonding orbitals formed by the lower d cluster with a configuration in cylindrical symmetry of $(\sigma)^2(\delta)^4(\sigma^*)^2(\delta^*)^4$. (Note the principal z-axis changed once two $[CpCo]^-$ are put together; see Fig. 13b.) And the two sets of e_1 orbitals from two fragments combine to form Co–Co π and π* MOs, e'_1 and e''_1, respectively. Thus, the frontier orbital layout of $[(CpCo)_2]^{2-}$ is nearly the same as $[Co(CO)_3]_2$. The differences in the electronic structures of the final molecules occur on addition of the bridging ligands. If local C_{2v} symmetry is assumed for $Co_2(XO)_2$ (X=C/N) ring and two bridging ligands are placed on the

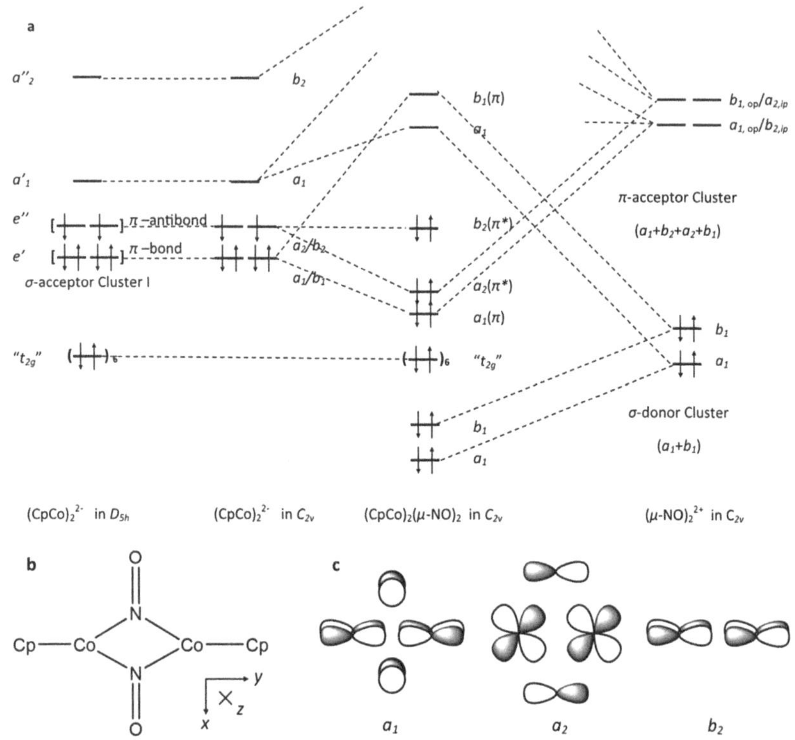

Fig. 13 (**a**) MO diagram of $(CpCo)_2(NO)_2$. (**b**) The sketch of the Co_2C_2 ring and Cartesian coordinates (z-axis is assigned such that it remains the principal axis even for the puckered (folded) structure, where the point symmetry reduces from pseudo-D_{2h} to C_{2v}). (**c**) The sketches of HOMO-2–HOMO and the interactions between bridging NO and two metals. (*Round parentheses* and the *number* in the subscript are used to indicate multiple nondegenerate orbitals when they are grouped for simplicity. For degenerate orbitals, *brackets* are used)

xy-plane and aligned to the x-axis (Fig. 13b), then the bridging ligands are exactly on the horizontal nodal plane of the b_2 orbital of $(CoCp)_2$ (antibonding combination of d_{yz}) and no orbital of bridging ligand can overlap with such an orbital, while a_1 (bonding combination of d_{yz}) and a_2 orbitals (antibonding combination of d_{xz}) are all stabilized by the backbonding from bridging ligands. The configuration for $(CpCo)_2(\mu\text{-NO})_2$ is still $(\pi)^2(\pi^*)^4$ with one excess antibonding orbital as that in $Co_2(CO)_8$, while the neutral $(CpCo)_2(\mu\text{-CO})_2$ has two fewer electrons, $(\pi)^2(\pi^*)^2$, such that there is just simple cancelation of any net Co–Co bonding. The MO diagram presented above is based on the singlet state. Benard et al. [50] predicted the ground state of $(CpCo)_2(\mu\text{-NO})_2$ may be triplet with MC-SCF calculations, but this was disapproved by its diamagnetism [45] and photoelectron spectroscopy [49].

$(CpFe)_2(\mu\text{-NO})_2$ and $(CpNi)_2(\mu\text{-CO})_2$ are isoelectronic species of $(CpCo)_2(\mu\text{-CO})_2$ and $(CpCo)_2(\mu\text{-NO})_2$, respectively, and they share similar orbital layout and configurations. The relative stability of $[(CpCo)_2(\mu\text{-}CO)_2]^-$ compared to $(CpCo)_2(\mu\text{-}CO)_2$ is attributed to the increased backbonding to bridging CO because the occupancy of a metal–metal antibonding orbital raises the donor level and finally stabilizes the whole molecule [48]. The replacement of Cp with Cp* has similar effects. Another related species $(CpFeCO)_2(\mu\text{-}CO)_2$ [53], which has four fewer *d* electrons and four more σ-electrons donated from the two additional terminal carbonyls, still has one excess metal–metal antibonding orbital as it has two fewer bonding electrons and two fewer antibonding electrons compared to $[(CpCo)_2(\mu\text{-}NO)_2]$ [54, 55]. The number of metal–metal bonds in this family, predicted by 18-*e* rule along with metric data [40, 44, 56], is generally not reliable. For all four members of the $(CpM)_2(\mu\text{-}CO)_{2\text{-}x}(\mu\text{-}NO)_x$ family, only a (3,1) ring critical point was detected instead of a (3,−1) bond critical point, by topology analysis of computational electron density, supporting the absence of net metal–metal bonds [52].

The next question is about the conditions that tend to pucker (fold) the M_2A_2 ring, where the Co_2A_2 (A=C/N) ring in $[(CpCo)_2(\mu\text{-}AO)_2]$ is an example. Generally, the filled Co–Co bonding and antibonding orbitals formed by the lower *d* cluster with a configuration in cylindrical symmetry of $(\sigma)^2(\delta)^4(\sigma^*)^2(\delta^*)^4$ are relatively insensitive to the puckering angle (the puckering angle ω is defined as the dihedral angle $C_1(N_1)$–M_1–M_2–$C_2(N_2)$)as the three bonds are canceled by three antibonds. A key factor to the puckering angle is the number of electrons in the upper *d* cluster, and those with six electrons, the ${a_1}^2{a_2}^2{b_2}^2$ in the planar structure as shown in Fig. 13, tend to be puckered while those with two fewer electrons tend to be planar. This difference arises because in the planar structure, the b_2 is completely nonbonding, but as the bridging ligands move to pucker the ring, the b_2 begins to be stabilized by π-backbonding with these ligands. Thus, if only the b_2 is doubly occupied, the system tends to pucker, but with fewer electrons, this driving force is diminished. Although $[(CpCo)_2(\mu\text{-}CO)_2]^{2-}$ might be expected to have a low-energy structure with a pucker angle of 180° [47], initial deviation from $\omega = 180°$ actually stabilizes both a_2 and b_2 until the critical point is reached and then successive deviation starts to destabilize a_2 such that a pucker angle of 127° is predicted [47]. Likewise, calculations on $(CpCo)_2(NO)_2$ predict an optimal ω around 120°, while those on $(CpCo)_2(\mu\text{-}CO)_2$ predict an optimal ω of 180° [51]. The application of one-pair generalized valence bond (GVB) is able to significantly lower the barrier but does not necessarily change the minimum structure and reveals the non-innocent interaction between cobalt and nitrosyls, particularly in the planar structure [51].

2.4 Comments on the Symmetric Bridging Carbonyls

While direct M–M interactions would stabilize bonding orbitals below antibonding ones, bridging ligands like CO have π-acceptor orbitals that stabilize the antibonding M–M orbital and σ-donor orbitals that destabilize M–M bonding orbitals. Since the first-row transition metals have more compact *d*-orbitals, they have weaker direct M–M interaction than the heavier metals. Thus, bridging carbonyls stabilize clusters by replacing weak M–M bonding with stronger M–C bonding [57] which results in a shuffling of bonds and antibonds, causing the difficulties identifying a formal metal–metal bond [12]. A similar antibonding M–M would be too repulsive in the complexes containing heavier metals for the carbonyl to stabilize sufficiently. With excess antibonds, the bond orders between two metals in $Fe_2(CO)_9$ and $Co_2(CO)_8$ can be recognized to be negative. The 18-electron rule never considers the idea of occupied antibonding orbitals in its scheme and it (wrongly) treats the excess antibonding as net bonding and predicts a (nonexisting) metal–metal bond in these carbonyl-bridged complexes. The readers are further referred to a previous Structure and Bonding chapter for the discussion of bond orders of metal–metal bonds [58].

3 Transition from Terminal to Bridging in a Compensating Carbonyl Pair

Unsymmetrically bridging carbonyls have different M–C bond lengths and M–C–O angles, and when the two metals are equivalent, they usually occur in symmetry-related pairs, so that each metal feels the same effects. Figure 14a depicts a carbonyl pair along a pseudoreaction coordinate from symmetrical to unsymmetrical and finally to terminal.

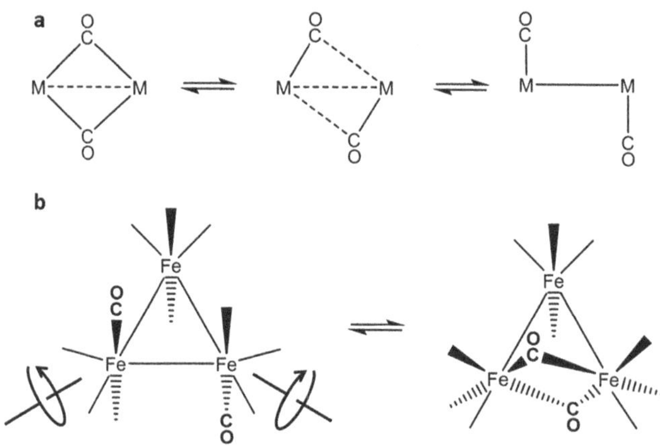

Fig. 14 (**a**) A general unsymmetrical bridging carbonyl pair (*middle*). It may convert into two symmetric bridging carbonyls (*left*) or two terminal carbonyls (*right*) depending on the C–M bond lengths. (**b**) Terminal and bridging $Fe_3(CO)_{12}$. Only two carbonyls are drawn explicitly. The motions involved in the conversion are drawn with *arrows*

3.1 A Representative Example: $Fe_3(CO)_{12}$

A good example of an unsymmetrical bridging carbonyl pair is $Fe_3(CO)_{12}$, whose structure [59, 60] is related to $Fe_2(CO)_9$ by replacing one of three bridging carbonyls of $Fe_2(CO)_9$ by a $Fe(CO)_4$ unit (i.e., bridging structure in Fig. 14b). The history surrounding the structural determination of $Fe_3(CO)_{12}$ was covered by a very recent review [61]. The existence of the bridging carbonyls was first rationalized by the ability of metal clusters to fit into carbonyl clusters arranged in certain polyhedrons. The carbonyls in $Fe_3(CO)_{12}$ form a "tightly packed" icosahedron, but the carbonyls in $Ru_3(CO)_{12}$ or $Os_3(CO)_{12}$ do not; thus, the geometry was believed to be controlled by steric effects [62, 63]. However, at lower temperature, these bridging carbonyls become more symmetric, and such behavior is attributed to decreased molecular motions [64]. Regardless of their "resting-state" structure, $Fe_3(CO)_{12}$ can be a good platform to investigate the relationship between the bridging carbonyls and the electronic structure, in relation to the D_{3h} all-terminal $Fe_3(CO)_{12}$ featuring three $Fe(CO)_4$ units (Fig. 14b).

The construction of molecular orbital diagram of $Fe_3(CO)_{12}$ starts from two trigonal pyramidal $Fe(CO)_3$ and the $Fe(CO)_4$ fragment (see the SCF and DFT calculations [54, 65] for additional details). The fragment orbitals of $Fe(CO)_3$ were already presented in Figs. 8 and 9. As we did for $Fe(CO)_3$, we apply d^2sp^3 hybridization to the Fe of $Fe(CO)_4$ in C_{2v} symmetry, but now, four of six hybrid orbitals are used to accept σ-donation from four terminal carbonyls and only two of them are left for interactions with other fragments (Fig. 15). As emphasized in previous cases, the lower *d*-orbitals (the *d*-orbital cluster that would correspond to the t_{2g} orbitals in O_h symmetry) are all doubly occupied in late transition metal complexes and are quite inert with respect to the metal–metal bonding as occupation of both the in-phase (bonding) and out-of-phase (antibonding) pairs always cancels most of the bonding. Therefore, only the frontier orbitals, σ-acceptor orbital cluster I (the d^2sp^3 hybrids not consumed by the ligands), will be used in this analysis. As shown in Fig. 15, the HOMO of $Fe(CO)_4$ is the asymmetric combination of the two unused hybrid orbital, while the LUMO is formed by symmetric combination. Although this order may seem unintuitive, the asymmetric combination of the two hybrid orbitals has more 3*d* orbital character, while the symmetric combination has more 4*s* and 4*p* character (orbitals higher in energy in the free atom).

The all-terminal $Fe_3(CO)_{12}$ is easily built from three $Fe(CO)_4$ fragments [66]. Its HOMO and HOMO-1 orbitals represent the symmetry-adapted form of three equivalent Fe–Fe bonds as expected by the 18-*e* rule (Fig. 16). The construction of the bridging $Fe_3(CO)_{12}$ needs two $Fe(CO)_3$ units, and a $Fe(CO)_4$ unit and a $(\mu\text{-}CO)_2$ unit will parallel the previous work [65, 67]. As mentioned above, we will concentrate our description on the formation of the cluster bonding using the frontier orbitals (Fig. 17). Each $Fe(CO)_4$ unit and $Fe(CO)_3$ has two or three d^2sp^3 orbitals not used by σ-donation from terminal COs. To assemble $Fe_3(CO)_{10}$ fragment in the absence of the two bridging carbonyls, $Fe(CO)_4$ unit contributes two

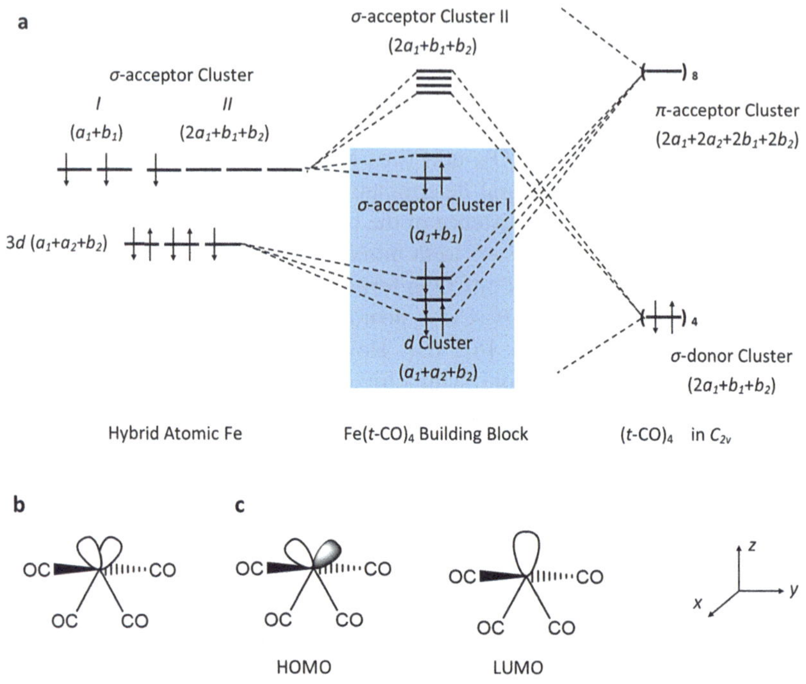

Fig. 15 (**a**) The molecular orbital diagram of $Fe(CO)_4$ fragment. (**b**) The two hybrid d^2sp^3 orbitals in the σ-acceptance orbital cluster I. (**c**) The molecular orbitals reorganized from the two hybrid orbitals

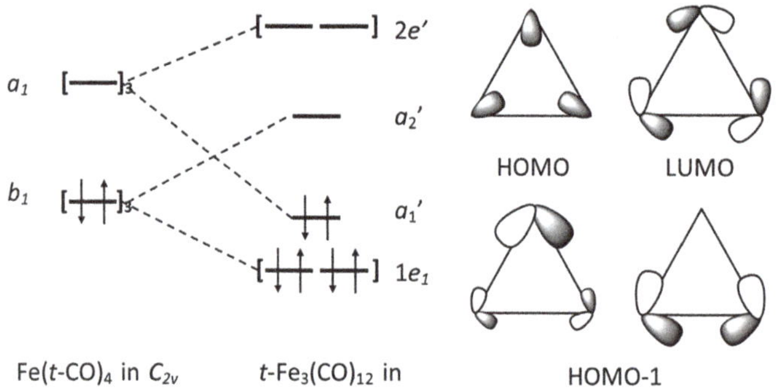

Fig. 16 The molecular orbital diagram of the all-terminal $Fe_3(CO)_{12}$

in-plane hybrid orbitals (#1 and #2 on Fe3; Fig. 17a), which overlap with the in-plane orbitals of the two $Fe(CO)_3$ units (#1 on both Fe1 and Fe2; Fig. 17a) to form two σ bonds (Fe1–Fe3 and Fe2–Fe3). The remaining hybrid orbitals on the Fe $(CO)_3$ units deviate from the Fe_3 plane and form one set of π/π* MOs and one set of σ/σ* MOs, among which the π MO (b_1) is doubly occupied as it has more

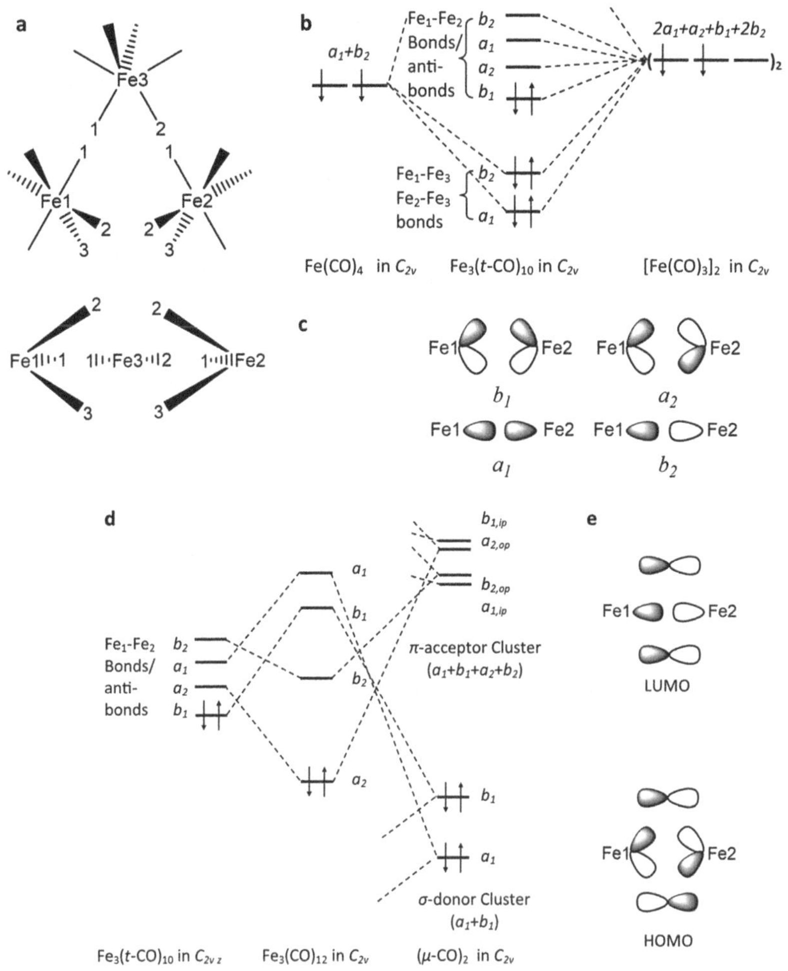

Fig. 17 (**a**) The *top* and *front* views $Fe_3(t\text{-}CO)_{10}$ (i.e., bridging $Fe_3(CO)_{12}$ less the two bridging carbonyls). Those unused hybrid orbitals are marked with numbers. The carbonyls are omitted. (**b**) The construction of the molecular orbitals by hybrid orbitals and (**c**) the sketches of HOMO–LUMO+2 and the bonding between Fe_1 and Fe_2; the a_1 is less stable than the a_2 because the latter has more *d*-orbital character. (**c**) Sketches of four Fe_1–Fe_2 bonding/antibonding orbitals of the $Fe_3(CO)_{10}$ fragment. (**d**) The MO diagram of the $Fe_3(CO)_{10}$ fragment interacting with the bridging COs to form $Fe_3(CO)_{12}$ (**e**) with sketches of its HOMO and LUMO

d character. This b_1 MO resembles the banana-shaped B–B orbital bridged by hydrogens in the description of the electron-deficient bonding of B_2H_6.

The final step puts the two bridging carbonyls into the $Fe_3(CO)_{10}$ unit. As expected, in the $Fe_2(\mu\text{-}C)_2$ pseudo-plane, the asymmetric combination of 2π orbitals of the bridging carbonyls significantly lowers the energy of the π* (a_2) Fe1–Fe2 MO and transfers the two electrons from the Fe1–Fe2 π bonding MO to the Fe1–Fe2 π* MO (Fig. 17d). Such backbonding stabilization prevents the formation of the

Fe1–Fe2 bond and no such bond is detected by the AIM analysis as no bond critical point exists between Fe1 and Fe2, while the analysis of electron localization function (ELF) suggests a delocalized bonding description for the appearance of bonds between the bridging Cs and both irons. The ELF analysis also indicates the charge transferred from irons to one bridging carbonyl is twice that to one terminal carbonyl [68].

Interestingly, there is an analogy between a third bridging CO and the $Fe(CO)_4$ unit, as the $Fe(CO)_4$ unit replaces one of the three bridging CO in $Fe_2(CO)_9$ to yield $Fe_3(CO)_{12}$. Further, the $Fe(CO)_4$ moiety has frontier orbitals like those of CO, namely, a σ orbital (the symmetric combination of the two unused hybrid orbitals), an in-plane (Fe_3 plane) π orbital, and two electrons, though the π orbital is expected to be lower than the σ orbital. Although the energetic order of these orbitals differs from that in CO, they interact with the hybrids on Fe1 and Fe2 (one from each) to form the two Fe–Fe bonds, an analogue of a ketonic CO.

The HOMOs of the terminal and bridging $Fe_3(CO)_{12}$ structures have symmetries of $a_{1'}$ (in D_{3h}) and a_2 (in C_{2v}), which are symmetric and asymmetric with respect to the mirror plane bisecting Fe1–Fe3–Fe2 angle, while the LUMOs are asymmetric and symmetric, respectively. However, the four orbitals share the same irreducible representation of a (in C_2) during the conversion from the terminal structure to the bridging structure. Therefore, the nonadiabatic HOMO energy plot on the reaction coordinate from D_{3h} $Fe_3(CO)_{12}$ to C_{2v} $Fe_3(CO)_{12}$, though smooth, indeed reflects the adiabatic occupancy transfer from the symmetric orbital ($a_{1'}/a_1$) to the asymmetric orbital ($a_{2'}/a_2$), i.e., the swap of HOMO and LUMO in two isomers. The Walsh diagram showed the net rise in orbital energy of HOMO after the transition from D_{3h} to C_{2v} [67]. The original authors concluded such electron transfer during geometric transformation is driven by the rise in energy of the symmetric orbital rather than the fall in energy of the asymmetric orbital [67]. And the HOMO itself in C_{2v} actually is less stable. However, the bridging $Fe_3(CO)_{12}$ is still more stable as one M_1–M_2 δ^* antibonding orbital in lower "t_{2g}" region can also be stabilized by the bridging carbonyls [67].

3.2 Analogues of $Fe_3(CO)_{12}$: $Ru_3(CO)_{12}$ and $Ru_3(CO)_{10}(NO)_2$

The heavier analogues of $Fe_3(CO)_{12}$, namely, $Ru_3(CO)_{12}$ and $Os_3(CO)_{12}$, were determined to have terminal CO structures with longer average metal–metal distances of ca. 2.85 and 2.88 Å, respectively [69, 70]. Their MO diagrams and occupancies are similar to that of terminal $Fe_3(CO)_{12}$ (Fig. 16) though the energetic order of the MOs varies depending on the method applied [65, 71, 72]. However, the NO variant $Ru_3(CO)_{10}(\mu\text{-}NO)_2$ does have a pair of bridging nitrosyls and a structure similar to bridging $Fe_3(CO)_{12}$ with extended bridged Ru–Ru distance, 3.150 Å [73] (versus the bridged Fe–Fe distance in bridging $Fe_3(CO)_{12}$, 2.558 Å

[60]). Such a long distance intuitively suggests the absence of a metal–metal bond [74]. A recent computational paper managed to reproduce the experimental metric data of $Ru_3(CO)_{10}(\mu\text{-}NO)_2$ to a good extent, but the original authors did not investigate the electronic structures in details [75]. No other theoretical work addresses this problem.

The MO diagram developed for $Fe_3(CO)_{10}$ fragment (Fig. 17b), whose HOMO (b_1) and LUMO (a_2) are the metal–metal bonding and antibonding orbitals between Fe_1 and Fe_2, respectively, is assumed to be applicable to the $Ru_3(CO)_{10}$ fragment as well [65]. To form an imaginary bridged $Ru_3(CO)_{12}$ cluster, two bridging carbonyls are added to the $Ru_3(CO)_{10}$ fragment and they stabilize the a_2 and destabilize the b_1, so that the a_2 becomes the HOMO of the bridged $Ru_3(CO)_{12}$ cluster (Fig. 17d). This is not the observed structure of $Ru_3(CO)_{12}$ because the heavier transition metals generally have stronger metal–metal bonds with more diffuse *d*-orbitals than first-transition-row metals, and backbonding from CO may fail to stabilize the antibonding orbital to a satisfactory extent so that the overall stability of the bridging $Ru_3(CO)_{12}$ structure underperforms that of the terminal $Ru_3(CO)_{12}$ one [65]. In contrast, the addition of two bridging nitrosyls brings two more 2π electrons to the $Ru_3(CO)_{10}(\mu\text{-}NO)_2$. These two additional electrons would be accommodated by the LUMO of the bridged $Ru_3(CO)_{12}$ structure (the antibonding b_2 in Fig. 17d) [65, 67]. Now, both HOMO (b_2) and HOMO-1 (a_2) of $Ru_3(CO)_{10}(\mu\text{-}NO)_2$ are bridging Rh–Rh antibonding orbitals and are stabilized by NO backbonding interaction; hence, a very longer Ru–Ru distance is expected for this bridged framework. From a different perspective, the all-terminal $Ru_3(CO)_{10}(NO)_2$ also has two more electrons to be accommodated and they are likely to fill the antibonding $a_{2'}$ (Fig. 16) if NO ligand is recognized as NO^+. This antibonding orbital does not receive any backbonding stabilization. It lowers the average bond order of three Rh–Rh bonds to 0.67 and may further cause the rupture of one Ru–Ru bond to accommodate single bonds to the other Ru. Thus, the added electrons from NO destabilize the terminal structures and help maximize the backbonding capacity of bridging NO in the bridging $Ru_3(CO)_{10}(\mu\text{-}NO)_2$ structure. As a result, the bridging structure is more stable than the terminal structure. Such analysis leads to an interesting question whether $Ru_3(CO)_{12}{}^{2-}$, the reduced form of $Ru_3(CO)_{12}$, has a bridging structure. In addition, does the electronegativity of N also contribute to the stability of the bridging structure? Albeit, $Ru_3(CO)_{12}{}^{2-}$ was only probed by cyclic voltammetry, was reduced from terminal $Ru_3(CO)_{12}$ with an ECE mechanism, and is apparently unstable [76, 77].

It is clear in the previous cases studied that the distance between the metals bridged by carbonyl(s) and nitrosyl(s) is rarely relevant to the formal bond order. The metals are held together or separated by the bridging ligands rather than the pure metal–metal interactions. The long Ru–Ru distance in the $Ru_2(\mu\text{-}NO)_2$ framework correctly attests to the absence of the metal–metal bond, but further, its length arises from occupation of two Ru–Ru antibonding orbitals, which produce repulsive interactions between two Ru with a formal bond order of negative two.

4 Semibridging Carbonyls

The semibridging carbonyl is the most general case of bridging carbonyls [5, 10], where in the simplest case, one carbonyl bridges two metal atoms with two different C–M bond lengths. Depending on the M_1–C–O bond angle θ (defined in the previous section), the semibridging CO can be classified as bent ($\theta < 170° \sim 165°$) or linear ($\theta > 170° \sim 165°$). The bent semibridging carbonyl is more common and historically, the first example was $[C_4(CH_3)_2(OH)_2]Fe_2(CO)_6$ [78] as shown in Fig. 18a. As summarized by Cotton [4, 79], such semibridging carbonyls provide a mechanism to delocalize the excess negative charge on the metal that the CO does not terminally bind to and is usually accompanied by a highly polarized metal–metal bond or dative bond. Thus, the semibridging carbonyls help maintain the neutrality of the two metals bridged by the carbonyls [4]. In $[C_4(CH_3)_2(OH)_2]$ $Fe_2(CO)_6$ (Fig. 18a), M_1, with the shorter M–C bond length, has an electron count of 18, while the other metal core M_2 has an electron count of 16, if the metal–metal bond is neglected. It is natural to recognize the M_1–M_2 bond as a dative bond to satisfy 18-*e* rule, but the imbalance of electron densities is also created as M_1 donates to M_2. The surplus electron density on M_2 can be reduced by back-donating to the semibridging CO which primarily resides on M_1. The backbonding is achieved by the overlap of metal's occupied *d*-orbital and CO's empty 2π orbital in a fashion that is similar to a terminal CO though the relative orientations of the orbitals are different.

The linear semibridging was first observed in 1975 on $Mn_2(CO)_5(PH_2PCH_2PPh_2)_2$ by Curtis et al. [80] followed by the examples like $[MoCp(CO)_2]_2$ which has been thoroughly investigated by theoretical studies [81, 82]. In $[MoCp(CO)_2]_2$ (Fig. 18b), each molybdenum atom has 15 electrons if all the carbonyls are treated as terminal ligands; thus, metal–metal triple bonds are needed by the 18-*e* rule. Furthermore, the acute M_2–M_1–C bond angle $\varphi = 67.4°$ and lower IR frequencies were attributed to additional bonding interaction [82]. (C–M_1 bond length is shorter than C–M_2, as defined in the introductory section.) In addition to the usual σ-donation to M_1 as a terminal CO, the linear semibridging CO was thought to donate electrons from its 1π orbitals to the second metal M_2. That both C and O from the carbonyl are involved in the second bond was supported by extended Huckel overlap population[82] between O and M_2. In this case, the semibridging CO was recognized as a "four-electron donor"; Curtis further proposed that if a semibridging carbonyl is a four-electron donor, its "asymmetric parameter" α ($\alpha = (d_2\text{-}d_1)/d_1$, in which d_1 and d_2 refer to C–M_1 and C–M_2 distances, respectively) is not relevant to the M_1–C–O angle θ, while other semibridging COs that accept charge from M_2 through their 2π orbital should have positive correlation between θ and α [83].

Crabtree et al. created a very different classification scheme in which he systematically classifies the linear semibridging carbonyls into four subclasses depending on M_2 and structures [5]. Type I complexes consist of early transition metals (up to group 6; Certain complexes with group 6 metals may also be classified

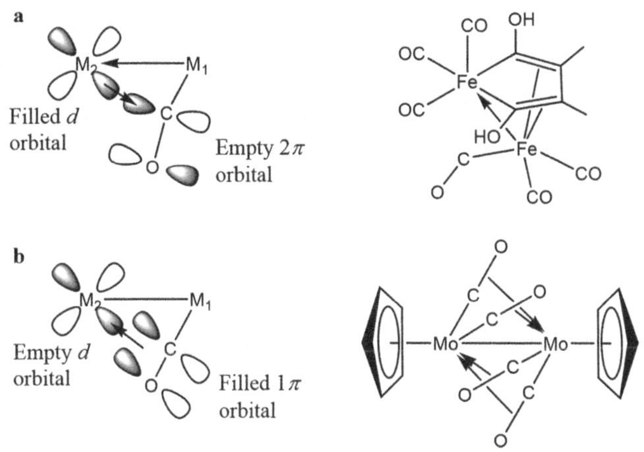

Fig. 18 Two typical scenarios of semibridging carbonyls. (**a**) The rebalancing of electron density with semibridging carbonyls as an electron acceptor with the example of $[C_4(CH_3)_2(OH)_2]Fe_2(CO)_6$; (**b**) the four-electron donor scheme of semibridging carbonyls, with the example of $[CpMo(CO)_2]_2$ [81, 82]; however, such a scheme was not really supported by calculations (see text)

as type II). This type of linear semibridging carbonyls has two structural characteristics: (1) the M_2–M_1–C angle φ is extraordinarily small, < ~50° (compared to ca. 70° in complexes of other types); (2) though M_1–C–O angle θ is close to 180°, it is still slightly bent toward M_2; types II–IV linear semibridging carbonyls are all bent away from M_2. Type II is the most general case, usually having middle or late transition metals; type III complexes contain metals that are known not to have carbonyl complexes or are not expected to provide electrons for backbonding, including very late transition metals Au, Cu, or even Zn and Ga (usually group 11/12) which can be described as if they were main group elements; type IV complexes are special because they have a special geometry with a backbone eight-member ring $M_2(PCP)_2$ (e.g., $Mn_2(t\text{-}CO)_4(\mu\text{-}CO)(\mu\text{-}dmmp)_2$).

4.1 *Bridging Carbonyl vs. Semibridging Carbonyl I: Mn_2(t-CO)$_4$(μ-CO)(μ-dmmp)$_2$, Fe_2(t-CO)$_4$(μ-CO)(μ-dmmp)$_2$, and Mn_2(t-CO)$_6$(μ-dmmp)$_2$*

$Mn_2(t\text{-}CO)_4(\mu\text{-}CO)(\mu\text{-}dmmp)_2$ (Crabtree's type IV) is the earliest example featuring a linear semibridging carbonyl, though the exact Mn-(μ-C)-O angle was not originally reported [80]. In comparison, the bridging carbonyl in $Rh_2(t\text{-}CO)_2(\mu\text{-}CO)(\mu\text{-}dmmp)_2$ (Fig. 19) is not so linear with a (Rh)–(μ-C)–O angle of 158.2°. Both of them look similar to (but are not necessarily classified as) so-called A-frame, M_2L_2L', in which L is the bidentate bridging dppm or similar ligand and L′ is the ligand at the apex of the letter "Λ" with mirror symmetry. The general electronic structure of "A-frame" complexes was investigated by Hoffmann [84], though the

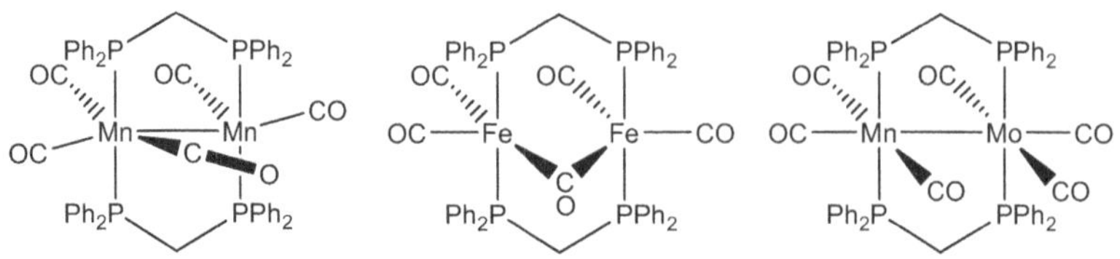

Fig. 19 The molecular structures of $Mn_2(t\text{-}CO)_4(\mu\text{-}CO)(\mu\text{-}dmmp)_2$, $Fe_2(t\text{-}CO)_4(\mu\text{-}CO)(\mu\text{-}dmmp)_2$, and $Mn_2(t\text{-}CO)_6(\mu\text{-}dmmp)_2$

general solution may not be applied to these two cases as the bridging carbonyls do not really stay on the apex with the ideal symmetry. The discussions of electronic structures in details for both of them are presented in the following two sections. More complexes featuring a linear bridging carbonyl can be found in the review [10].

The first theoretical investigation of this type of linear semibridging carbonyl was carried out by Benard et al. on the model of $[Mn(t\text{-}CO)_2(PH_3)_2]_2(\mu\text{-}CO)$ to simulate $Mn_2(t\text{-}CO)_4(\mu\text{-}CO)(\mu\text{-}dmmp)_2$ [85]. The molecule can be assembled with two $M(t\text{-}CO)_2(PH_3)_2$ fragments and one bridging carbonyl. Currently the reader should be very familiar with orbital diagram of ML_4 or $M(CO)_4$ (Fig. 15), i.e., the 3 +2 pattern, where three is lower d-orbitals and two is higher-energy unused d^2sp^3 hybrid orbitals. The replacement of two of four carbonyls by phosphines does not qualitatively change the 3+2 pattern of orbitals. Each Mn(0) atom has 7 d electrons in these ML_4 fragments, respectively. Like the previous examples, one can neglect the three lower d-orbitals as they are all doubly occupied and will not contribute to the overall M–M bond order. The two unused d^2sp^3 orbitals on each MnL_4 form two bonding orbitals and two antibonding orbitals, but only the lowest of the four is doubly occupied in $Mn_2(t\text{-}CO)_4(dmmp)_2$ with absence of the bridging carbonyl (middle panel of Fig. 20). Due to the dissymmetry introduced by the semibridging carbonyl, the orbitals are not as perfectly overlapped as shown in Fig. 20.

The molecular orbital diagram for $Mn_2(t\text{-}CO)_4(\mu\text{-}CO)(dmmp)_2$ with the semibridging carbonyl added is presented in the left part of Fig. 20. The semibridging carbonyl is oriented in such a way that its in-plane $1\pi/2\pi$ orbitals may interact heavily with those metal–metal orbitals. Especially the in-plane p-contribution from the carbon is pointing to the midpoint of the vector connecting two metal atoms and has overlapped with d-orbitals from the metals. Though 1π and 2π all contain major contributions from the p orbitals of the carbon and the oxygen, they interact with metal orbitals in different ways. The CO's 1π orbital is stabilized by the metal orbitals to a very limited extent due to the large energy difference between these orbitals. In contrast, the carbon's p-contribution in the 2π orbital can mix strongly with HOMO (i.e., the metal–metal π bond) of $Mn_2(t\text{-}CO)_4(PH_3)_4$, forming a de facto three-center two-electron bond; such delocalization results in the net stabilization of bonding interaction between metals. In other words, the stability of the M–M bond is reinforced by the involvement of the 2π orbital through the carbon's p-contribution. Population analysis also indicates the major charge transfer from the metal to the in-plane 2π orbital of the semibridging carbonyl. On the

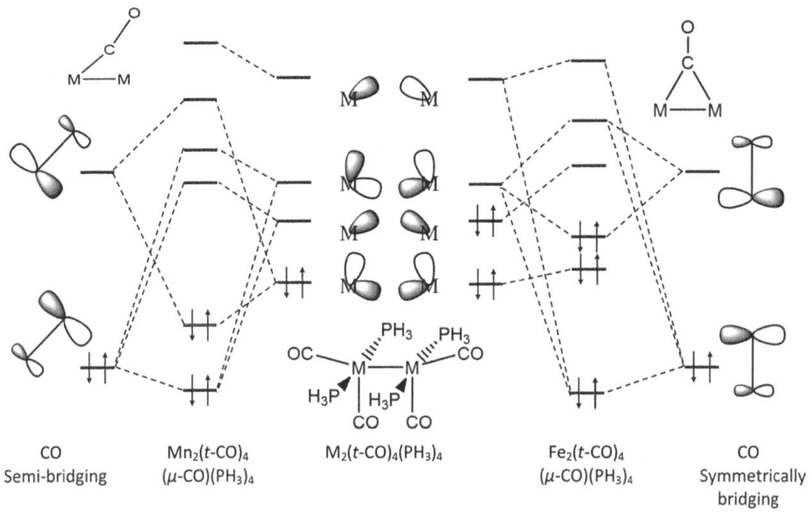

Fig. 20 Molecular orbital diagrams of $Mn_2(t\text{-}CO)_4(\mu\text{-}CO)(dmmp)_2$, and $Fe_2(t\text{-}CO)_4(\mu\text{-}CO)(dmmp)_2$, which differ by two electrons [85]. The σ-donation from the bridging carbonyl was not drawn explicitly

other hand, the deformation density map does not show the buildup of electron density between O of the semibridging carbonyl and the second metal M_2, revealing the absence of orbital interaction between *p* orbitals of O and *d*-orbitals of M_2. Such evidence together contradicts the four-electron donor proposal, in which the carbonyl keeps a linear binding mode to donate its 1π electrons to the metal as if it was an olefin. Interestingly, the Mn–Mn distance is 2.934 Å, longer than an ordinary Mn–Mn bond (~2.5 Å) though the formal bond order is 1 as indicated by the previous orbital analysis. The separation does not destabilize the metal–metal bond as the extra space is made for the insertion of the *p*-contribution of the carbon as the "relay" of the bonding interaction. The linearity of the semibridging carbonyl may be better rationalized by maximizing such a "bonding relay."

For comparison, the molecular orbital diagram of $Fe_2(t\text{-}CO)_4(\mu\text{-}CO)(dmmp)_2$ is presented in the right panel of Fig. 20. The iron complex has two more electrons and the semibridging carbonyl shifts to a symmetric bridging position. The interaction between the bridging CO and the metal dimer is similar to previous examples: metal–metal σ bonding orbital is destabilized by ligand 5σ-donation and the metal–metal π-antibonding orbital is stabilized by ligand 2π-acceptance. The stabilization also swaps the order of HOMO (σ bonding) and LUMO (π-antibonding) of $Fe_2(t\text{-}CO)_4(PH_3)_4$, reducing the formal bond order from 2 of $Fe_2(t\text{-}CO)_4(PH_3)_4$ to 0 of $Fe_2(t\text{-}CO)_4(\mu\text{-}CO)(PH_3)_4$. In such a bridging pattern, the carbonyl tends to hold two metals together to optimize the overlap between its 2π orbital and *d*-orbitals from both metal atoms. Therefore, the Fe–Fe distance is relatively short, 2.709 Å, regardless of the formal nonbonding status between two metal atoms [85].

The dynamic NMR experiment indicates that the semibridging CO migrates between the two Mn atoms in $Mn_2(t\text{-}CO)_4(\mu\text{-}CO)(\mu\text{-}dmmp)_2$ with a barrier >15.6 kcal/mol, presumably through a transition state with a symmetrically bridging carbonyl [86]. Carbonylation of $Mn_2(t\text{-}CO)_4(\mu\text{-}CO)(\mu\text{-}dmmp)_2$ into $Mn_2(t\text{-}CO)_6(\mu\text{-}dmmp)_2$ happens easily upon the addition of CO under pressure [86, 87]. Unlike $Mn_2(t\text{-}CO)_4(\mu\text{-}CO)(\mu\text{-}dmmp)_2$, $Mn_2(t\text{-}CO)_6(\mu\text{-}dmmp)_2$ has a global CO scrambling which makes the carbonyls indistinguishable at 30°C by ^{13}C NMR. The broad carbonyl peak splits into two peaks with intensity ratio of 2:1 at −75°C confirming that the two dppm ligands indeed bridge two metal atoms [87]. DFT calculations reported an adjusted barrier of 13 kcal/mol of the global CO scrambling, through a transition state which features two symmetrically bridging carbonyls, but the original authors thought this value is still overestimated by 5–10 kcal/mol compared to dynamic NMR investigation [88]. The general gist is that it is easier to have two symmetrically bridging carbonyls than one in the rearrangement mechanism of $Mn_2(t\text{-}CO)_x(\mu\text{-}dmmp)_2$. The synergic effects of backbonding from two carbonyls are more capable of flipping the order of higher *d*-orbitals and stabilizing the system overall.

4.2 *Bridging Carbonyl vs. Semibridging Carbonyl II: $[Rh_2(t\text{-}CO)_2(\mu\text{-}CO)(\mu\text{-}dmmp)_2]^{0/2+}$*

Compared to $Mn_2(t\text{-}CO)_4(\mu\text{-}CO)(dmmp)_2$, each Rh in $Rh_2(t\text{-}CO)_2(\mu\text{-}CO)(dmmp)_2$ (Fig. 21) has two more *d* electrons, but two fewer σ-electrons because of the loss of terminal COs. The electronic structure of $Rh_2(t\text{-}CO)_2(\mu\text{-}CO)(dmmp)_2$ may be understood by assembling a planar $Rh_2(t\text{-}CO)_2(dmmp)_2$ fragment and then adding the bridging carbonyl [89]. The nearly planar $Rh_2(t\text{-}CO)_2(dmmp)_2$ fragment features two Rh(0, d^9) atoms and a single metal–metal bond, with the configuration (in cylindrical symmetry) of $(\sigma)^2(\pi)^4(\pi^*)^4(\delta)^4(\delta^*)^4$; the only vacant metal–metal orbital is the σ* orbital and it is relatively high in energy [89]. The semibridging carbonyl is not symmetric, but if a symmetric bridging CO is applied to the $Rh_2(t\text{-}CO)_2(dmmp)_2$ fragment, the transient complex would have a C_{2v} symmetry. Such C_{2v} complex can be treated as the transition state for shifting the unsymmetrical bridging carbonyl from one Rh to the other, as observed by VT-NMR [90]. The symmetrically bridging carbonyl can destabilize the metal–metal σ bond and stabilize the metal–metal π* bond in $Rh_2(\mu\text{-}CO)$ plane, through its σ-donation and π-acceptance in the manner discussed in previous sections. The σ* antibond, although it shares the same symmetry as the in-plane ($Rh_2(\mu\text{-}CO)$ plane) π* bond, is too high in energy to interact with the bridging carbonyl. One symmetric bridging carbonyl is not strong enough to change the order of metal–metal σ and σ* orbitals and presumably does not provide much net stabilization. In comparison, the oxidized form, $[Rh_2(t\text{-}CO)_2(\mu\text{-}CO)(dmmp)_2]^{2+}$, which has vacant metal–metal σ orbital[90], has a nearly symmetrically bridging carbonyl as the symmetrically

Fig. 21 The sketches of the $Rh_2(t\text{-}CO)_2(\mu\text{-}CO)(\mu\text{-}dmmp)_2$ and $[Rh_2(t\text{-}CO)_2(\mu\text{-}CO)(\mu\text{-}dmmp)_2]^{2+}$

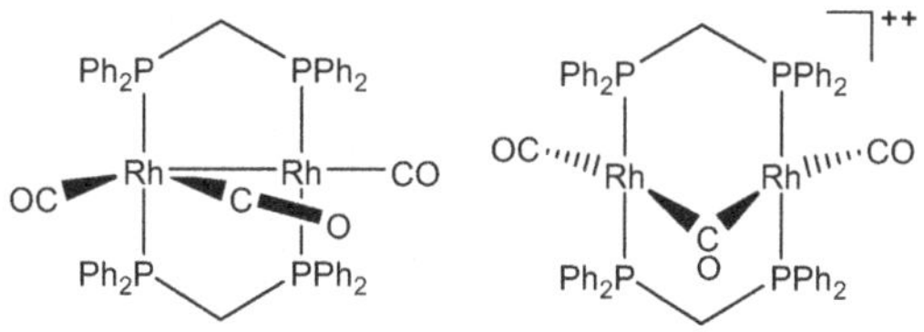

bridging carbonyl is able to provide net stabilization to the system through backbonding [89].

4.3 *Comments on the Occurrence of the Asymmetric Bridging Carbonyl*

At this moment, one may already have a general feeling about the orientation of the bridging carbonyl. A symmetric bridging carbonyl is actually harmful to the (occupied) metal–metal bond, especially for σ-type metal–metal bond, because the symmetric bridging carbonyl donates its σ-electron to the midpoint of metal–metal σ bond, where a critical point of electron density is expected and such donation has to overcome large electronic repulsion. In order to keep a symmetric bridging position, the carbonyl (or nitrosyl) needs either a vacant metal–metal σ-orbital to reduce the mentioned repulsion or one or more occupied metal–metal antibonding orbital with proper symmetry to stabilize the bond between the metal dimer and the carbonyl through backbonding.

On the one hand, each carbonyl is a bifunctional ligand as it is an L ligand with its σ-donation orbital but can also be treated as somewhat a Z ligand with its π* orbitals; on the other hand, these orbitals are very directional and hard to polarize so that the metal–ligand interactions through those orbitals are dependent on the orientation of the carbonyl. Besides, the occupancies of metal-based orbitals may also be altered accompanying CO's twofold effects on the orbital energies.

When the two metal atoms have a vacant metal–metal σ bonding MO, the carbonyl may occupy the symmetric bridging position to donate its σ-electrons into this bonding MO. However, if the two-metal system features a net metal–metal bond, the carbonyl can either choose to have a semibridging orientation, pointing its vacant 2π to the center of the metal–metal σ bond to stabilize the σ bond through three-center interaction, or alternatively, the carbonyl may keep a symmetrically bridging position, which alters the order of the metal–metal bonding and antibonding MOs and pushes the electrons into the antibonding MO, which is heavily stabilized by the π-backbonding. One carbonyl may not be strong enough to alter these orbitals, but multiple bridging carbonyls may work in a synergic way to reinforce each other. This is why symmetric bridging carbonyls usually appear on systems where a metal–metal σ bond is expected by 18-*e* rule, explaining Cotton's early prediction, which was at first proposed to explain the short M–M distance [24].

The next section concerns a system with expected metal–metal multiple bonds.

4.4 Linear Semibridging Carbonyls I: $[CpMo(CO)_2]_2$

The structure of $[CpMo(CO)_2]_2$ (Crabtree's type II) is illustrated in Fig. 18, the two cyclopentadienyls are almost collinear, and the carbonyl pairs attached to each molybdenum atom are oriented over the other molybdenum atom, but the Mo–C–O angle is quite linear, around 176°. As mentioned earlier, those semibridging carbonyls were proposed to be four-electron donors by experimental chemists. The observation by Curtis et al. that [82] the IR frequencies of these semibridging CO (~1,900 cm^{-1}) are lower than those of conventional terminal CO was thought to be indicative of the depopulation of CO 1π orbital. The linearity of the semibridging carbonyl was the result of both its carbon and oxygen interacting with the second metal atom as the charge is donated from CO 1π orbital.

Extended Huckel calculations and molecular orbital analyses from Hoffmann, building the system from ML_5 ($Mo(CO)_5$) fragments, give an alternative explanation to the electronic structure of $[CpMo(CO)_2]_2$ [91]. A related scheme built directly from $CpMo(CO)_2$ moieties was presented by Hall et al. [92], where the order of orbitals is slightly different due to the different methods applied. Like M $(CO)_3$ in Fig. 9 and $M(CO)_4$ in Fig. 15, a $M(CO)_5$ fragment needs five of the six d^2sp^3 hybrid orbitals as incoming donor orbitals, leaving only unused, energetically above the three "pure" d-orbitals as in previous examples. According to the symmetry, the unused hybrid orbital should be dominated by d_{z^2} from the metal (Fig. 22). The imaginary compound $[Mo(I, d^5)_2(CO)_{10}]^{2+}$ (the second column of Fig. 22) has a structure similar to $Mn(d^7)_2(CO)_{10}$ but could have three metal–metal bonds: one σ, two π bonds, and a pair of δ/δ^* bonds canceling each other if the Mo–Mo bond were short enough that the degenerate π^* orbitals shown as the LUMO are pushed above the σ bond (note that this situation was dismissed in earlier descriptions of the placement of the pure d and d^2sp^3 hybrid orbitals.) Nevertheless, the net metal–metal bond order is three if the order is as shown in Fig. 22, and adding four more electrons to make a molecule such as $Mn(d^7)_2(CO)_{10}$ reduces the bond order to a single M–M bond.

The cyclopentadienyl anion (Cp^-) can be treated as a tridentate L_3 (in ionic counting) ligand and replace three carbonyls on the $[Mo(CO)_5]^+$ moiety. The cyclopentadienyl is a much poorer π-acceptor; thus, the splitting between pseudo-"e_g" and "t_{2g}" in $CpM(CO)_2$ is smaller and such replacement may alter the order of orbitals even though the other two terminal carbonyls are kept (i.e., $[CpMo(t\text{-}CO)_2]_2$, the third column in Fig. 22). Finally, each $CpMo(CO)_2$, which still features terminal carbonyls, must rotate to bring the terminal carbonyls into their semibridging positions and the two Cp rings collinear with two Mo atoms. The molecular orbitals after the rotation (as depicted in Fig. 23a) are drawn in the rightmost column of Fig. 22. Note

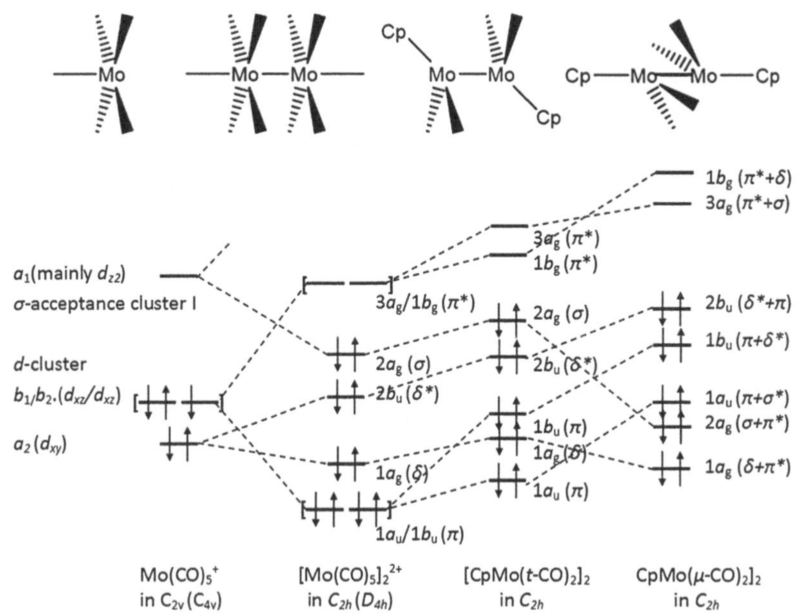

Fig. 22 The molecular orbital diagram of $[CpMo(\mu\text{-}CO)_2]_2$, reproduced from Hoffmann's paper [91] with modifications. The carbonyls are omitted in these drawings

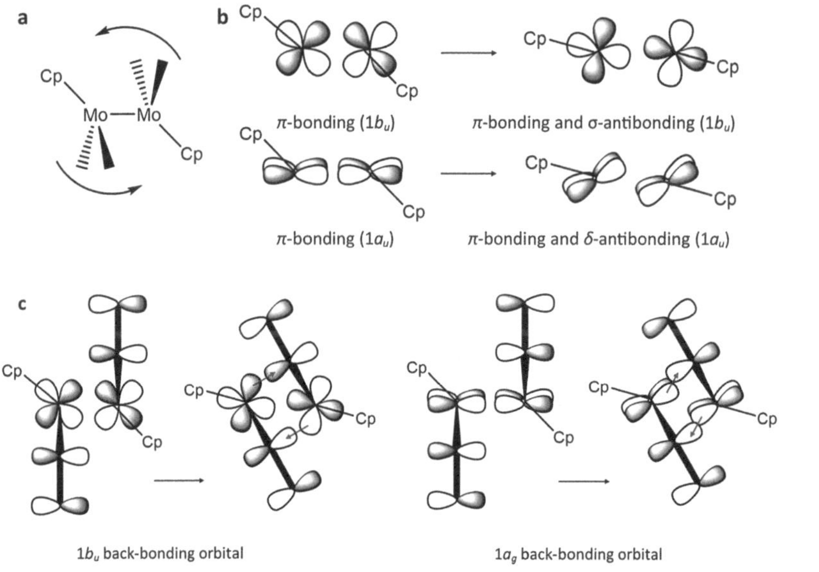

Fig. 23 (**a**) The rotation of Mo moieties to the experimental structure. (**b**) The change of metal–metal bonds after the rotation. (**c**) The additional π-backbonding is established after rotation. The 2π orbital, from only one carbonyl that goes out of the plane of each carbonyl pair, is drawn for clarity

that cylindrical symmetry assignments $\sigma/\pi/\delta$ along the Mo–Mo axis no longer exist in the rotated C_{2h} geometry, but these designations were kept to help visualize these orbitals. During the rotation, the bonding properties of these orbital may change dramatically (see Fig. 23b) if it is assumed fragment orbitals of each CpMo(t-CO)$_2$ moiety are kept the same; designations are provided in the parentheses in Fig. 22 to indicate original and newly gained bonding/antibonding characters before and after the rotation. However, one should note that molecular orbitals are free to reorganize to keep the lowest energy and the final configuration between two molybdenum atoms might be written as $(\sigma)^2(\pi)^4(\delta)^2(\delta^*)^2$ or $(\pi)^2(\pi^*)^4(\delta^*)^4$ or something in between depending on the extent of orbital reorganization introduced by rotation. A fractional metal–metal bond is lost as the overlap between d-orbitals is not optimal.

The rotation transforms the terminal carbonyls into the semibridging carbonyls and creates additional contact between CO and the other Mo as it tilts the 2π orbital of CO, whose lobes were parallel with the Mo–Mo vector, toward the other Mo, to form "banana" bonds (Fig. 23c). This is the key factor that compensates the inferior metal–metal contact (Fig. 23b) in the bonding orbitals. The stabilization gained from four semibridging carbonyls overwhelmed the destabilization of any metal–metal antibonding character and explains the preference over the typical semibridging orientations, which stabilize M–M antibonding orbitals.

It was proposed [82] that the linearity of the semibridging carbonyls can help maximize the overlap between O and Mo in this case and helps in the donation from 1π orbital to the d-orbital. However, population analysis [91] of [CpMo(μ-CO)$_2$]$_2$ gives small, negative overlap between O and the other Mo and large, positive overlap between C and the other Mo. Additionally, each linear semibridging CO in [CpMo(μ-CO)$_2$]$_2$ has ca. 0.09 more electrons than the terminal CO in [CpMo(t-CO)$_2$]$_2$ and each Mo atom has ca. 0.18 fewer electrons. These two pieces of information indicate that the additional contact is π-backbonding rather than π-donation, a description that offers a better explanation of the linearity of the carbonyls.

Hall et al. [92] reported Fenske–Hall calculations and argued that the linearity of these semibridging carbonyls is related to the extreme unsaturation of metal cores which require multiple metal–metal bonds to fulfill 18-e rules. Previous examples, like $Fe_2(CO)_9$ and $(CpCO)_2(CO)_2$, only require one additional "formal" metal–metal bond to reach the 18-e rule, and effectively, one of the symmetrically bridging carbonyls donates its 5σ-electron pair to both metals, while all the bridging COs stabilize M–M antibonding orbitals with their 2π orbitals. The carbonyl's 2π is able to stabilize metal–metal bonding and antibonding orbitals depending on its orientation as shown in Fig. 24. The linear semibridging carbonyl is able to provide the C p orbital of its 2π to the metals to form a three-center interaction that resembles the electron-deficient (banana) bond in B_2H_6 (Fig. 24). In such a manner, the multiple metal–metal bonds can be kept and stabilized. On the other hand, a bent bridging carbonyl inserts both lobes (+ and −) of the C p orbital between two metals and cannot stabilize the bonding orbitals but does stabilize antibonding ones (Fig. 24). In addition, the 5σ-donor orbital of a linear semibridging CO points to the center of one metal atom rather than the midpoint of the two metals and further destabilizes the metal–metal σ bond. Thus, for metals having net multiple bonds,

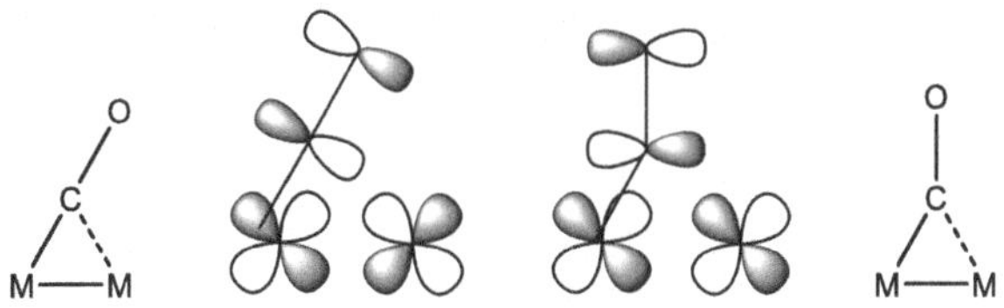

Fig. 24 The sketches of π-backbonding modes of linear and bent semibridging carbonyl connecting two metal atoms. The former stabilizes the M–M bonding, while the latter stabilizes the M–M antibonding

it is very energetically inefficient to have symmetric bridging or bent semibridging carbonyls.

Further, Fenske–Hall calculations were able to show that Crabtree's type III complexes (which contain unsaturated very late transition metals that traditionally are not expected to be good at donating backbonding electrons, e.g., $Cu(NR_3)_2$ or Au (PR_3)) may be able to interact more strongly with the linear semibridging carbonyls [93, 94].

Later, Hartree–Fock (HF) calculations were performed on a 6-atom model system $M_2(CO)_2$ (M=Mn or Sc), in which all six atoms are coplanar and maintain C_{2v} symmetry, while the carbonyls were varied from terminal to bridging (as in the unsymmetrical bridging carbonyl pair illustrated in Fig. 14a) [95]. The result predicted that configurations with different occupied metal–metal orbitals show different preferences: generally, configurations with occupied $\sigma(a_g)$, $\pi_{ip}(b_u)$, and $\pi_{ip}{}^*(a_g)$ prefer a terminal carbonyl but configurations with filled $\sigma^*(b_u)$, $\delta_{ip}(a_g)$, and $\delta^*{}_{ip}(b_u)$ prefer a semibridging carbonyl (note: ip means in-plane). Furthermore, the calculations revealed that when a terminal carbonyl is preferred, but the M_2–M_1–C angle φ is forced to reduce to yield a bridging carbonyl, then the bridging carbonyl itself is always linear. However, for all of the electronic configurations that yielded bridging carbonyls naturally, they were always bent. The configuration of [CpMo $(CO)_2]_2$ (which has two nearly orthogonal $M_2(CO)_2$ planes) can be written as $(\sigma)^2(\pi_{ip})^4(\delta_{op})^2(\delta^*{}_{op})^2$, which would appear to have a preference for terminal carbonyls; although the COs are semibridging, they are linear as predicted. It would appear that on purely electronic configuration arguments, the COs should be terminal, but they are "forced" to adopt the semibridging orientation due to the steric crowding and the metal need to form bonds and thus finally appear as linear semibridging carbonyls. In this regard, the linear semibridging carbonyls can be understood as a special case of the conventional terminal carbonyl as the major electronic interaction, including σ-donation and π-acceptance between CO and its closest M, remains the same. The stabilization by the new interaction between the other M and CO, as presented in Fig. 24, is verified by calculations at SCF level [96]; but it may be secondary, in part, and a response to the carbonyls being forced to be bridging by steric constrains. The same steric-driven, π-interaction-stabilized explanation also applies the crowded $[Ta(CO)_4]_2$ dimer, which features 3 or 4 linear semibridging carbonyls depending on the structures of isomers [97].

4.5 Linear Semibridging Carbonyls II: [HB(pz)$_3$](CO)$_2$W (μ-CS)MO(CO)$_2$(Cp)

None of the examples presented above support the idea that the carbonyl can be a four-electron donor, as 1π orbitals of CO are generally very low in energy and will be poor donors. The final example here discusses if any possibilities exist that a carbonyl or similar ligand may act as a four-electron donor and the best candidate is the Crabtree's type I complex. For comparison, the thiocarbonyl-bridged species is introduced here, as the thiocarbonyl is both a better π-donor and a better π-acceptor; hopefully, this difference may "magnify" the ligand–metal π-donor interaction (Fig. 25).

The real-world molecules investigated are [HB(pz)$_3$](CO)$_2$W(μ-CS)Mo (CO)$_2$(In) and its carbonyl analogue [HB(pz)$_3$](CO)$_2$W(μ-CO)Mo(CO)$_2$(In) [98], in which $HB(pz)_3^- = HB(N_2C_3H_3)_3^-$ and In-=η^5-$C_9H_7^-$. The structure of the former species, which features a Crabtree's type I linear semibridging carbonyl with W–C–S angle of 170.8°, was determined by X-ray diffraction and the latter is only stable in solution. The In^- ring was replaced by a simpler Cp^- ring in the theoretical analysis [94].

Fenske–Hall calculations were done with this system [94]. The complex [HB (pz)$_3$](CO)$_2$W(μ-CS)Mo(CO)$_2$(Cp) is arbitrarily divided into two ion fragments [[HB(pz)$_3$](CO)$_2$W^{II}]$^+$ and [(μ-CS)Mo^0(CO)$_2$(Cp)]$^-$. Such an assignment of charge helps keep the closed-shell wave function for both fragments, but the initial charge assignment does not necessarily change the final partitioning of the molecular charges. The former W ion is an ML_5 fragment and the later Mo ML_6, which are expected to have 3+1 and 3+0 orbital patterns just like the 3+3 and 3+2 fragment frontier orbital patterns of ML_3 and ML_4, respectively. Interestingly, the unused d^2sp^3 hybrid orbital of the W fragment points toward the semibridging thiocarbonyl. (Again, one should note those hybridizations are used as indications of the directions of the orbital lobes rather than the implication of orbital contributions. This unused d^2sp^3 hybrid orbital actually should be dominated by metal's d contribution.)

Upon combination of two fragments, the lower three d-orbitals from each fragment interact to form three bonding orbitals and three antibonding orbitals (Fig. 26). It is difficult to assign those bonds to σ/π/δ-type due to the distortion and mixing of orbitals, but it can be clearly seen that three bonding orbitals and two antibonding orbitals are filled by a total of 10 d electrons and the only unoccupied one of six orbitals is left as LUMO. Such bonding scheme results in a net flow of 0.9

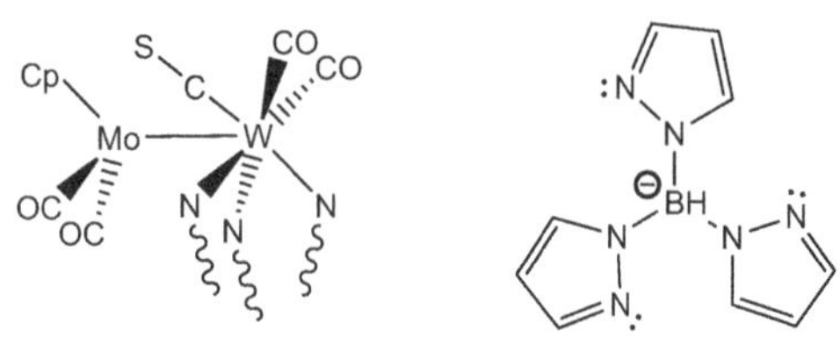

Fig. 25 The molecular structures of [HB(pz)$_3$] (CO)$_2$W(μ-CS)Mo (CO)$_2$(Cp) and its ligand [HB(pz)$_3$]$^-$

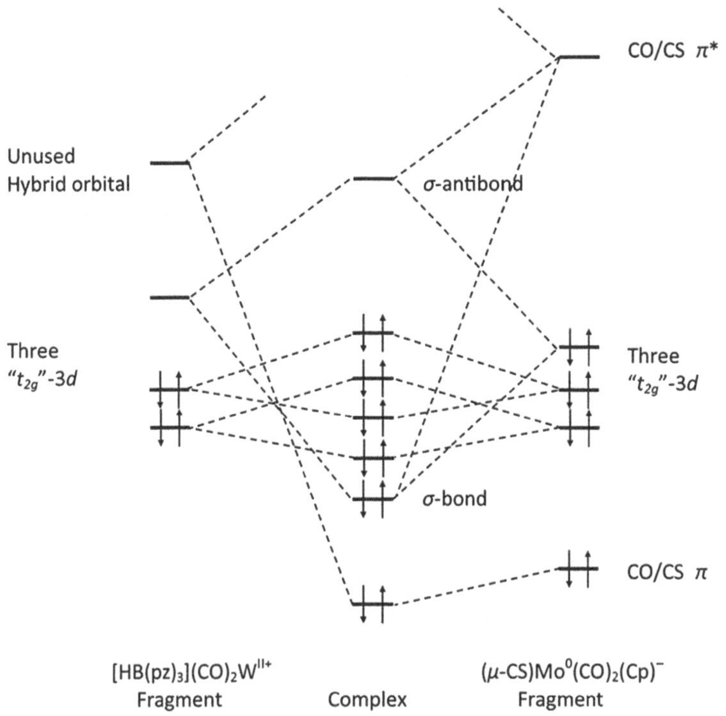

Fig. 26 The molecular orbital diagram of $[HB(pz)_3](CO)_2W(\mu\text{-}CS/CO)Mo(CO)_2(Cp)$

electron from W fragment to Mo fragment and gives in a de facto d^5-d^5 configuration regardless of the initial charge assignment. The single bond between W and Mo is reinforced by the insertion of the π* orbital of the bridging thiocarbonyl as seen in previous case of linear carbonyls. The most interesting new interaction is between the unused hybrid orbital on the Mo fragment and the π orbital of the bridging ligand. The vacant hybrid orbital on the isolated Mo fragment gains 0.246 electron through interaction with the W fragment and the bridging thiocarbonyl donates 0.208 electron from its π orbital. As the plot of relevant orbitals also shows moderate contact between these two orbitals, it is reasonable to assume the electron transfer occurs directly between them. On the other hand, the π* orbital accepts only 0.038 electron from the W unit. In conclusion, the thiocarbonyl's intra-fragment π-acceptance is overwhelmed by its inter-fragment π-donation; thus, a four-electron donor model should be applicable here. For the carbonyl-bridged analogue, $[HB(pz)_3](CO)_2W(\mu\text{-}CO)Mo(CO)_2(Cp)$, the analysis is qualitatively the same, though both π-donation (0.093 *e*) and π-acceptance (0.025 *e*) become less significant. Thus, CS provides better π-donation as its π is energetically closer to the *d*-orbitals of the metals.

Crabtree [5] classified linear bridging carbonyls by metals' positions in the periodic table and suggested different electronic structures for them. The examples presented above indicate that type II–type IV linear semibridging carbonyls actually have similar interaction with the metals, i.e., the carbonyl accepts donation from second M through its π^* orbital. Only type I linear semibridging carbonyl actually realizes the four-electron donor model which was originally proposed for $[CpMo(CO)_2]_2(\mu\text{-}CO)$. However, it turns out that the bridging CO on $[CpMo(CO)_2]_2(\mu\text{-}CO)$ is type II [81, 82]. Another example, the analysis on the complex $CpCo(\mu\text{-}CO)_2ZrCp_2$ [99], which contains one symmetric bridging carbonyl and one type I linear semibridging carbonyl, still supports the idea of four-electron donor, as Zr(II) is even more electron deficient: the semibridging carbonyl donates ca. 0.1 electron through π-donation and retrieves ca. 0.074 electron through π-acceptance. The small net π-donation may be related to the symmetric bridging carbonyl that coexists on the complex. Generally speaking, the second M has to be an electron-deficient early transition metal to accept π-donation from the weakly donating carbonyl.

5 Face Bridging Carbonyl

5.1 Classification of the Face Bridging Carbonyls

All the previous examples are about the carbonyls bridging two metal atoms, or edge bridging. A carbonyl may also bridge three atoms at the same time, face bridging, and can be classified as various types [100]. A type I face bridging carbonyl binds to three metal atoms equally and forms a pseudo-C_3 local symmetry. A type II face bridging carbonyl binds to M_1 and M_3 symmetrically and leans its M_1M_3CO plane toward M_2. Finally, a type III carbonyl binds to M_1 terminally and tilts its M_1–C–O linear bond to the midpoint of M_2–M_3 vector (Fig. 27).

5.2 Type I Face Bridging Carbonyl: $(CpRh)_3(\mu_3\text{-}CO)_2$

Complexes like $(CpNi)_3(\mu\text{-}CO)_2$[101, 102] and $[(C_6H_6)Fe]_3(\mu\text{-}CO)_2$ [103] containing type I face bridging carbonyls were initially synthesized as early as 1958 though the structures were solved later. The electronic structure of type I face bridging carbonyls is illustrated here by the hypothetical $(CpRh)_3(\mu_3\text{-}CO)_2$ [104]

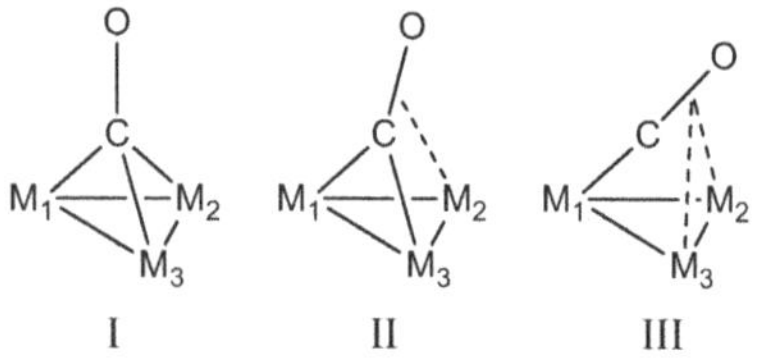

Fig. 27 Face bridging carbonyls. The lines connecting atoms do not necessarily indicate bonds between them

because of its structural similarity to the next example bearing type II face bridging carbonyl. The structure is expected to be close to the crystal structure of $(Cp^{*}Rh)_3(\mu_3\text{-}CO)_2$ as well [105]. The construction of the MO diagram follows Hoffmann's publications [66, 106]. $(CpRh)_3(\mu\text{-}CO)_2$ has a pseudosymmetry of D_{3h} if Cp rings are treated as points. Three CpRh fragments and one $(\mu\text{-}CO)_2$ fragment are needed to assemble the molecule. Each CpRh fragment is an ML_3 fragment and is expected to have 3+3 frontier orbital pattern with eight *d* electrons (six in the three lower *d*-orbitals and two in unused hybrid orbitals). By anticipating the difference between in-plane and out-of-plane interactions in the formation of $(CpRh)_3$, the three unused hybrid orbitals of the CpRh fragment are reorganize to give fragment orbitals with irreducible representations a_1 and e (in pseudo-C_{3v}) (Fig. 28). Once three CpRh fragments interact (center of Fig. 28), the six higher *d* electrons occupy the three bonding orbitals originating from the *e* orbitals of the CpRh fragments and result in single bonds between Rh atoms.

The addition of the two face bridging carbonyls alters the order of metal orbitals in two ways: on one hand, it destabilizes two bonding orbitals $a_{1'}$ and $a_{2''}$ through σ-donation and evacuates the electrons from the latter; on the other hand, it stabilizes the bonding $1e_{'}$ and antibonding $e_{''}$ through π-acceptance, putting electrons in the degenerate e'' orbitals. At this stage, one may realize that the addition of the bridging carbonyls has reduced the average bond order between metals from 1 to 0.33. Green et al. used three-center two-electron bonds to explain the special case of 18-*e* rule in $Fe_2(CO)_9$. Here, the theory may be extended to a four-center two-electron bond as the 5σ orbitals of the bridging carbonyls are interacting with orbitals contributed by three metal atoms equally. In other words, every carbonyl

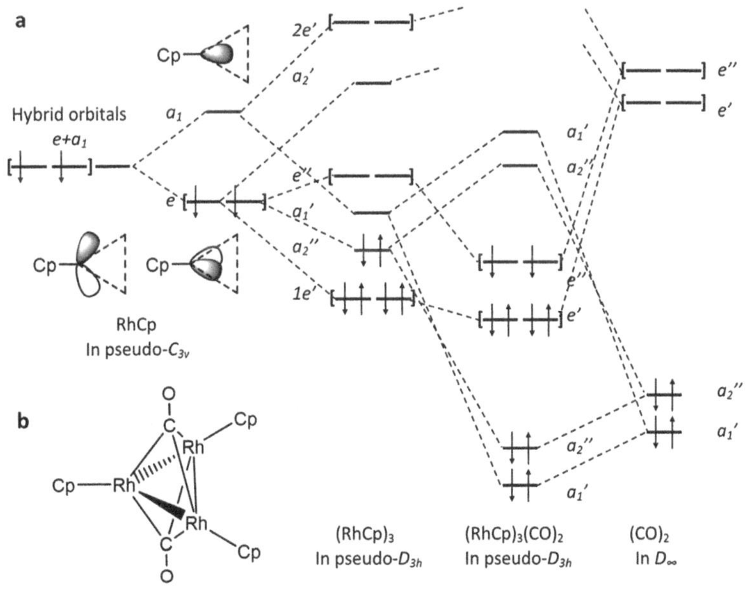

Fig. 28 (**a**) The molecular orbital diagram of $(CpRh)_3(\mu\text{-}CO)_2$ and (**b**) its structure

donates the same two electrons to every metal atom so that each Rh atom can have 18 electrons (6 from Cp^-, 8 from Rh^+, and 4 from CO), if the small (0.33 bond order per Rh–Rh) net metal–metal bonds are neglected (alternatively, two net Rh–Rh bonds and each carbonyl donation to two Rh will also satisfy the 18-*e* rule). A very similar electronic structure [66] was given to the realistic molecule $(CpCo)_3(\mu\text{-}CO)_2$, whose paramagnetic property was confirmed by solution magnetic susceptibility measurement [107]. Theoretical work is also available on $(CpNi)_3(\mu\text{-}CO)_2$ by Fenske–Hall method and the results are essentially the same [108]; of course, here, the Ni–Ni bond order is smaller because of the additional electrons.

5.3 Type II Face Bridging Carbonyl: $[(CpRh)_2(\mu\text{-}CO)_2[Rh(t\text{-}CO)_2]]^-$

The type II face bridging carbonyl was discovered on diamagnetic $[(CpRh)_2(\mu\text{-}CO)_2[Rh(t\text{-}CO)_2]]^-$ [109], followed by other examples like $[(CpCo)_2(\mu\text{-}CO)_2[Pt(PPh_3)_2]]$ [110] and $(CpCo)_2(\mu\text{-}CO)_2(IrCp)$ [111]. In the Rh dimer, the two carbonyls symmetrically bridge the two CpRh fragments with a small puckering (or hinge) angle (C–Rh–Rh–C dihedral angle) and then they bind the $Rh(t\text{-}CO)_2$ fragment like tweezers (Fig. 29a). Hoffmann et al. calculated this system with extended Huckel method and proposed a molecular orbital diagram [106]. However, his diagram is based on an idealized structure, in which the two Cp rings are collinear and two bridging COs, Rh1 and Rh2, are all in the same plane, i.e., the $(CpRh)_2(\mu\text{-}CO)_2$ fragment has exactly the same structure as Fig. 13b with a C–Rh–Rh–C of 180°. Although Hoffmann addresses these differences in this discussion, here, we propose a molecular orbital diagram based on the highly puckered structure with the knowledge gained from $Cp_2Co_2(CO)_2$ and $Fe_3(CO)_{12}$ discussed in the previous sections.

$[Cp_2Rh_3(CO)_4]^-$ can be viewed as an adduct of $[Rh(t\text{-}CO)_2]^-$ and $Cp_2Rh_2(\mu\text{-}CO)_2$, and it may also be assembled by replacing the third CpRh fragment in $(CpRh)_3(\mu\text{-}CO)_2$ with a $[Rh(t\text{-}CO)_2]^-$ fragment. In addition, the fragment $Cp_2Rh_2(\mu\text{-}CO)_2$ may be further dissembled into two Rh^ICp fragments (denoted as Rh1 and Rh2) and one fragment of two face bridging carbonyls $(\mu\text{-}CO)_2$. Each RhCp fragment is an ML_3 fragment and the "super" fragment, $(CpRh)_2(\mu\text{-}CO)_2$, is expected to resemble $(CpCo)_2(CO)_2$ less two electrons. It should be noted that the two Cp rings are not collinear in $[Cp_2Rh_3(CO)_4]^-$, as each CpRh fragment needs to tilt to aim one of three unused hybrid orbitals to Rh of the $[Rh^{-I}(t\text{-}CO)_2]^-$ fragment (denoted as Rh3) to form Rh1–Rh3/Rh2–Rh3 bonds, where one electron on each CpRh fragment is consumed by these two Rh–Rh bonds, as in the bonding of $Fe_3(CO)_{12}$. At this moment, each CpRh has one electron and two unused hybrid orbitals to interact with the bridging carbonyls; the totally four orbitals from two fragments give linear combinations following a_1, a_2, b_1, b_2 irreducible

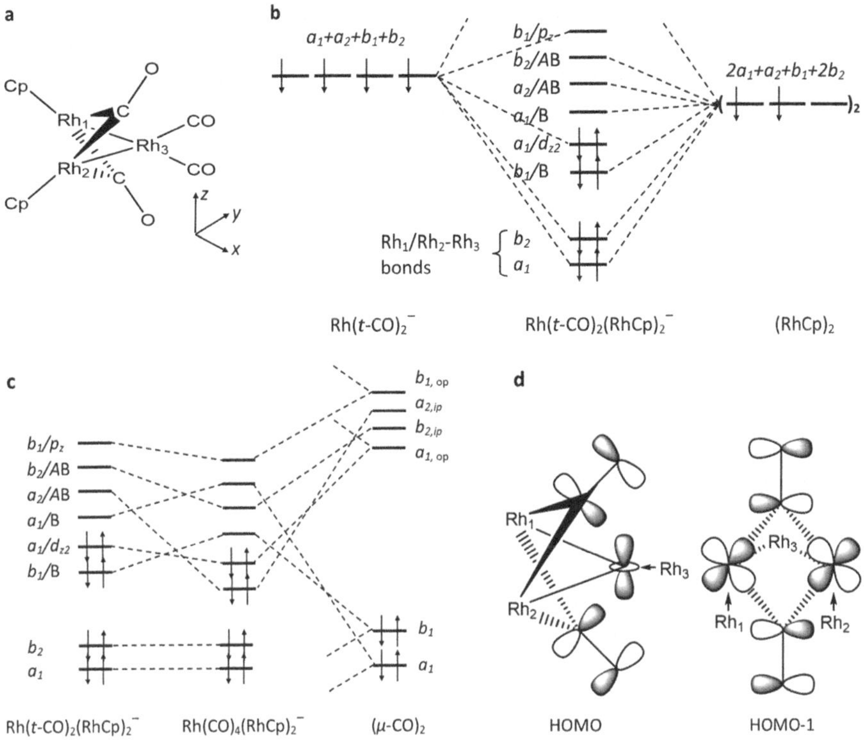

Fig. 29 (**a**) The molecular structure of $[Cp_2Rh_3(CO)_4]^-$. The electronic structures of (**b**) $[Rh(t\text{-}CO)_2(RhCp)_2]^-$ and (**c**) $[Cp_2Rh_3(CO)_4]^-$. Only unused hybrid orbitals are presented and the exact order of occupied orbitals could be different. (**d**) Visualized HOMO and HOMO-1 of $[Cp_2Rh_3(CO)_4]^-$. The notations AB and B refer to bonding and antibonding interactions between Rh_1 and Rh_2. The notations p_z and d_{z2} refer to the orbitals dominated by the respective atomic orbitals of Rh_3. Ip (in-plane) and op (out-of-plane) orbitals are w.r.t. Rh_1Rh_2CO plane

representations (see Fig. 29b). On the other hand, the $[Rh(CO)_2]^-$ fragment is an ML_2 fragment which is expected to have a 3+4 frontier orbital pattern with ten *d* electrons (six on three lower d-orbitals and four on the hybrid orbitals). Two in-plane ($Rh3(t\text{-}CO)_2$ plane) hybrid orbitals, with one electron on each, are used for Rh–Rh bonding and two out-of-plane ($Rh3(t\text{-}CO)_2$ fragment plane) hybrid orbitals, which can be recombined to give a_1 and b_1 orbitals, are left to interact with bridging carbonyls (Fig. 29b).

Finally, the $(\mu\text{-}CO)_2$ fragment is added, and its 5σ orbitals interact with the a_1/b_1 bonding MOs between Rh1 and Rh2 and raise their energy. The $(\mu\text{-}CO)_2$ fragment's in-plane (Rh1Rh2CO planes) 2π orbitals stabilize the a_2/b_2 metal–metal antibonding MOs (Fig. 29c, d). The last two electrons belonging to Rh1 and Rh2 occupy the a_2 Rh–Rh antibonding orbital as it is well stabilized due to the overlap with the 2π orbitals. The a_1 symmetric combination of the out-of-plane 2π orbitals of the COs stabilizes the a_1 symmetric combination of Rh3's hybrid orbitals with the same symmetry, which is dominated by d_{z2} contribution from Rh3 (Fig. 29d). The overall gist from the above analysis is that the Rh1–Rh2 bonding MO (before interaction with the (μ-CO)$_2$ fragment) is replaced by an Rh1–Rh2 antibonding MO (stabilized by the in-plane 2π orbitals of the symmetrically (edge) bridged COs), while the bonding to Rh3 is stabilized by the linear semibridging CO by its out-of-plane 2π orbitals.

Here, it seems unlikely for the linear semibridging COs to donate electrons from the 1π orbitals to the electron-rich $[Rh(t\text{-}CO)_2]^-$ fragment. However, Hoffmann's extended Huckel calculations, which leads to a qualitatively similar MO diagram, predicted a net flow of electrons from these COs to the 16-*e* $Rh(t\text{-}CO)_2$ fragment [106] such that the bridging carbonyls become a relay of electrons from the Cp_2Rh_2 to the $Rh(t\text{-}CO)_2$. Calculations with modern DFT methods with better basis set are needed to verify this proposed π-donation.

Alternatively, one might argue that $[Cp_2Rh_3(CO)_4]^-$ can be broken down into $[Cp_2Rh_2(\mu\text{-}CO)_2]^{2-}$ and $[Rh(t\text{-}CO)_2]^+$. In this case, the d_{z2} orbital of $Rh(t\text{-}CO)_2$ is vacant and may be able to accept electrons from the carbonyl's out-of-plane π orbital. Meanwhile, two additional electrons need to be accommodated on Rh1–Rh2 MOs, so the small hinge angle (C–Rh–Rh–C) must allow these COs to stabilize two antibonding Rh1–Rh2 orbitals. However, these COs actually bend away from Rh3 so that they present more of their 2π rather than their 1π to the Rh3. This structural feature seems to support the idea of these COs as π-acceptors [109]. For comparison, Crabtree's types II–IV linear edge semibridging carbonyls all have such behavior and are all determined to be π-acceptor (see previous sections).

5.4 Type III Face Bridging Carbonyl: $Cp_3Nb_3(CO)_6(\mu_3\text{-}CO)$

Type III face bridging carbonyls are relatively rare and the idea of treating the linear semibridging carbonyl as a four-electron donor can be extended to a six-electron donor. The example is the stable complex $Cp_3Nb_3(CO)_6(\mu_3\text{-}CO)$ [100, 112]. (Refer to Fig. 27 for the orientation of the CO.) Structurally, the CO bonds as a typical σ-donor to Nb1 and is linearly semibridging to Nb2 and Nb3, the C-Nb2 and C-Nb3 distances are close to the sum of covalent Pauling radii, and the CO shows a very long C–O bond length (1.30 Å) and an extremely low IR frequency (1,330 cm^{-1}). Thus, the C–O must have a very low bond order by either π-acceptance or π-donation. The face bridging carbonyl on this complex may migrate from one niobium to another with the activation energy of 17 kcal/mol through a transition state featuring either a type I symmetric face bridging carbonyl or type II face bridging carbonyl [113]. This system also satisfies the following conditions: (1) The Nb1–C–O θ is almost linear (169.6°) and it bends toward the Nb2–Nb3 vector (which is the signature characteristic of Crabtree's type I linear semibridging carbonyl.); (2) the two $CpNb(CO)_2$ recipient fragments contain electron-deficient, early transition metal. Computational work by Schaefer et al. has predicted that the μ_3-CO containing isomer is most stable for $Cp_3Nb_3(CO)_6(\mu_3\text{-}CO)$ [114] and for another model $[(CO)_2CSFe]_2(\mu_3\text{-}CS)[(CO)_2Fe]$, which has the same metal electron count [115]. Both cases of the CO were claimed to be a six-electron donor, but further analysis of the electronic structure is needed to confirm the role of the carbonyl since both π-donation and π-acceptance can lower the C–O bond order.

6 Bridging Isocarbonyl

The bridging isocarbonyl connects two or more metal atoms by terminally bonding to a metal at both C and O, while the carbon terminal of a carbonyl may still connect to multiple metals. Although one might think of naming this structure a linear bridging carbonyl, to reduce ambiguity with the term linear semibridging carbonyl, the term "bridging isocarbonyl" is selected to stress that the oxygen end participates in its own distinct dative bond. Generally speaking, the carbon terminal of the carbonyl is a π-acid ligand and prefers electron-rich metals; on the other hand, the O terminal interacts with electron-poor metals in the Lewis acid–base fashion. The adduct of main group metal aluminum in the form of AlR_3 was the first to be discovered, and this Lewis acid has been shown to bond to the oxygen of a variety of transition metal carbonyl complexes [116]. The coordination of such Lewis acids to the O terminal of the carbonyls bound to transition metals depletes the electron density on the carbonyl and dramatically increases the backbonding between the transition metal and the carbonyl, which are reflected by IR and UV–Vis spectra [116].

The first example of an all-transition-metal bridging isocarbonyl with available structural information is $(V(CO)_5(\mu\text{-}CO))_2(V(THF)_4)$, which has a structure like a sandwich with $V(THF)_4$ in the middle and connected to the other two $(V(CO)_6)$ moieties through the O terminal of the carbonyl [117]. $[Cp^*_2(CH_3)Ti](\mu\text{-}OC)[Mo(t\text{-}CO)_2Cp]$ is the first example of a d^0 transition metal coordinated to the oxygen of a carbonyl [118]. Fenske–Hall calculations on this system revealed strong electron redistribution upon the coordination of electron-deficient Ti core. Unexpectedly, the bridging isocarbonyl has a more negative Mulliken charge than the terminal carbonyls, which seems odd as the Ti atom is supposed to remove electron density from this carbonyl. An orbital-by-orbital population analysis found that the bridging isocarbonyl is a three-way donor and one-way acceptor. On the carbon end, the 5σ and 2π orbitals interact with electron-rich Mo center as a regular π-acid ligand; on the other end, the 4σ orbitals and 1π (Fig. 1) donate their electrons to the d^0 Ti metal atom. The bent Ti–O–C angle of 144.3° increases the overlap between the Ti *d*-orbital and $1\pi_{ip}$ (in Ti–O–C–Mo plane). The Ti^{IV} cannot backbond with CO 2π because it is d^0, so it only depletes the CO electron density. However, this depletion enhances the backbonding between the carbonyl and Mo and the CO gains more electron density from Mo than what it loses to Ti [118]. Norton et al. explained such a bent M_2–O–C angle by referring to the re-hybridization of the O atom of the carbonyl as the strong backbonding altered the C–O bond to an extent that the triple bond no longer exists (Fig. 4). The O of the bridging carbonyl then has sp^2 or sp^3 hybrid orbitals and one of those orbitals donates its electrons to the electron-poor M_2 [119]. The C of the bridging isocarbonyl may bridge more than one metal atom but the oxygen end may only connect to one electron-poor cores, as the 4σ and 1π orbitals themselves are low in energy and do not have good capacity for electron donation [116].

A complete linear bridging isocarbonyl (i.e., both M_1–C–O and C–O–M_2 angles are linear) was discovered on $[V(CO)_5(\mu\text{-}CO)]$ $[V(Cp^*)_2]$ by Trogler et al. in which $V(Cp^*)_2$ connects the oxygen terminal of one of six carbonyls on the $V(CO)_6$ moiety [120]. Its MOs are presented in Fig. 30 as calculated by the SCC-Xα-DV method (though only closed-shell calculation was done at that moment). Magnetic susceptibility reveals this molecule has two unpaired electrons and those two electrons are assigned to the nearly degenerate α-HOMO and α-HOMO-1. The formal oxidation states of the complex is written as $[V^{-I}(CO)_5(\mu\text{-}CO)]$ $[V^{III}(Cp^*)_2]$ [120].

Here, the major interaction between the $[V(CO)_5(\mu\text{-}CO)]^-$ fragment and the $[V(Cp^*)_2]^+$ is the mixing of the d_{xz} orbitals of two vanadium atoms, while the detailed discussion of the electronic structure of $[V(Cp^*)_2]^+$ and compounds with the common structure $M(Cp)_2L_x$ is available from Hoffmann et al. [121]. In this system, the d_{xz} orbital in the $[V(Cp^*)_2]^+$ is stabilized by the δ vacant orbitals of the Cp ring, while that in the $[V(CO)_6]^+$ is stabilized by 2π *orbitals* of CO. Such *d–d* coupling is conducted through the bridging isocarbonyl's 2π orbital, creating one bonding orbital $1b_1$ and one antibonding orbital $2b_1$ between CO and $V(Cp^*)_2$, and the latter is almost degenerate with $2a_1(d_{x2}$ of $[V(Cp^*)_2]^+)$ and the two unpaired electrons reside on those two orbitals. Thus, such a configuration $(1b_1)^2(2b_1)^1$ provides net π-backbonding between the $[V(Cp^*)_2]^+$ and the bridging carbonyl. On the other hand, the $2a_1$ orbital (mainly d_{x2} of $[V(Cp^*)_2]^+$ moiety), which has the correct symmetry to accept 4σ-donation from the carbonyl, only has 1% atomic contribution from C/O of the bridging carbonyl. In other words, in this case, the carbonyl, though bound to the $[V(Cp^*)_2]^+$ through its O terminal, is still a π-acid ligand rather than the π-donor in the case of d^0 $Cp^*_2(CH_3)Ti$ adduct. It may be

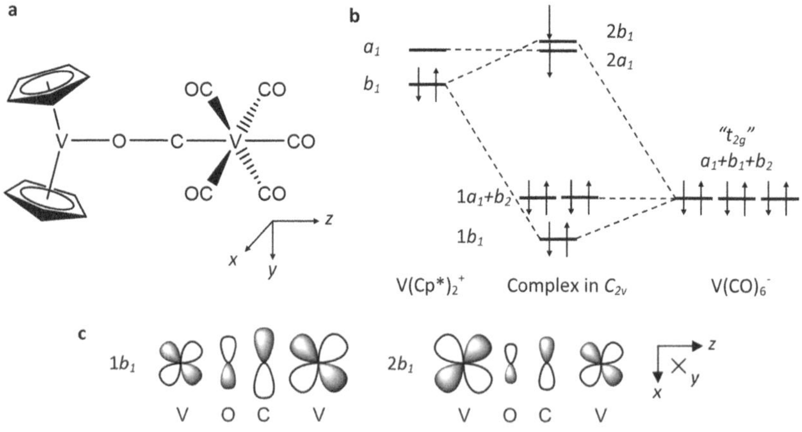

Fig. 30 (**a**) The structure and (**b**) the molecular orbital diagram of $[V(CO)_5(\mu\text{-}CO)]$ $[V(Cp^*)_2]$. (**c**) The sketches of two important orbitals reflecting the interaction between two fragments (note the change of coordinates)

related to the *d* electron count, as only d^0 metals may accept π-donation through bridging isocarbonyl. The bridging isocarbonyl connecting two V atoms also has a much higher negative charge than other terminal carbonyls, leading Trogler et al. to conclude the interaction may be somewhat ionic [120]. Though the calculations rendered here did not explicitly explain the linearity of M′–O–C angle as caused by the π-interaction, it could be a plausible explanation. Such complete linear bridging isocarbonyl was reproduced in the imaginary $[Cr(CO)_5](\mu\text{-}CO)[Cr(CO)_5]$ complex by Schaefer et al. [122]. The calculations also verified the backbonding from both $Cr(CO)_5$ moieties connected to the oxygen and the carbonyl.

7 Final Comments

The analysis in the previous sections gives some general rules of the behavior of bridging carbonyls: the bridging carbonyl is generally a π-acid ligand beyond its conventional σ-donation capacity on the C end. For carbonyl bridging two metals, it may bind with either a bent M1–C–O (symmetrical bridging or bent semibridging carbonyls) or linear M1–C–O (linear semibridging carbonyl). In the bent M1–C–O structures, the axial nodal plane in CO's 2π orbital means that the backbonding stabilizes the antibonds between two metals, while the CO's 5σ-donation destabilizes the metal–metal bonding orbitals, such that the occupancies of the orbitals along the metal–metal vector may be altered from those that would have been occupied in the absence of the bridging COs. For linear semibridging carbonyl, at least linear semibridging carbonyl on electron-rich mid- to late transition metals, the carbonyl can insert one lobe of its vacant 2π between the bonding orbital of two metal cores, stabilizing the metal–metal bond. The many linear semibridging carbonyls may actually be a terminal CO distorted by the steric pressure. For rare cases, if M_2 is an electron-deficient early transition metal, the carbonyl may donate its 1π electrons to M_2 and a signature structural feature is that the linear semibridging carbonyl leans toward M_2 to maximize the contact of both C and O with M_2. For face bridging carbonyl and bridging isocarbonyl, it is still generally determined to be a π-acid ligand unless an electron-deficient metal core is involved.

As an expansion to the Lewis bonding theory, an electron-deficient three-center (or four-center, etc.), two-electron bond is sometimes needed [25] to maintain the 18-electron rule for polynuclear complexes such as those described here. However, the orbital-by-orbital analysis described here is easily accomplished only on relatively small clusters, and the orbital mixing becomes overwhelming with an increasing number of metals in the cluster, even when high symmetry may help reduce this complexity. For larger clusters containing four or more metals, there is a tendency to analyze the bonding situation inside the cluster as a whole, without detailing the bonding between individual metal–ligand fragments, by the polyhedral skeletal electron pair theory or the Wade/Mingos rule [123, 124], as well as Stone's symmetry classification of cluster orbitals [125]. In such a manner, the 18-*e* rule is either skipped or reversely validated (often conditionally) after the MO diagram is derived, rather than being used as a priori tools to predict the electronic structures. Nevertheless, molecular orbital theory offers a powerful tool to achieve a deeper understanding of the exact nature of the metal–ligand, metal–metal bonding and serves as the final judge of disputes as there are still some justifications for not assuming that all of the electrons in the lower *d*-orbitals (the *d*-orbital cluster that would correspond to the t_{2g} orbitals in O_h symmetry) are uninvolved in the cluster bonding [21, 22, 25, 126].

References

1. Pauling L (1960) The nature of the chemical bond, and the structure of molecules and crystals; an introduction to modern structural chemistry. Cornell University Press, New York, p 332
2. Dewar JS (1951) Bull Soc Chim Fr 18:C71
3. Chatt J, Duncanson LA (1953) J Chem Soc 2939
4. Cotton FA (1976) Prog Inorg Chem 21:1
5. Crabtree RH, Lavin M (1986) Inorg Chem 25:805
6. de la Cruz C, Sheppard N (1990) J Mol Struct 224:141
7. Bruce King R, Zhang Z, Li Q-S, Schaefer HF III (2012) Phys Chem Chem Phys 14:14743
8. Braterman PS (1972) Struct Bond 10:57
9. Horwitz CP, Shriver DF (1984) Adv Organomet Chem 23:219
10. Sappa E, Tiripicchio A, Braunstein P (1985) Coord Chem Rev 65:219
11. Braterman PS (1976) Struct Bond 26:1
12. Macchi P, Sironi A (2003) Coord Chem Rev 238–239:383
13. Haaland A (1989) Angew Chem Int Ed 28:992
14. Green MLH (1995) J Organomet Chem 500:127
15. Green JC, Green MLH, Parkin G (2012) Chem Commun 48:11481
16. Freund H-J, Hohlneicher G (1979) Theor Chim Acta 51:145
17. Summerville RH, Hoffmann R (1979) J Am Chem Soc 101:3821
18. Freund H-J, Dick B, Hohlneicher G (1980) Theor Chim Acta 57:181
19. Mealli C, Proserpio DM (1990) J Organomet Chem 386:203
20. Rosa A, Baerends EJ (1991) New J Chem 15:815
21. Reinhold J, Hunstock E, Mealli C (1994) New J Chem 18:465
22. Reinhold J, Kluge O, Mealli C (2007) Inorg Chem 46:7142
23. Cotton FA, Troup JM (1974) J Chem Soc Dalton Trans 800
24. Cotton FA, Hunter DL (1974) Inorg Chem 13:2044
25. Landis CR, Weinhold F (2015) J Comput Chem. doi:10.1002/jcc.24001
26. Orgel LE (1966) An introduction to transition-metal chemistry: ligand-field theory, 2nd edn. Wiley, New York, p 146
27. Ponec R, Gatti C (2009) Inorg Chem 48:11024
28. Ponec R, Lendvay G, Chaves J (2008) J Comput Chem 29:1387
29. Ponec R (2015) Comp Theor Chem 1053:195
30. Bo C, Sarasa JP, Poblet JM (1993) J Phys Chem 97:6362
31. Sumner GG, Klug HP, Alexander LE (1964) Acta Crystallogr 17:732
32. Hall MB, Fenske RF, Dahl LF (1975) Inorg Chem 14:3103
33. Thorn DL, Hoffmann R (1978) Inorg Chem 17:126
34. Low AA, Kunze KL, MacDougall PJ, Hall MB (1991) Inorg Chem 30:1079
35. Tiana D, Francisco E, Macchi P, Sironi A, Martín Pendás A (2015) J Phys Chem A 119:2153
36. Finger M, Reinhold J (2003) Inorg Chem 42:8128

37. Kluge O, Finger M, Reinhold J (2005) Inorg Chem 44:6494
38. Swart I, de Groot FMF, Weckhuysen BM, Rayner DM, Meijer G, Fielicke A (2008) J Am Chem Soc 130:2126
39. Ignatyev IS, Schaefer HF, King RB, Brown ST (2000) J Am Chem Soc 122:1989
40. Calderón JL, Fontana S, Frauendorfer E, Day VW, Iske SDA (1974) J Organomet Chem 64: C16
41. Brunner H (1968) J Organomet Chem 14:173
42. Schore NE, Ilenda CS, Bergman RG (1976) J Am Chem Soc 98:256
43. Cirjak LM, Ginsburg RE, Dahl LF (1982) Inorg Chem 21:940
44. Bernal I, Korp JD, Reisner GM, Herrmann WA (1977) J Organomet Chem 139:321
45. Brunner H (1968) J Organomet Chem 12:517
46. Byers LR, Dahl LF (1980) Inorg Chem 19:680
47. Pinhas AR, Hoffmann R (1979) Inorg Chem 18:654
48. Schugart KA, Fenske RF (1986) J Am Chem Soc 108:5100
49. Griewe GL, Hall MB (1988) Organometallics 7:1923
50. Demuynck J, Mougenot P, Benard M (1987) J Am Chem Soc 109:2265
51. Low AA, Hall MB (1993) Inorg Chem 32:3880
52. Hall MB. Unpublished Work
53. Mitschler A, Rees B, Lehmann MS (1978) J Am Chem Soc 100:3390
54. Benard M (1979) Inorg Chem 18:2782
55. Benard M (1978) J Am Chem Soc 100:7740
56. Schore NE, Ilenda CS, Bergman RG (1977) J Am Chem Soc 99:1781
57. Evans DG (1983) J Chem Soc Chem Commun 675
58. Farrugia L, Macchi P (2012) Struct Bond 146:127
59. Wei CH, Dahl LF (1969) J Am Chem Soc 91:1351
60. Cotton FA, Troup JM (1974) J Am Chem Soc 96:4155
61. Campana C, Guzei I, Mednikov E, Dahl L (2014) J Cluster Sci 25:205
62. Johnson BFG (1976) J Chem Soc Chem Commun 211
63. Benfield RE, Johnson BFG (1980) J Chem Soc Dalton Trans 1743
64. Braga D, Grepioni F, Farrugia LJ, Johnson BFG (1994) J Chem Soc Dalton Trans 2911
65. Hunstock E, Mealli C, Calhorda MJ, Reinhold J (1999) Inorg Chem 38:5053
66. Schilling BER, Hoffmann R (1979) J Am Chem Soc 101:3456
67. Braga D, Grepioni F, Tedesco E, Calhorda MJ, Lopes PEM (1995) J Chem Soc Dalton Trans 3297
68. Chevreau H, Martinsky C, Sevin A, Minot C, Silvi B (2003) New J Chem 27:1049
69. Churchill MR, DeBoer BG (1977) Inorg Chem 16:878
70. Churchill MR, Hollander FJ, Hutchinson JP (1977) Inorg Chem 16:2655
71. Tyler DR, Levenson RA, Gray HB (1978) J Am Chem Soc 100:7888
72. Delley B, Manning MC, Ellis DE, Berkowitz J, Trogler WC (1982) Inorg Chem 21:2247
73. Norton JR, Collman JP, Dolcetti G, Robinson WT (1972) Inorg Chem 11:382
74. Mingos DMP (2014) Struct Bond 153:1
75. Bag A, Ghorai PK (2015) RSC Adv 5:31575
76. Cyr JE, DeGray JA, Gosser DK, Lee ES, Rieger PH (1985) Organometallics 4:950
77. Downard AJ, Robinson BH, Simpson J, Bond AM (1987) J Organomet Chem 320:363
78. Hock AA, Mills OS (1961) Acta Crystallogr 14:139
79. Cotton FA, Troup JM (1974) J Am Chem Soc 96:1233
80. Colton R, Commons CJ, Hoskins BF (1975) J Chem Soc Chem Commun 363
81. Klingler RJ, Butler W, Curtis MD (1975) J Am Chem Soc 97:3535
82. Klingler RJ, Butler WM, Curtis MD (1978) J Am Chem Soc 100:5034
83. Curtis MD, Han KR, Butler WM (1980) Inorg Chem 19:2096
84. Hoffman DM, Hoffmann R (1981) Inorg Chem 20:3543
85. Benard M, Dedieu A, Nakamura S (1984) New J Chem 8:149
86. Caulton KG, Adair P (1976) J Organomet Chem 114:C11

87. Marsella JA, Caulton KG (1982) Organometallics 1:274
88. Decker SA, Klobukowski M (1999) Can J Chem 77:65
89. Bo C, Costas M, Poblet JM, Rohmer M-M, Benard M (1996) Inorg Chem 35:3298
90. Woodcock C, Eisenberg R (1985) Inorg Chem 24:1285
91. Jemmis ED, Pinhas AR, Hoffmann R (1980) J Am Chem Soc 102:2576
92. Morris-Sherwood BJ, Powell CB, Hall MB (1984) J Am Chem Soc 106:5079
93. Sargent AL, Hall MB (1989) J Am Chem Soc 111:1563
94. Sargent AL, Hall MB (1990) Polyhedron 9:1799
95. Simpson CQ, Hall MB (1992) J Am Chem Soc 114:1641
96. Williamson RL, Low AA, Hall MB, Guest MF (1992) NATO Sci Ser 378:155
97. Miller TF, Strout DL, Hall MB (1998) Organometallics 17:4164
98. Doyle RA, Daniels LM, Angelici RJ, Stone FGA (1989) J Am Chem Soc 111:4995
99. Barger PT, Bercaw JE (1980) J Organomet Chem 201:C39
100. Herrmann WA, Ziegler ML, Weidenhammer K, Biersack H (1979) Angew Chem Int Ed 18:960
101. Fischer EO, Palm C (1958) Chem Ber 91:1725
102. Byers LR, Uchtman VA, Dahl LF (1981) J Am Chem Soc 103:1942
103. Sinn H, Lundborg C, Kirchner K (1958) Angew Chem 70:744
104. Green M, Hankey DR, Howard JAK, Louca P, Stone FGA (1983) J Chem Soc Chem Commun 757
105. Bray AC, Green M, Hankey DR, Howard JAK, Johnson O, Gordon F, Stone A (1985) J Organomet Chem 281:c12
106. Pinhas AR, Albright TA, Hofmann P, Hoffmann R (1980) Helv Chim Acta 63:29
107. Barnes CE, Orvis JA, Staley DL, Rheingold AL, Johnson DC (1989) J Am Chem Soc 111:4992
108. Rives AB, You XZ, Fenske RF (1982) Inorg Chem 21:2286
109. Jones WD, White MA, Bergman RG (1978) J Am Chem Soc 100:6770
110. Macklin PD, Mirkin CA, Viswanathan N, Williams GD, Geoffroy GL, Rheingold AL (1987) J Organomet Chem 334:117
111. Herrmann WA, Barnes CE, Zahn T, Ziegler ML (1985) Organometallics 4:172
112. Herrmann WA, Biersack H, Ziegler ML, Weidenhammer K, Siegel R, Rehder D (1981) J Am Chem Soc 103:1692
113. Lewis LN, Caulton KG (1980) Inorg Chem 19:3201
114. Peng B, Li Q-S, Xie Y, King RB, Schaefer HF III (2009) Dalton Trans 3748
115. Zhang Z, Li Q-s, Xie Y, King RB, Schaefer HF (2009) Inorg Chem 48: 6167
116. Shriver DF, Alich A (1972) Coord Chem Rev 8:15
117. Schneider M, Weiss E (1976) J Organomet Chem 121:365
118. Hamilton DM, Willis WS, Stucky GD (1981) J Am Chem Soc 103:4255
119. Marsella JA, Huffman JC, Caulton KG, Longato B, Norton JR (1982) J Am Chem Soc 104:6360
120. Osborne JH, Rheingold AL, Trogler WC (1985) J Am Chem Soc 107:6292
121. Lauher JW, Hoffmann R (1976) J Am Chem Soc 98:1729
122. Zhang Z, Li Q-s, Xie Y, King RB, Schaefer HF (2010) J Phys Chem A 114:4672
123. Wade K (1976) Adv Inorg Chem Radiochem 18:1
124. Mingos DMP (1984) Acc Chem Res 17:311
125. Stone AJ (1981) Inorg Chem 20:563
126. Hall MB (1990) In: Fackler J Jr (eds) Metal–metal bonds and clusters in chemistry and catalysis. Springer, New York, p 265

Index

The manufacturer's authorised representative in the EU is Springer Nature Customer Service Centre GmbH, Europaplatz 3, 69115 Heidelberg, Germany. If you have any concerns regarding our products, please contact ProductSafety@springernature.com

Printed and bound by CPI Group (UK) Ltd, Croydon, CR0 4YY

30/09/2026

02225689-0009